THE STORY OF EARTH'S CLIMATE IN 25 DISCOVERIES

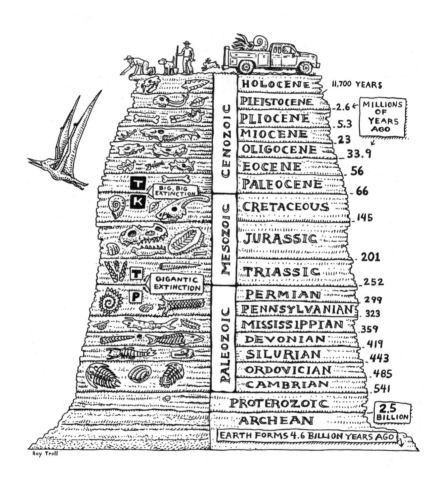

	HOLOCENE	11,700 YEARS	
CENOZOIC	PLEISTOCENE	2.6	MILLIONS OF YEARS AGO
	PLIOCENE	5.3	
	MIOCENE	23	
	OLIGOCENE	33.9	
	EOCENE	56	
	PALEOCENE	66	
MESOZOIC	CRETACEOUS	145	
	JURASSIC	201	
	TRIASSIC	252	
PALEOZOIC	PERMIAN	299	
	PENNSYLVANIAN	323	
	MISSISSIPPIAN	359	
	DEVONIAN	419	
	SILURIAN	443	
	ORDOVICIAN	485	
	CAMBRIAN	541	
	PROTEROZOIC		
	ARCHEAN	2.5 BILLION	

BIG, BIG EXTINCTION

GIGANTIC EXTINCTION

EARTH FORMS 4.6 BILLION YEARS AGO

Ray Troll

THE STORY OF EARTH'S CLIMATE

in 25 DISCOVERIES

HOW SCIENTISTS FOUND THE CONNECTIONS BETWEEN CLIMATE AND LIFE

DONALD R. PROTHERO

COLUMBIA UNIVERSITY PRESS   NEW YORK

COLUMBIA UNIVERSITY PRESS

Publishers Since 1893

New York   Chichester, West Sussex

cup.columbia.edu

Library of Congress Cataloging-in-Publication Data

Title: The story of Earth's climate in 25 discoveries : how scientists discovered
the connections between climate and life / Donald R. Prothero.

Description: New York : Columbia University Press, [2023] |
Includes bibliographical references and index.

Identifiers: LCCN 2023030945 | ISBN 9780231203586 (hardback) | ISBN 9780231555135 (ebook)

Subjects: LCSH: Climatic changes—History. | Mass extinctions.

Classification: LCC QC903 .P97 2023 | DDC 551.609—dc23/eng/20231016

LC record available at https://lccn.loc.gov/2023030945

Printed in the United States of America

Cover design: Julia Kushnirsky

Cover images: Shutterstock

Frontispiece: Ray Troll

THIS BOOK IS DEDICATED TO MY SONS
ERIK, ZACHARY, AND GABRIEL
AND TO MY STUDENTS
MAY THEY FIND A WAY TO SOLVE FUTURE CLIMATE
PROBLEMS THAT EARLIER GENERATIONS COULD NOT.

CONTENTS

PREFACE

This project came about because of my recent experience teaching about Earth systems and climate change at the college level. My original training was in paleontology and geology, but most paleontologists must be versatile and teach a wide array of subjects, from Earth history to sedimentary geology, and many other topics as well. Over my research career, I have spent a lot of time publishing on climate changes of the Cenozoic, especially the Eocene-Oligocene transition, so I count myself a climate scientist. Now I find myself not only teaching meteorology and oceanography but also a course called "Studies of a Blue Planet," which covers atmospheres and oceans and geology, the history of the Earth's atmospheres and oceans, and the history of life on this planet as well. In this course, I emphasize the integration of climate and life and how these systems are all part of one great "Earth system." That idea has become very popular in the in recent years, and many geology courses now integrate an understanding of oceans and atmospheres along with the history of rocks and of living organisms. In addition, the "Earth system" includes the formation of the stars and solar systems, the Sun and the Moon, as well as the formation and early history of the Earth and its oceans, so I have included chapters on those topics, even if they are not about "climate" in the narrow sense.

Lots of books discuss the current situation with regard to climate change, but few of them put it in the context of 4.5 billion years of Earth's climate history. Nor are they written for the intelligent lay reader without formal scientific training in climate science. This book is intended to address that gap, giving scientists and nonscientists alike a broad background in the history of Earth's climate and how life and climate interact with each other. I discuss the ice ages and the current climate situation, of course, but it's important to know how Earth's atmosphere works, how it has changed, and how the planet has gone from one climate extreme to another—from a "greenhouse" Earth with warm temperatures, high sea levels, and no ice, to a "snowball" Earth when the planet was completely frozen right down to the equator. People often dismiss the current rapid changes in climate with comments such as "Climate changes all the time, so why worry?" But that is precisely the point: scientists know *when* and *how* and *why* climate has changed in the past, and the way it is changing now is so utterly different that climate scientists recoil in alarm as each year breaks previous global temperature records and extreme climate and weather events become more and more common. By looking at the EPICA-1 ice core drilled in the thickest part of the Antarctic ice sheet, we can sample air bubbles of our atmosphere going back 780,000 years. From this, and from the detailed record of ice cores, we know that "natural climate variability" never produced carbon dioxide values higher than 280 ppm even during the warmest interglacials of the last 780,000 years, but as I write this, global carbon dioxide levels are already more than 415 ppm and still climbing. Yes, climate has fluctuated in the past—but scientists know and understand why and how much it has changed. They are insulted and alarmed when climate deniers imply that current climate changes are just "normal fluctuations."

I feel that a book like this is timely and is essential. Everyone needs a background in our own climate history to provide a context for our current climate situation. I hope you will find this story interesting and entertaining. I have incorporated tales of some of the amazing scientists who discovered many of these facts about Earth's past climates and highlighted the cultural context of many of those discoveries. This book is intended not only to educate the intelligent lay reader about climate and Earth's history but also to entertain and inform people as well. I hope you enjoy learning a bit more about the interplay between climate and life on this planet that we all call home.

ACKNOWLEDGMENTS

I thank Miranda Martin at Columbia University Press for supporting this book and for giving me lots of encouragement while I was writing it. I thank Bruce Lieberman and Greg Retallack for scholarly reviews, and I thank Brian Smith at Columbia University Press for his help with the project. I also thank Kat Jorge, Ben Kolstad, and Kay Mikel for their work in production and editing.

I especially thank the many generations of my students over 40 years of teaching at Vassar, Knox, and Occidental colleges, and at Cal Poly Pomona, who forced me to learn how to explain and describe the events of Earth's history clearly and in an exciting and interesting way. In many ways, my students have taught me as much as I have taught them.

Finally, I thank my family, especially my sons Erik, Zachary, and Gabriel, and my wonderful wife, Dr. Teresa LeVelle, for support and love and encouragement over the years it took to complete this book.

THE STORY OF EARTH'S CLIMATE IN 25 DISCOVERIES

RARE EARTH

We live on a hunk of rock and metal that circles a humdrum star that is one of 400 billion other stars that make up the Milky Way Galaxy which is one of billions of other galaxies which make up a universe which may be one of a very large number, perhaps an infinite number, of other universes. That is a perspective on human life and our culture that is well worth pondering.

—CARL SAGAN, *COSMOS*

WHY WON'T ET PHONE HOME?

This striking bit of dialogue between Ellie Arroway (played by Jodie Foster) and Palmer Joss (played by Matthew McConaughey) is from the 1997 film *Contact*, based on the novel by the late Carl Sagan:

ELLIE ARROWAY: You know, there are four hundred billion stars out there, just in our galaxy alone. If only one out of a million of those had planets, and just one out of a million of those had life, and just one out of a million of those had intelligent life; there would be literally millions of civilizations out there.

PALMER JOSS: [looking at the night sky] Well, if there wasn't, it'll be an awful waste of space.

Unfortunately, the screenwriter who wrote Ellie's lines was a bit math-challenged. One millionth of 400 billion is only 400,000, and one millionth of that is only 0.4, and one millionth of that is only 0.0000004 civilizations out there. Estimates of the number of stars in the Milky Way, our galaxy, vary from 100 to 400 billion. Of course, most of those stars probably do not have planets or planets in a zone and of a mass conducive to the development of either intelligent life or any life at all. In fact, it is possible that in our entire galaxy, no star system but ours harbors intelligent life.

THE NUMBERS GAME

In 1961, Frank Drake, the founder of SETI (Search for Extra-Terrestrial Intelligence), proposed an equation for calculating the number of extraterrestrial civilizations with whom radio communication might be possible. The Drake Equation is written as $N = R_* \cdot f_p \cdot n_e \cdot f_l \cdot f_i \cdot f_c \cdot L$. The values expressed are as follows:

N = number of extraterrestrial civilizations with whom radio communication is possible

R_* = rate of star formation in the galaxy

f_p = fraction of stars in the galaxy with planets

n_e = number of planets capable of supporting life

f_l = fraction of planets actually harboring life

f_i = fraction of planets harboring intelligent life

f_c = fraction of planets harboring technologically advanced life

L = length of time during which technologically advanced civilizations broadcast radio waves

Most, if not all, of these quantities are unknown, so the Drake Equation isn't really a guide upon which to base calculations. Rather, it is a tool for philosophical contemplation. Is there any way to actually calculate or at least to get a rough idea of how many extraterrestrial civilizations might exist in our galaxy?

Perhaps. Following Ellie Arroway's example but with a bit better math, we could make a 1 percent rule. If we assume a low estimate—that there are 100,000,000,000 stars in our galaxy and only 1 percent have planets— then 1,000,000,000 stars could have planets. If only 1 percent of those stars have planets in what we might call the "Goldilocks zone"—where temperatures are neither too hot nor too cold so water on the surface is mostly in liquid form—we are down to 10,000,000 stars. If only 1 percent of those stars have planets that actually harbor life, we are down to 100,000 such planets in the Milky Way. If only 1 percent of these planets have higher life-forms, such as vertebrates, we are down to 1,000 planets. If intelligent life exists on only 1 percent of these, there would be only 10 planets with intelligent life-forms in the entire Milky Way. However, this 1 percent rule is arbitrary and could easily be off by many orders of magnitude, giving us as many as 10,000 planets or fewer than one (that is, none) planet with intelligent life-forms in our galaxy.

A better way of calculating how many intelligent life-forms exist in our galaxy might be to consider the Morgan-Keenan (MK) and Harvard classification systems, which state that our sun is a type G-2 star. These constitute 7.5 percent of the main sequence stars in the Milky Way, or 7,500,000,000 stars, if we assume there are 100,000,000,000 stars in our galaxy. If only 1 percent of these have had time to develop intelligent life, there would be 75,000,000 intelligent life-forms in the Milky Way. Another way to calculate the abundance of life in the galaxy is to consider the number of stars with planets. An estimate based on observations made by NASA's Kepler Space Telescope is that 22 percent of sun-like stars have planets, so even this estimate might be too low.

THE FERMI PARADOX: WHERE IS EVERYBODY?

We now come up against the Fermi Paradox. Following a casual conversation at Los Alamos in 1950, Enrico Fermi, one of the main architects of the Manhattan Project, along with the physicist Michael H. Hart came up with what is called the Fermi Paradox. Its main assumptions are: The sun is a typical star; billions of stars are far older than the sun; there's a high probability that some of these might harbor life, even intelligent life; even with starships taking hundreds of years to traverse the stars, intelligent beings should have colonized the entire galaxy by now. But they haven't. Fermi's question is: "Where is everybody?"

Somewhat related to the Fermi Paradox are the Dyson sphere and the Kardashev scale. In 1960, the American theoretical physicist and mathematician Freeman Dyson suggested that as civilizations advance to the point of colonizing their solar systems their energy demands would rise to the point that they would require all the energy output of their sun. Dyson proposed that such civilizations would be capable of building a megastructure that would entirely enclose their star to collect all of its energy. This came to be known as a Dyson sphere. The enclosed star would then radiate only in the infrared wavelengths, the waste heat given off by the Dyson sphere's energy consumption.

In 1964, the Russian astronomer Nikolai Kardashev proposed three levels of highly advanced civilizations classified by their energy use. On the Kardashev scale, a Type I civilization utilizes all the energy available on its planet. A Type II civilization utilizes all the energy of its sun. A civilization that has enclosed its sun in a Dyson sphere is Type II on the Kardashev scale. Type III civilizations utilize all the energy output of their home galaxy. A galaxy hosting such a civilization would not emit any light in our visible spectrum. In fact, it would only, or at least chiefly, emit infrared radiation, i.e., what we perceive as heat. Recently, researchers using the Wide-field Infrared Survey Explorer (WISE) surveyed approximately 100,000 galaxies and found no sources that emitted only or chiefly infrared radiation, indicating that none of them host a civilization that is Type III on the Kardashev scale. One possible explanation for this lack of evidence of Type III civilizations is that their heat signature simply hasn't reached us yet. Our nearest galactic neighbor is the Andromeda Galaxy, which is about 2.5 million light years away from us, a light year being the distance light travels in a year. Had a Type III civilization begun to enclose all the stars in the Andromeda Galaxy about 2,000,000 years ago, we wouldn't begin to see its stars winking out and giving off only infrared radiation for another 500,000 years.

Of course, both Dyson spheres and Kardashev-type civilizations are highly theoretical. We don't really know if interstellar civilizations will require such vast amounts of energy. Limiting ourselves to only the 100 to 400 billion stars in our own galaxy, however, Fermi's question remains unanswered. We have not found convincing evidence that Earth has been visited in the past by extraterrestrials, nor have we found evidence of radio waves broadcast from other parts of the galaxy. So here we are, like the stereotypical wallflower at the senior prom sitting alone on the sidelines of the dance floor. Why hasn't anyone asked us to dance?

One answer to the Fermi Paradox is that our sun, a G-2 star, isn't that typical. Another answer to the paradox is the sheer size of the galaxy, even if we accept the low estimate of its total number of stars as being 100 billion. If we assume that 1 percent of the G-2 stars in our galaxy harbor intelligent life, giving us a striking number of 75 million intelligent species, they must still navigate through 100 billion stars. That is, only 75 out of every 100,000 stars would harbor intelligent life. To put it another way, intelligent life would be found on the planets of only 0.075 percent of the stars in the Milky Way, so our potential dance partners simply may not have found us yet. It's also possible we haven't found them yet. According to the SETI website:

> The failure so far to find a signal is hardly evidence that none is to be found. All searches to date have been limited in one respect or another. These include limits on sensitivity, frequency coverage, types of signals the equipment could detect, and the number of stars or the directions in the sky observed. For example, while there are hundreds of billions of stars in our galaxy, only a few thousand have been scrutinized with high sensitivity and for those, only over a small fraction of the available frequency range.[1]

If we take "a few thousand" as about 10,000 and assume a low-end estimate of 100,000,000,000 stars in the Milky Way, then SETI has only managed to scan one out of every 10,000,000 stars or 0.0001 percent of the stars in our galaxy. Likewise, if other civilizations are looking for our signal, they might also have a very hard time finding it.

Another barrier posed by the vastness of our galaxy is that of interstellar space and what we might call "Einstein's speed limit." A button and T-shirt sold at science fiction conventions reads, "186,000 miles per second: It's not just a good idea. It's the law!" The velocity of light in a vacuum is 186,000 miles per second. Any particle with a measurable rest mass (the rest mass of photons and neutrinos is zero) is subject to Fitzgerald-Lorentz contractions. Independently proposed by George Fitzgerald in 1889 and Hendrik Antoon Lorentz in 1892, these contractions were borne out by Einstein's theory of relativity. As particles with a measurable rest mass are accelerated, their dimensions, including time, decrease, and their masses increase. At the speed of light, the dimensions of any object become zero, and its mass becomes infinite, meaning in essence that it ceases to exist. The nearest star to our sun is Alpha Centauri, which is about four light years away from us. If aliens from Alpha Centauri could manage to harness vast amounts of energy and go even one-quarter of the speed of light, it would take them 16 years to get here and another 16 years to get back home. Groombridge 1618, the fiftieth closest star to our sun, is nearly

16 light years away. At one-quarter of the speed of light, a one-way trip from there to Earth would take 64 years.

Our potential dance partners might be hampered by these great distances and the time it takes to traverse them, which might even keep them from reaching us at all. When dealing with travel through interstellar space, science fiction authors have either used the concept of generational ships or resorted to such literary conventions as wormholes or *Star Trek*'s "warp drive" to get around Einstein's speed limit. The problem with the latter two means is that they are, in essence, nothing more than literary conventions. Nobody really knows if such a thing as warp drive is even possible. Generational ships are at least theoretically possible. These are envisioned as huge vessels with long-term life support systems sufficient to last for generations. People aboard these floating colonies when they are launched eventually die off and are replaced by the next generation, and several generations are born and die before the ship finally reaches its distant goal. Thus those aboard a generational ship will have severed their ties with their own species.

The velocity of light in a vacuum also affects radio signals broadcast by neighbors seeking to make contact. A signal broadcast from a star 100 light years away would take a century to reach us, and our response to it would take another century to reach them. Communications over such vast reaches of time would be strained to say the least. This was one of the points of the movie and novel *Contact*. When Ellie Arroway and her colleagues finally break the code of the signal from the aliens, it reveals old newsreel footage of Adolf Hitler giving a speech. The powers that be in the room are shocked until it is pointed out that Hitler's broadcasts were among the first signals humans ever sent out into space, and the aliens were simply recording the first messages they received and sending them back to us.

Assuming there is a way around the light-speed limit, another, darker, question arises: Why haven't we been colonized and either subjugated or even annihilated by space aliens with a vastly superior technology? This frightening possibility is sometimes voiced by celebrity scientists on the lecture circuit to elicit an emotional response from the audience. However, it may well be based on false assumptions. First of all, consider that any civilization capable of sending an invasion fleet across interstellar space would first have to solve its basic energy and material needs. In other words, the space aliens wouldn't really need to plunder our planet. This is one of the great weaknesses in the 2009 movie *Avatar*. If Earth's resources were so badly stripped that they had to plunder the planet Pandora to obtain the unobtainable "unobtanium"—the heavy-handed name for the material Earth needed so badly—they probably wouldn't have been able to get to Pandora in the first place. The same problem plagues *Independence Day* and other movies that explain the motives of the alien invasion as exploiting Earth's resources.

In fact, planets already harboring life might be hazardous to colonize, having their own rich supply of nasty microbes. Much like the scenario of H. G. Wells's *War of the Worlds*, beings from another star system invading Earth might be wiped out by diseases against which they have no immunity. Conversely, they might infect us with some of their microbes, a scenario Michael Crichton explored in his 1969 novel *The Andromeda Strain*. Some have argued that our microbes, which have evolved to attack earthly hosts, wouldn't be able to infect extraterrestrials. This would certainly be the case with the *Plasmodium falciparum* parasite, the causative agent of malaria. It might require not only the *Anopheles* mosquito as a vector but human beings as hosts. The *Anopheles* mosquito might not even bite a space alien. Other diseases communicated by mosquitoes are yellow fever and West Nile virus. Other insect and arachnid vectors include fleas, which carry bubonic plague; ticks, which transmit Lyme disease and Rocky Mountain spotted fever; and lice, which transmit typhus. It is quite possible that visiting or invading extraterrestrials would be safe from all of these. However, at least two genera of our bacteria are extremely opportunistic, infecting a wide range of organisms with a variety of diseases. These are *Pseudomonas* and *Staphylococcus*. At least one species of *Pseudomonas*, *Pseudomonas fluorescens*, can even use hydrocarbons as nutrients. Such a microbe could easily adapt to using the flesh of visiting aliens as food. Both our would-be conquerors and potential dance partners might be leery of landing on what the poet A. E. Housman characterized in his poem "Terence, This Is Stupid Stuff" as "the many-venomed earth."

We should also consider the possibility that nobody really wants to interact with us—at least not yet. Perhaps the space aliens most likely to contact us are 2,000 years beyond us in cultural as well as technological development. We can put ourselves in their place by remembering that 2,000 years ago one of the most advanced societies on Earth was the Roman Empire, a culture that accepted slavery as the norm and whose chief form of public entertainment was to make a spectacle of violent death in the arena. If we didn't absolutely have to, would we really care to associate with such people? Perhaps advanced extraterrestrials would view us in much the same way, finding our problems of endemic racism and our penitentiary systems barbaric.

However, perhaps the extraterrestrials are only 200 years ahead of us. Two hundred years ago, prior to 1824, race-based slavery was practiced in the United States. Native Americans were being driven off their lands. Women didn't have the right to vote. Child labor was common, and most adult laborers worked 12-hour days, six days a week. Today we would regard such a society as criminally backward. Of course, advanced space aliens could land on Earth, take over, and forcibly civilize us for our own good, as did the Overlords in Arthur C. Clarke's 1953 novel *Childhood's End*. However, individual species might be regarded as important in providing new ways of thinking and problem solving. Finding us repugnant in our present state, ethically superior aliens might have a noninterference policy similar to that of the Prime Directive of *Star Trek*. Thus they'll leave us to our own devices and only ask us to dance once our manners have improved.

Finally, let us consider a possibility somewhere between the extremes of our galaxy teeming with thousands of civilizations and us being its sole form of intelligent life. If intelligent life is out there but is quite rare, we might be one of only 30 intelligent species in the galaxy. We might also be the foremost intelligent species. If we are one of 75,000 intelligent forms, it is doubtful that we are the first to have reached high technological status. However, if we are one out of 30, there is a 3.33 percent chance that we are in first place. Another way to look at this is to assume that most species lie in the middle of a probability curve. With 30 species, we have 10 chances out of 30 (or a fraction over 33.3 percent) of being in the middle of the curve, a bit over 33.3 percent chance of being somewhere on the high side of the curve, and bit over 33.3 percent chance of being on the low side of the curve. Maybe we haven't heard from anyone else because we are the first species in our galaxy to broadcast radio waves. Even as we scan for radio signals from elsewhere in the galaxy, perhaps we are the first to walk out onto the dance floor.

WHY IS LIFE UNLIKELY ELSEWHERE IN THE UNIVERSE?

In recent years, two new sets of discoveries have changed the perspective on the possibility of life on other planets and the probability of extraterrestrial life. One is the evidence of more and more planetary bodies in other parts of the universe that may have the right set of properties (not too hot or cold for liquid water, not too extreme in other aspects) for life to evolve there. Current estimates suggest the possibility of millions of such smaller bodies that might fit the requirements for a planet not hostile to life.

The other discovery is that microbial life occurs in far more places than we ever imagined. Very primitive organisms (mostly Archaebacteria, an ancient group of bacteria with the most primitive genetic code of any living thing) are known to live in superheated hot springs on land, in the superheated "black smoker" springs in the midocean ridges deep on the floor of the ocean, and even deep in rocks far below the earth's surface. From the carbonaceous chondritic meteorites formed in the early solar system, we know that amino acids are widespread in the universe, so the possibility that simple life has originated elsewhere is not so implausible now. As a consequence, most scientists would no longer rule out the possibility that very simple bacteria or other primitive microbes occur on other planets.

It's one thing to say that there may be microbes on lots of other planetary bodies. But it's quite another to imagine that they could have evolved into complex life-forms communicating with us electronically, and it is even less likely that they have been secretly invading us with their flying saucers and interacting with humans in weird ways. As the paleontologist Peter Ward and the astronomer Donald Brownlee point out in *Rare Earth: Why Complex Life Is Uncommon in the Universe*, the astounding coincidence of events that caused Earth to be hospitable to the evolution of complex organisms was extremely unlikely—and it is even less likely to have occurred anywhere else in the universe. Here are some factors these authors list as being required for the evolution of complex organisms.

1. **Right kind of galaxy:** Most regions in our galaxy are "galactic dead zones" that could never support complex life because they are too close or too far from the galactic center. If we are too close to the center, metallic planets do not form, and having the right metals (such as iron and nickel) are essential. So a planet with life can't be in the inner galaxy. Also, if it is located in the inner galaxy, a planet is more likely to be struck by objects from space. But the planet must be the right distance from dangerous black holes in the galactic center with its damaging X-rays and gamma radiation. Thus a planet with life can't be too far into the outer galaxy either.

2. **The "Goldilocks zone":** Earth lies in the Goldilocks zone, just far enough from the sun, but not too far, for surface temperatures that are not too hot for liquid water on the surface (like Venus) and not too cold (like Mars), but "just right" (like Goldilocks's porridge). That zone is very narrow, so life could never have arisen on Venus (even hotter than when it started due to its super greenhouse atmosphere, hot enough to melt lead). Mars once had liquid water on its surface, but now it is a deeply frozen dead planet, and there is no evidence so far that any kind of life has ever evolved there.

3. **Solar system dynamics:** Our sun is the right size and had the right composition of elements to give Earth what it needed—most other stars don't have the right composition or size. Most other planets orbiting stars are too big or too small or are poor in the elements necessary for our iron-nickel core, which gives us the shielding effects of our magnetic field. Many are in a perilous part of the universe such as in a globular cluster or near an active gamma-ray source, in an unstable binary or multiple-star system, near a pulsar, or near a star likely to go supernova.

4. **Jupiter's gravity:** Most catastrophic large impacts of asteroids or comets have been avoided because Jupiter diverts them away from Earth.

5. **Moon's gravity:** Thanks to the gravitational effects of our extraordinarily large moon, Earth has a stable rotation with only a slight tilt of its axis, unlike the wildly unstable wobbles and tilting motions of Neptune (currently spinning on its side with respect to the plane of its orbit around the sun). For the same reasons, our planet has a nearly circular orbit (not an extremely elliptical one). This means that Earth's surface temperatures are fairly stable and constant over millions of years and do not fluctuate wildly from freezing to boiling as they do on many other planets.

6. **Core and plate tectonics:** Earth was also the right size and composition to have a liquid iron-nickel core and a mantle that produces plate tectonics, which constantly remodels and shapes Earth's surface and helps maintain the range of temperatures and the balance of atmospheric gases in a fairly stable, narrow range.

7. **Magnetic field:** The liquid iron-nickel core of Earth generates our magnetic field. The field, in turn, deflects most of the bombardment of ionized plasma from the sun known as the "solar wind," diverting most of its effects away from the planet. If we did not have that protection, Earth's surface would be inhospitable to life.

8. **Oxygen:** Even more important, life itself has made Earth supremely habitable. It's possible to imagine many other planets with simple microbes, but not so easy to imagine that they evolved the crucial system of photosynthesis that earthly blue-green bacteria (cyanobacteria) first developed 3.5 billion years ago. Without photosynthesis, there is no free oxygen. (Lots of other planets have oxygen, but it is all bound up into other molecules and is not in the form of free oxygen, or O_2). In fact, most solar system planets have an atmosphere of components such as nitrogen and carbon dioxide (the major gases in the atmospheres of Mars and Venus) but no free oxygen. As you will see in chapter 6, for 80 percent of life's history on Earth, there was not enough free oxygen for even the simplest animals to evolve. It wasn't until about 2.4 billion years ago that oxygen first began to accumulate in our oceans and atmosphere, and it was not until the Late Cambrian or Ordovician (about 500 million years ago) that free oxygen probably reached the current level of 21 percent. Without free oxygen, the only life possible is simple microbes, and possibly simple multicellular plants—but no animals of any kind, let alone space aliens that look like us. As far as we can tell from spectral analysis of other similar-sized planetary bodies elsewhere in the universe, *not one of them has abundant free oxygen in the atmosphere*—and that rules out the idea that they have animal life as we know it in any form. In 2004, the Hubble Space Telescope found a planet called Osiris (about 150 light years from Earth) with a small amount of oxygen and carbon in its atmosphere, but it was so hot that the atmosphere was boiling off and constantly being lost to space. Recent research describes other pathways for forming free oxygen on planets, but so far there has been no physical evidence of another planet with an atmosphere rich in free oxygen.

9. **Ozone:** Without a layer of free oxygen in our atmosphere, there would be no ozone either. The ozone layer is formed in our stratosphere when free oxygen molecules (O_2) are split into O-oxygen radicals, each of which then latches onto another O_2 molecule to form ozone (O_3). Without ozone, the harmful UV radiation from the sun would cause severe sunburns and skin cancer that would destroy any land life not protected by being in the water.

10. **Carbon cycle:** Our planet walks a narrow tightrope to keep enough carbon dioxide in the atmosphere to warm the planet but not too much (or a runaway greenhouse will kill everything) or too little (and the planet will freeze over). Chapter 8 describes several "Snowball Earth" events from 2 billion years ago to 600 million years ago that affected life on planet Earth. Excess

carbon is trapped in crustal rocks such as coal (made of the tissues of extinct plants turned to rock) and limestone (made of the calcium carbonate shells of sea creatures). Several times in the past billion years the excess carbon dioxide in the atmosphere was transferred into the earth's crust by huge volumes of coal or limestone trapping the carbon in rocks. Indeed, one of the main reasons for our current climatic crisis of global warming is because we have been burning the coal and oil that once trapped all that carbon millions of years ago and returning it to the atmosphere as carbon dioxide. Without advanced life-forms such as land plants or shell-building animals, a planet has no mechanism for regulating the carbon dioxide in its atmosphere and can swing wildly from one extreme too warm for life (like Venus) to a frozen planet too cold for life (like Mars). Life itself is a global thermostat, as well as the source of all of our oxygen.

11. **Contingency:** Throughout most of human history we have taken the anthropocentric view that humans are the most important beings in the universe. More recently, these views have evolved so that we now think of ourselves as the inevitable by-product of billions of years of the evolution of life. But as the late Stephen Jay Gould pointed out in his 1989 book *Wonderful Life: The Burgess Shale and the Nature of History*, this view is completely false. Gould argued that chance, accident, and contingency are important factors in evolution, and that evolution is not predestined to follow a certain path that—as the anthropocentric bias dictates—eventually leads to us. The history of life could have been played out in a near infinite number of alternate scenarios if one or two events early in the sequence had been different. Gould made this contingency argument most strongly in his book, which discussed the amazing soft-bodied Cambrian fauna from the Burgess Shale of Canada. With so many bizarre experimental body forms (none of which survive today), he argued that there was no way of predicting what the future shape of life would be like by examining this early experimental phase. (The title of the book refers to the famous Frank Capra movie *It's a Wonderful Life*, in which Jimmy Stewart gets to view an alternate future predicated on a very slight difference in initial conditions—in this case, that his character had never been born.) Our ancestors include the Burgess Shale fossil known as *Pikaia*, which was just a tiny wormlike lancelet relative back then and a very minor part of the fauna. A Cambrian biologist would have no way of guessing that eventually the vertebrate body plan would be a great success and most of the other types would die out. If the video of life had been rewound

and replayed, perhaps some other group would have dominated and we would not be sitting here speculating about it! Or, as I discuss in chapter 18, if the Cretaceous extinctions had not happened, dinosaurs would still be ruling Earth 66 million years later, and we would still be tiny shrew-like mammals hiding in the underbrush from our reptilian overlords (as mammals did for the first 130 million years when they overlapped in time with dinosaurs). Human evolution would never have happened, let alone humans developing intelligence and science that could communicate with other planets. Humanity is just a lucky accident, and if we reran the video-tape of life's history again from the start, there is no possibility that it would repeat the events of the first run.

All of these factors came together on Earth, and there is good reason to believe that Earth is indeed an unusual and exceptional planet. The odds are extremely slim that any of the many new planetary bodies being dis-covered every year will have all of these right factors in the right range. No other planet we have found so far has the right set of conditions for complex multicellular life (let alone aliens like you see on TV). Simple microbial life is much more likely, but it's not like ET ready to phone home. Despite the optimistic ideas of science fiction authors, we are almost certainly alone in the universe.

NOTE

1. SETI Institute, "Is There an 'Eerie Silence?,'" FAQ, accessed May 30, 2023, www.seti.org/faq#obs19.

FOR FURTHER READING

Comins, Neil F. *What If the Moon Didn't Exist? Voyages to Earths That Might Have Been.* New York: HarperCollins, 1993.

Gribbin, John. *Alone in the Universe: Why Our Planet Is Unique.* New York: Wiley, 2011.

Lane, Nick. "Life: Is It Inevitable or Just a Fluke?" *New Scientist* 214, no. 2870 (2012): 32–37.

Lissauer, Jack J. "How Common Are Habitable Planets?" *Nature* 402, no. 6761 Suppl. (1999): C11–14.

Prantzos, Nikos. "On the Galactic Habitable Zone." *Space Science Reviews* 135 (2008): 313–322.

Simpson, George Gaylord. "The Nonprevalence of Humanoids." *Science* 143, no. 3608 (1964): 768–775.

Taylor, Stuart Ross. *Destiny or Chance: Our Solar System and Its Place in the Cosmos.* Cambridge: Cambridge University Press, 1998.

Ward, Peter D., and Donald Brownlee. *Rare Earth: Why Complex Life Is Uncommon in the Universe.* New York: Copernicus Books (Springer Verlag), 2000.

Webb, Stephen. *If the Universe Is Teeming with Aliens, Where Is Everybody? Fifty Solutions to the Fermi Paradox and the Problem of Extraterrestrial Life.* New York: Copernicus Books (Springer Verlag), 2002.

Waltham, David. *Lucky Planet.* New York: Basic Books, 2013.

IN THE BEGINNING

We are stardust, we are golden
We are billion year old carbon
And we got to get ourselves back to the garden
—JONI MITCHELL, "WOODSTOCK"

ORIGINS

Our view of the universe and the solar system has changed dramatically in the last 500 years. Before 1543, almost all humans thought the earth was flat and was the center of the universe and that the stars were tiny points of light on the dome of the heavens. In 1543, Copernicus showed that the sun, not Earth, was at the center of our world, and that Earth was a planet in orbit around the sun. In 1611, Galileo used one of the newly invented devices called telescopes to discover that the stars were beyond counting and were not scattered on a big dome over our head (figure 2.1). He also confirmed that Jupiter has its own moons that could move completely around it, showing that the planet was not surrounded by a perfect celestial sphere. He debunked the notion that the planetary bodies were perfect and unsullied when his telescope revealed that Earth's moon is covered in craters and is not a perfect celestial sphere. Finally, by discovering that Venus has phases like the moon ("full Venus," "half Venus," etc.), he showed that Venus was moving around the sun in an orbit inside our own orbit. Most important, he confirmed Copernicus's idea that Earth was just another planet orbiting the sun. By the early 1700s, Isaac

Figure 2.1 ▲

A famous 1888 woodcut by Camille Flammarion showing the old notion that Earth is flat, the stars are on a dome in the heavens, and the sun, moon, and planets are on great wheels outside the dome. (Courtesy of Wikimedia Commons)

Newton had worked out the laws of motion and gravitation, showing how the entire system could be explained by basic physics.

Today we look at the amazing images from space coming from both land-based telescopes and the Hubble and Webb space telescopes and see what no one could have possibly imagined even 30 years ago. We can see the different stages of how other stars are born and die and how other planets and solar systems have formed. These images, and the astrophysical calculations and models that explain them, give us a new view of the origin of the solar system and enable us to explain much of what was simply guesswork before this century.

As scientists, we must not only describe what we know about the world but also ask the question: *How* do we know this? In a book like this, it is not enough to just list scientific facts about a topic; I want the reader to *understand how things work*, and especially *how we know* about these events that took place in the past. So a fair question is: How do we know about the formation of the universe during the "Big Bang," which took place more than 13 billion years ago, and about the formation of the solar system, which took place more than 4.5 billion years ago?

By examining multiple lines of evidence, we can answer these seemingly impossible questions. This list of discoveries informs our search for the answers we seek.

1. Astronomical observations have shown that almost all of the bodies in space—stars, quasars, galaxies, and so on—are rapidly moving away from us, with the most distant objects moving fastest. This observation led to the realization that the universe is expanding and that this expansion must have resulted from a single event in the past, now nicknamed the "Big Bang."

2. The accidental discovery of the previously predicted cosmic background radiation confirmed that the Big Bang must have occurred.

3. The motions of all the planets in the solar system are in the same plane, moving in the same direction. This suggests that they all started out together as part of large disk of matter that was moving around the sun.

4. Another line of evidence comes from examining the rest of the universe and finding other solar systems that are in earlier stages of formation. The Hubble and Webb space telescopes and extremely high-resolution telescopic arrays now on land—such as the Atacama Large Millimeter Array in Chile showing solar systems such as HL Tauri (figure 2.2) and Beta Pictoris—provide images of what our early solar system might have looked like. In 2018, we obtained the first good image of a new planet forming.

5. Numerous space probes have flown by all the planets in our solar system, and some have even landed on Mars and Venus. They have given us a clear idea of the composition of the planets and what lies in the space between the planets. We need to explain the chemical composition of the solar system to understand why the planets are different.

6. We have actual samples of rocks from space. These include meteorites that came from the earliest stages of solar system formation as well as those from planets that formed and then broke up or were blasted off the surface of Mars and reached Earth.

7. Last but not least, we have actual pieces of rock brought back from the moon during the Apollo missions from 1969 to 1976, and our Mars rovers have been collecting samples and analyzing the composition of the surface of Mars in detail.

Putting all of this together, a remarkably consistent and well-supported model of the origin of the solar system emerges. Throughout the following description of how the solar system was formed, evidence is provided explaining *why* scientists accept certain aspects of the model.

Figure 2.2 ▲

A forming solar nebula in HL Tauri, part of the constellation Taurus. The protoplanetary disk is condensing around the center, and the circular tracks are already being formed by small planetary objects sweeping up all the dust in their tracks by the pull of gravity. This image was taken by the Atacama Large Millimeter Array, a single telescope with 66 high-precision antennae that is located in Chile. (Courtesy of Wikimedia Commons)

THE BIG BANG AND THE ORIGIN OF THE UNIVERSE

Where did we come from? When and where did it all begin? These questions have fascinated and troubled people for thousands of years. For millennia, the explanations came from a wide variety of religious myths and stories from every culture on Earth. Early in the twentieth century it became possible to go beyond myth and speculation, and we now use the methods of science to understand what really happened.

The first breakthrough came from a number of women astronomers working at Harvard College Observatory under Edward C. Pickering. They were known as the "Harvard Computers" (figure 2.3) because they were talented mathematicians who could make fast calculations in their heads and on paper and do measurements by hand. (Only much later did the word "computer" come to mean the electronic devices we all use.) Pickering

Figure 2.3 ▲

Harvard Computers at work, circa 1890, including Henrietta Swan Leavitt seated, third from left, with magnifying glass (1868–1921), Annie Jump Cannon (1863–1941), Williamina Fleming standing at center (1857–1911), and Antonia Maury (1866–1952). (Courtesy of Wikimedia Commons)

hired them because they were not only good at math but also careful and meticulous; the women studied and analyzed thousands of photographic glass plates of the night sky shot by different telescopes. They were also paid less than male assistants (25 cents an hour, less than a secretary) and worked hard without complaining six days a week.

During this time, most women were barred from scientific careers completely, and those who tried to get advanced educations or do research in science met huge barriers every step of the way. However, the women's talents were soon recognized, and they each made discoveries that revolutionized astronomy and outshone most male astronomers of their time. The most famous was Annie Jump Cannon, who cataloged the stars of the night sky and proposed our modern system of star classification—from red giants to white dwarfs. She built her system using the first complete star classification system by Antonia Maury.

For this story, however, the key woman was Henrietta Swan Leavitt. She was assigned to study classes of stars known as "variable stars" because their brightness fluctuated from one night to the next. She soon realized that their brightness variations had a regular period of fluctuation, with the brightest stars (most luminous stars) having the longest periods of brightness variation. She found many variable stars in a cluster in the constellation Cepheus (known as "Cepheids") that were all the same distance away from Earth. This allowed her to calibrate the brightness spectrum. In 1913, after studying some 1,777 variable stars, she worked out the relationship between the period of brightness fluctuation and the luminosity and showed that we could determine how far away a star was from us by measuring its luminosity and its period of fluctuation. Thanks to Leavitt, astronomers now had a reliable tool to tell how far away a star or galaxy was from Earth.

The next step was made by a legendary astronomer, Edwin Hubble. In 1919, he was assigned to work at the newly completed Mt. Wilson Observatory (figure 2.4) in the mountains above Pasadena, California. He had free use of what was then the world's most powerful telescope,

Figure 2.4 ▶

Mount Wilson observatory. (A) Old aerial image of the largest of the three telescope domes on Mt. Wilson. (B) The original 100-inch Hooker reflecting telescope in the main dome at Mt. Wilson. (C) Edwin Hubble at the sighting device on the main telescope at Mt. Wilson. (Courtesy of Wikimedia Commons)

THE ONE HUNDRED INCH HOOKER TELESCOPE.

[Fig. 1.]

Figure 2.4 ▲
(*continued*)

a reflecting telescope with a 100-inch mirror. His first major discovery in 1924 used Leavitt's Cepheid variable stars to show that the spiral nebulae were in fact galaxies outside our own Milky Way Galaxy, and that the Milky Way was just one of many galaxies. This expanded our understanding of the size of the universe far beyond what people once thought was possible.

Hubble then used the telescope to systematically study as many stars and galaxies and other large celestial objects as he could. He not only measured their distance using the Cepheid variable method but also analyzed the spectrum of light from the star. Like taking a prism and splitting sunlight into its major colors, the light from the stars can be split into a spectrum of colors as well (figure 2.5A). The spectrum, however, has dark "bands" across the color scale, caused by the absorption of certain elements. We can find these same bands when we analyze the spectrum of burning sodium or other metals in the lab, so each set of bands tells us what elements we are seeing.

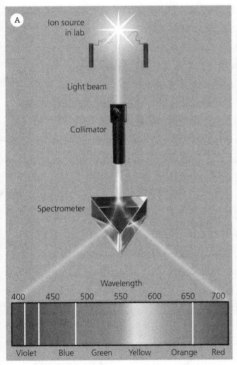
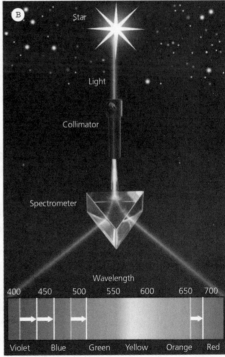

Figure 2.5 ▲

(A) A prism breaks the rays of white light into their constituent colors, and black absorption lines occur at wavelengths characteristic of certain elements. (B) The spectrograph on a telescope breaks starlight into their own spectra, and the black lines are shifted to the red end of the spectrum. This is a Doppler shift, which indicates that all the stars and galaxies are receding from us. (Courtesy of Wikimedia Commons)

Hubble's major collaborator in this effort was Milton Humason, who had no education past age 14 but was eager to prove himself. Humason originally drove the mules that hauled the telescope and other materials up that steep mountain. He then became a janitor during the night shift when the astronomers were at work, so Hubble got to know him. He found that Humason had unexpected talents and promoted him to be his assistant. Hubble admired Humason's quiet determination to study the difficult photographs and make careful measurements of the spectrum on thousands of photographic plates from the telescope. After measuring dozens of stars and galaxies, Hubble and Humason noticed something very peculiar. The nearest stars had absorption lines in their spectra that resembled the same spectrum for that element on Earth. The further away the star or galaxy was, the more the dark absorption bands were shifted from their original positions toward the red direction of each spectrum (figure 2.5B).

Why do the absorption lines move toward the red end of the spectrum? This discovery had first been reported and explained for a few galaxies in 1912 by Vesto M. Slipher at Lowell Observatory in Flagstaff, Arizona. It is what is known as a Doppler shift, and it is caused by the Doppler effect. You have experienced the Doppler effect for sound many times. If you are standing on the street and a car or train rushes toward you blaring its horn, you will notice that the pitch of the sound gets slightly higher as it approaches. After the vehicle passes and rushes away from you, you will hear the sound of the horn drop in pitch again. The Doppler effect is caused by the fact that the sound waves approaching you are bunched up because their source is getting closer and closer. If the waves get bunched up, they go higher in pitch. Similarly, when the sound source travels away from you, the waves are stretched out because their source is retreating. Longer, more stretched out waves are lower in pitch than when they are not stretched out. The Doppler shift also applies to light waves (figure 2.6). If the source is moving very rapidly toward you, the light waves will get bunched up and have a shorter wavelength (which corresponds to the blue and violet end of the light spectrum). If the light source is rapidly moving away from you, its waves will be stretched to longer wavelengths, which correspond to the red end of the spectrum.

Slipher's first observations in 1912, and then Hubble and Humason's careful catalog of more than 46 galaxies and many stars in the 1920s, found that almost all of them showed the red shift; there were no blue-shifted

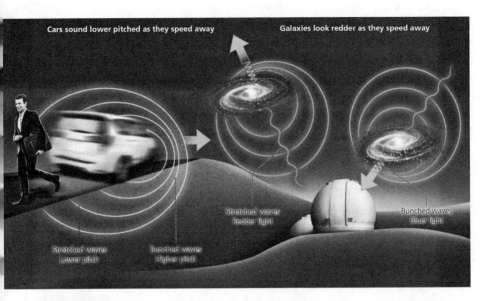

Cars sound lower pitched as they speed away

Galaxies look redder as they speed away

Stretched waves
Redder light

Bunched waves
Bluer light

Stretched waves
Lower pitch

Bunched waves
Higher pitch

Figure 2.6 ▲

The Doppler effect occurs when an object emitting waves (sound or light) is moving with respect to the observer. If it is moving away, its waves are stretched out by the moving source and the wavelengths get longer. If it is moving toward you, the waves are bunched up and the wavelengths are shorter. (Courtesy of Wikimedia Commons)

objects that might be moving toward us. More important, Hubble and Humason found that the furthest objects had the greatest red shifts, so they must be moving away from us the fastest. Hubble realized that this meant the universe must be expanding. It's analogous to making a loaf of raisin bread. When you start with the ball of dough, the raisins are all close together. But as the ball of dough expands, each raisin moves apart from every other raisin, and those raisins on the outer part of the ball of dough move the fastest.

The universe is expanding. This is a staggering thought, and at first most astronomers were not able to accept it. However, Hubble and Humason's data were reliable, and as time went on and more and more objects were analyzed, they all turned out to be red-shifted. In 1927, the Belgian astronomer Georges Lemaître postulated a model in which the universe expanded from a single point in the far distant past. Most astronomers did not like the idea that the universe had a beginning. They thought the universe was in a "steady state" of expansion, with new matter being created at the center all the time. One of these steady-state advocates, Fred Hoyle, coined the term "Big Bang" to mock Lemaître's model, and that name has stuck ever since.

The controversy of Big Bang versus steady state continued for about 30 years without any clear consensus. In the late 1950s, a crucial discovery was made purely by accident and not by astronomers but by two engineers: Arno Penzias and Robert W. Wilson. In 1964, they were employed by Bell Labs (the original research division of AT&T/ Bell Telephone) and were responsible for improving the technology of communication for "Ma Bell." They were working on improving the first antennas for receiving and transmitting signals by microwaves, primarily to enable communication with NASA's "Project Echo" (the first attempt to use satellites for global communication) and later with the Telstar satellite. As the chief electrical engineers on the project, their main job was to get the "bugs" out of the device and improve its efficiency. They found and eliminated many of the sources of noise from the antenna, but then they found a source of background hum that was 100 times stronger than they expected. It was detected day and night and evenly spread across the sky (so it was not coming from a single point source on Earth or in space). It was clearly coming from outside our own galaxy, and they could not explain it.

Luckily, just a year earlier and 37 miles away in Princeton, New Jersey, the physicists Robert Dicke, Jim Peebles, and David Wilkinson had predicted the existence of background "noise" left over from the Big Bang— when everything had exploded with a big blast of radiation. The Princeton scientists were just beginning experiments to detect this noise when a friend told Penzias that he'd seen a preprint of a paper by the Princeton group predicting the exact same background noise. The two groups got in touch, and Penzias and Wilson showed them what they had found. Lo and behold, the two Bell Lab engineers had accidentally discovered proof that the Big Bang had actually happened. For this discovery, Penzias and Wilson eventually received the 1978 Nobel Prize for Physics. And it was discovered entirely by accident!

Since this discovery, the Big Bang model has undergone many modifications as physicists use the properties of matter and the equations of physics to figure out how it all happened. The most recent methods date the Big Bang at about 13.8 billion years ago. At the very beginning, the universe was in a "singularity"—an infinitely small, high-energy region with an infinite density. Ten milliseconds after singularity, the universe was filled with high-energy particles at temperatures over 1 trillion degrees Kelvin and was expanding rapidly in all directions. It was so hot that it was only radiation, without matter. Space and time did not yet have meaning, but they were infinitely warped around this super dense region. Over the next billion years, the universe cooled enough to form subatomic particles and later matter in the form of atoms. Over the next 12 billion years the expansion continued, and random clumps of matter began to coalesce to form stars and galaxies and quasars. Some of these stars have already burned out and exploded, producing the heavier elements such as oxygen, silicon, carbon, and iron that make up most of the matter in the solar system. In that sense, we are all stardust.

THE SOLAR NEBULA HYPOTHESIS

The observation that all of the planets are moving in a flat plane in the same direction around the sun emerged as soon as Isaac Newton's laws helped explain planetary motions. In fact, the very idea of a solar system wasn't possible until this discovery was made. In addition, astronomers using the latest advances in telescopes had discovered large fuzzy blobs of gases out in space that they called "nebulae" (Latin for "clouds"). By the mid-1700s, the Swedish scientist Emanuel Swedenborg, the French mathematician Rene Descartes, and the German philosopher Immanuel Kant had all independently suggested that our solar system began as a cloud of dust, or nebula, that was spinning around a central axis. This became known as the solar nebula hypothesis. The detailed physics of how this system worked was deciphered in the late 1700s and early 1800s by the great French mathematician and astronomer Pierre-Simon Laplace.

According to the solar nebula hypothesis, the solar system started out as a ball of gases and matter that was about 90 percent hydrogen, 9 percent helium, and less than 1 percent all other elements (figure 2.7). This nebula was slowly spinning on its axis. Indeed, astronomers have found that most of the nebulae in space are spinning at various speeds. Eventually this loose mass of cosmic dust began to condense, possibly due to the shock of a supernova from a nearby star. As the early solar nebula began to condense, it increased the gravitational attraction of the matter toward its massive center, and it began to spin faster due to the law of conservation of angular momentum.

Figure 2.7 ▲

Artist's conception of the early solar nebula when it was just a disk of matter surrounding the condensing sun. (Courtesy of the National Aeronautics and Space Administration [NASA])

You are familiar with the law of conservation of angular momentum even if you may not recognize the name. When you watch a figure skater go into a spin, the skater spins slowly if their arms and legs are far out from their body, but the skater begins to spin faster as their arms and legs (and thus their mass) are pulled in toward their body. You can experience this by sitting on a stool, pushing off until you are spinning, and holding some small weights in each hand. As you pull your arms and the weights in toward your spin axis, you will spin faster; if you extend your arms holding the weights further out from your body, you will slow down.

The spinning solar nebula not only goes faster as it condenses but also begins to form a flattened disk. This is comparable to a glassblower spinning a blob of molten glass around and around on its axis until it flattens out. The simplest way for the mass of the nebula to resist spinning angular momentum is to spread out into a flat disk. As the disk continues to spin and condense and cool down, the small grains of cosmic dust collide with each other, clumped together by gravitational attraction, and get bigger and bigger. Eventually their mass is large enough that they have a significant gravitational attraction, which pulls even more tiny bits of matter toward them and enlarges them further, yet again increasing their gravitational pull. Once these growing clumps of matter reach a diameter of about 1 kilometer, they have enough gravity to behave like "gravitational vacuum cleaners," and they pull in all the cosmic dust in their path as they revolve around the sun.

These small bodies were first named in 1905 by the geologist Thomas C. Chamberlin and the astronomer Forest R. Moulton, who called them planetesimals (a mashup of the words "infinitesimal" and "planet") to describe a body that is a tiny fraction of a planet. As these planetesimals grow larger and larger, their gravitational attraction increases and they pull in more and more loose cosmic debris until they form a clear circular path or trackway within the solar disk. This is particularly true of the gas giant planets Jupiter and Saturn, which created huge cleared-out tracks as they orbited the central star. The same can be seen in young solar systems such as HL Tauri, part of the constellation Taurus, which have their own protoplanets creating tracks in their disks (figure 2.2).

Meanwhile the energy of the emerging sun in the center of the solar system (99.85 percent of the total mass of the solar system) changes everything. At first, the gravity of the protosun pulls in most of the matter in the original cosmic disk. As the solar nebula cools, it developed a temperature gradient from the center to the edge of the disk that begins to redistribute and shape the growing solar system. Closer to the protosun, the temperatures are above 2000°C and everything is vaporized. About 5 million miles from the protosun, the temperatures cool enough for rocky bodies to solidify. This is known as the "rock line," an area in which the smaller, inner, rocky planets (Mercury, Venus, Earth, and Mars) could eventually coalesce. Even further out is the "frost line," with temperatures of –375°C or lower. This is cold enough not only to freeze liquid water into ice but also to freeze carbon dioxide, methane (CH_4), and ammonia (NH_3). This distinctive composition is the most notable feature of the outer planets—Jupiter, Saturn, Neptune, and Uranus—which are giant frozen gas balls with very little rocky material.

In 2018, using the Very Large Telescope (VLT) in the European Southern Observatory in northern Chile, scientists recovered the first good image of a new planet in the process of formation. It was found in a star named PDS 70, more than 370 light years away. PDS b, as the planet is known, is still condensing in the inner part of the planetary disk at a distance of about 22 times the distance between our sun and planet Earth. The image clearly shows it cutting a circular track around the star in the center, just as would have happened during the early history of our solar system. Planet PDS b is a gas giant even larger than Jupiter, so it is not going to turn into a rocky earth-like planet like ours. In addition, its surface temperature is still 1200°K (1700°F), so it is a long way from cooling down to become a frozen gas giant like Jupiter.

After 3 million years, the planetesimals clumped together into bigger and bigger bodies, and when their diameters were about 100 kilometers or larger they became protoplanets. At this point, their gravity and internal heat were enough to mold them into a roughly spherical shape. Much of the loose cosmic dust of the early solar nebula was pulled in by the gravitational attraction of the growing protoplanets.

About 50 million years after the solar nebula first formed, a critical threshold was crossed. The protosun had accumulated enough heat and energy to collapse by gravity and trigger nuclear fusion of its hydrogen into helium. This is the same reaction that occurs in the hydrogen bomb and powers fusion reactors. This reaction enabled the sun to become a fully fledged star that can burn for more than 10 billion years, as it continues to power the solar system. The energy of this huge fusion reaction first came out as large bursts of solar wind, a huge flux of charged particles (known as a plasma) that poured continuously out from the sun. This first intense burst of solar wind blew away much of the remaining cosmic dust from the inner solar system, almost completely clearing out interplanetary space and preventing the inner planets from gaining much more mass.

THE EARTH DEVELOPS LAYERS

One of these protoplanets eventually became our Earth. It is about 149 million kilometers (93 million miles) away from the protosun, a distance sometimes referred to as the Goldilocks zone (see chapter 1). At this distance, it is neither "too hot" (like Venus, where the atmosphere is hot enough to melt lead) or "too cold" (like Mars, where solar heating is so weak that the planet is frozen solid).

At the temperatures of the Goldilocks zone, the swirling clumps of matter and planetesimals tend to concentrate, with large amounts of the common solid elements mixed with lighter gases, such as oxygen, nitrogen, carbon, and helium—and especially hydrogen, the element that makes up most of the sun and much of the rest of the solar system. When these elements first got together, they formed a well-mixed mass of uniform composition throughout the entire proto-Earth. Some of these elements (iron and nickel) would sink to the center of the protoplanet because of their greater density, whereas the lightest gases, especially the hugely abundant hydrogen and helium, largely escaped out into space because the gravitational pull of Earth was not strong enough to capture them. We use these gasses to float balloons today because they are less dense than air and our gravity cannot hold them back.

Gravitational settling alone is not sufficient to explain Earth's differentiation into discrete layers, with an iron-nickel core surrounded by a silicate-rich rocky mantle. To differentiate Earth into layers of completely different composition (the core and mantle), you need enough heat to melt the entire planet. This would allow almost all of the dense iron and nickel to sink to the center of gravitation (nicknamed the "iron catastrophe"), leaving the less dense silicates floating on top.

Where did all this heat come from? There are several possible sources.

1. Early proto-Earth was full of unstable elements that were undergoing radioactive decay. During radioactive decay, a tremendous amount of heat is released, and 50 percent of the decay and heat is released in the first half-life. All of the important radioactive elements on Earth today—uranium-238, uranium-235, rubidium-87, and potassium-40—were in their first half-life and were much more abundant as Earth began to form, so their heat production was at its maximum. However, meteorites, such as the Allende carbonaceous chondrite, also came from the earliest solar system just as Earth was forming. They suggest that another element might have provided most of the heat production. These meteorites had unusual quantities of the isotope magnesium-26, which is the daughter product of the radioactive decay of aluminum-26. Aluminum-26 decays very rapidly, with a half-life of only 700,000 years, so none of Earth's original aluminum-26 survives. However, it was apparently hugely abundant in the early condensing proto-Earth (judging from how much magnesium-26 was left behind), so it would have been the most important element of all in melting Earth.

2. In the early condensation of proto-Earth, a lot of cosmic debris was left over, and Earth was under constant bombardment from meteorites. A meteorite's impact represents a huge amount of kinetic energy (energy of motion) that converts to heat energy when it slams into Earth. Dating of the meteorite impact craters on the moon suggests that the early solar system was still under intense bombardment from space debris, which did not slow down until about 3.9 billion years ago. Among these impacts was the one that blasted off a chunk of the mantle to form the moon (see chapter 3). Earth would have had additional heat contributed from all the impacts of rocks from space.

3. As the blobs of iron and nickel sank to Earth's core, they released a lot of potential energy (comparable to the difference in energy between the rock at the top of a cliff and at the bottom). Like any other form of energy, potential energy cannot be destroyed; it had to be converted to another form of energy, namely, heat.

4. The interior of Earth is under intense gravitational forces (gravitational compression), which increase not only the pressure but also the temperature. This is enough to melt many of the materials that eventually became the mantle. It is also why the outer iron-nickel core of Earth is fluid.

5. Finally, as Earth's densest materials began to sink to the center, they changed the angular momentum of Earth. If Earth was as small as a figure skater, this would cause it to spin faster. However, Earth is too massive to respond to this small change in angular momentum. But the change of energy must go somewhere—so it is converted to heat.

By the end of this process some 4.5 billion years ago, Earth had discrete layers of an iron-nickel core and a magnesium-silicate-rich mantle. It was still too hot to allow a crust to cool on the outside, so Earth had only two primary layers.

FOR FURTHER READING

Bartusiak, Marcia. *The Day We Found the Universe*. New York: Pantheon, 2009.

Bembenek, Scott. *The Cosmic Machine: The Science That Runs Our Universe and the Story Behind It*. New York: Zoari Press, 2017.

Brockman, John, ed. *The Universe: Leading Scientists Explore the Origin, Mysteries, and Future of the Cosmos*. New York: Harper, 2014.

Canup, Robin M., and Kevin Righter, eds. *Origin of the Earth and Moon*. Tucson, AZ: University of Arizona Press, 2000.

Carroll, Sean. *The Big Picture: On the Origins of Life, Meaning, and the Universe Itself*. New York: Dutton, 2016.

Editors of Chartwell Books. *An Illustrated Guide to the Cosmos and All We Know About It*. London: Chartwell Books, 2017.

Hartmann, William K., and Ron Miller. *The Grand Tour: A Traveler's Guide to the Solar System*. New York: Workman, 2005.

——. *The History of Earth: An Illustrated Chronicle of an Evolving Planet*. New York: Workman, 1991.

Hazen, Robert. *The Story of Earth: The First 4.5 Billion Years from Stardust to Living Planet*. New York: Penguin, 2013.

Hawking, Stephen. *A Brief History of Time*. New York: Bantam, 1998.

Krauss, Lawrence. *The Greatest Story Ever Told—So Far: Why Are We Here?* New York: Atria, 2017.

McKenzie, Dana. *The Big Splat: Or How Our Moon Came to Be*. New York: Wiley, 2003.

Natarajan, Priyamvada. *Mapping the Heavens: The Radical Scientific Ideas That Reveal the Cosmos*. New Haven, CT: Yale University Press, 2016.

Perlov, Delia, and Alex Velenkin. *Cosmology for the Curious*. Berlin: Springer, 2017.

Ryden, Barbara. *Introduction to Cosmology*. Cambridge: Cambridge University Press, 2016.

Sagan, Carl. *Cosmos*. New York: Ballantine, 2013.

Saraceno, Pablo. *Beyond the Stars: Our Origins and the Search for Life in the Universe*. New York: World Scientific, 2012.

Silk, Joseph. *The Big Bang*. 3rd ed. New York: W. H. Freeman, 2001.

Singh, Simon. *Big Bang: The Origin of the Universe*. New York: Harper, 2005.

Sobel, Dava. *The Glass Universe: How the Ladies of the Harvard Observatory Took the Measure of the Stars*. New York: Viking, 2016.

Tyson, Neil de Grasse, and David Goldsmith. *Origins: Fourteen Billion Years of Cosmic Evolution*. New York: Norton, 2004.

Valley, John W., William H. Peck, Elizabeth M. King, and Simon A. Wilde. "A Cool Early Earth." *Geology* 30 (2002): 351–354.

Valley, John W., Aaron Cavosie, Takayuki Ushikubo, David A. Reinhard, Daniel F. Lawrence, David J. Larson, Peter H. Clifton, et al. "Hadean Age for a Post-Magma-Ocean Zircon Confirmed by Atom-Probe Tomography." *Nature Geoscience* 7 (2014): 219–223.

MOONSTRUCK

When the moon hits your eye like a big pizza pie, that's amore!

—LYRIC TO "THAT'S AMORE" BY HARRY WARREN AND JACK BROOKS

FLY ME TO THE MOON

Like many Americans older than 60, I was glued to the TV on July 20, 1969. I was visiting my cousins' ranch outside Hot Springs, South Dakota, and getting a few months' experience in ranch life: rising at dawn to collect eggs and feed the chickens, riding horses and tractors, taking care of the daily chores, and getting to know my extended family. We'd been hearing the buildup to the *Apollo 11* mission all week, but now we were about to witness something extraordinary: the first man was about to walk on the moon, and even more amazing for that time, the world would be able to watch it live on television! We all clustered around the TV in their small living room that afternoon as the networks began to broadcast preparations for the first moon walk. The magic moment finally arrived, and millions of people all over the world simultaneously saw one of the most stirring achievements in human history.

The race to the moon was launched by President John F. Kennedy in 1961, challenging the United States, and especially its space program, to land a man on the moon before the end of the decade. We had been trailing the Soviet Union in the space race ever since they first launched the Sputnik satellite in 1957, long before we could do so. The Soviets also launched the first animals, and then launched the first man into space, Yuri Gagarin, in 1961 (one month before the United States launched Alan Shepard into space). Between 1959 and 1963, the Mercury program launched the first Americans into space, and we were all riveted when John Glenn was the first American to orbit Earth in 1962. From 1965 to 1966, we moved on to the Gemini program, with two astronauts doing even more daring missions, including space walks and docking between spacecraft. From 1968 to 1972, the Apollo program with its three astronauts were building the expertise to land on the moon and then return, each mission flying longer times and farther distances around the moon than before.

Finally, on that fateful day in 1969, *Apollo 11* reached the moon. The third astronaut, Michael Collins, remained in orbit around the moon, and Neil Armstrong and Buzz Aldrin flew the lunar lander to the moon's surface and did a short moon walk (figure 3.1) before blasting off and returning to the command module *Columbia* for their voyage back to Earth. In each successive Apollo mission, *Apollo 12* through *Apollo 17* (except for the ill-fated *Apollo 13*, which had an explosion in space and barely returned its astronauts alive), the astronauts took longer and longer moon walks and brought back more and more samples. By the time Congress canceled the Apollo program in 1973, the six moon missions had landed 12 men on the moon, collected a huge amount of data about the moon, and returned with 381.7 kilograms (842 pounds) of lunar samples. The only scientist ever to walk on the moon was a geologist, Harrison Schmitt, who was on the last mission, *Apollo 17*, which spent several days on the moon in December 1972.

The space program generated a huge volume of research that produced enormous technological breakthroughs, not only in space but in all sorts of products and technologies as well. It jump-started the race for smaller and faster computers, hugely improved telephone communication, and especially satellites for communication and GPS navigation, among many other benefits. The robots assembled to build spacecraft eventually made our assembly lines for cars and many other products more efficient. A wide variety of products were developed based on NASA research, including artificial hearts, thermal blankets, strong but light metal alloys and lightweight composite materials, better drugs grown in zero gravity, smoke detectors, air purification systems, small practical lasers, high-capacity batteries, UV sunglasses, Teflon-coated fiberglass, better fire protection gear

Figure 3.1 ▲
Astronaut Buzz Aldrin walking on the surface of the moon during the *Apollo 11* moon walk, July 20, 1969. The reflection of the photographer Neil Armstrong can be seen in Aldrin's visor. (Courtesy of Wikimedia Commons)

for firefighters, solar power systems, artificial limbs, MRI and CAT scanning, LED technology, joysticks for video games, better golf balls, the TACS system that aircraft use to avoid collisions, virtual reality simulators, hydroponics, Direct TV, pacemakers, and even disposable diapers.

The space program was immensely important in other ways as well. The satellite images of Earth we recorded enabled us to study all sorts of processes happening on our planet. In addition, the humbling views of Earth we saw from space transformed the way we think about that "pale blue dot" we all depend on (figure 3.2). All of this and more was produced for less than 1 percent of the federal budget, a trivial amount compared to what we spend on other things that provide much fewer benefits.

GREEN CHEESE?

People have stared at the moon for thousands of years and have had all sorts of ideas about what it represents. The moon is a conspicuous object in the night sky, and its nightly rise and fall in the sky is such a powerful image that people could not help but focus on it. The moon's monthly phases—from new moon to first quarter moon to full moon to third quarter moon to new moon again—are helpful in marking off time. The moon's waxing and waning have made it a symbol of time, change, and repetitive cycles around the world. One such cycle is the constant alternation between birth and death, creation and destruction. People have linked the moon with both birth and death, and in some cultures the moon had a destructive aspect. The Aztecs of Mexico called it Mictecacuiatl and believed that it traveled through the night skies hunting out victims to consume. The Maori people of New Zealand referred to the moon as the "man eater." Africans and Semitic peoples of the ancient Near East also feared this terrifying aspect of the moon. In other cultures, the moon had a gentler association with death. Some ancient Greek sects thought that the moon was the home of the dead, and early Hindus believed that the souls of the dead returned to the moon to await rebirth. The moon could even symbolize birth and death at the same time. The Tartars of Central Asia called it the Queen of Life and Death.

In legends, a full moon caused werewolves to change out of their human form or caused real humans to act crazy. In astrology, the moon's position at the moment we are born supposedly affects our personality and future, but this idea is pure bunk. Many cultures blamed the moon for odd events or worshiped it as a god. Early science fiction imagined men on the moon or aliens from the moon invading Earth. Lots of cultures have looked up at the blotchy pattern of dark and light surfaces and imagined a "face" or seen the "man in the moon." In one of the earliest silent movies ever made, *A Trip to the Moon* (1902), the main characters are loaded into a cannon shell and shot to the moon, and the cannon shell sticks in the "eye" of the "man on the moon." This film was influenced by Jules Verne's 1865 science fiction

Figure 3.2 ▲

Earthrise photographed by astronaut William Anders on the *Apollo 8* mission, December 25, 1968. (Courtesy of Wikimedia Commons)

novel *From the Earth to the Moon,* which blasted the explorer to the moon with a cannon. None of these ideas makes sense anymore with our current understanding of the moon.

Almost every culture had some sort of deity that was associated with the moon or was represented by the moon. To the ancient Greeks, the moon was associated with the goddess Artemis, the huntress (Diana to the Romans), and the crescent moon was supposed to represent her bow (just as her brother Apollo was the main god of the sun). Another version of the Greek myth considered the moon to be the incarnation of the goddess Selene (Luna to the Romans). The Chinese had several moon deities, including Jie Lin, who carried the moon across the night sky; Chang'e and Wu Gang, immortals who live on the moon; and Chang Xi, the mother of the 12 moons that correspond to the 12 months of the year. Hindus worshipped Chandra, the moon god; the Japanese worshipped Tsukuyomi; the Koreans worshipped Myeongwoi; and the Turks worshipped Ay Ata. The Aztecs had several moon deities, including Coyolxauhqui, Metztli, and Tecciztecatl; and the Incas worshipped Mama Killa, Ka-Ala-Killa, and Coniraya. The Hopis described the moon as Muuya, and the moon was Hanwi to the Lakota and Pah to the Pawnee tribes.

People once believed that moonlight had a powerful effect on human behavior. Those who acted strangely were said to be "moonstruck," and *lunacy*, a term for madness, comes from Luna, the Latin name for the moon goddess. The Japanese believed that the moon was a god with powers to foretell the future. Priests studied the moon's reflection in a mirror, believing that gazing directly at the moon might drive them mad. Superstitions about the moon's evil influence made some people refuse to sleep in a place where moonbeams could touch them. In the 1200s, the English philosopher Roger Bacon wrote, "Many have died from not protecting themselves from the rays of the moon." Many cultures have implicitly linked the 29.5-day lunar cycle to women's menstrual cycles, which is evident in the shared linguistic roots of the words "menstruation" and "moon" in multiple language families. This identification was not universal, which is demonstrated by the fact that not all moon deities are female.

In Western culture, many people wondered what the moon was made of and how it was formed. There were all sorts of ridiculous or silly ideas, such as the "green cheese" notion, but only a few serious hypotheses were proposed by the scientific community. The astronomer George Darwin, son of Charles Darwin, proposed that the Pacific Ocean was the scar left when the moon spun off the earth. That idea has been debunked by our modern mapping of the Pacific Ocean floor and plate tectonics.

SCIENTIFIC IDEAS ABOUT THE MOON'S ORIGINS

The most scientifically credible ideas about the origins of the moon fall into three broad categories. Due to the all-male community of astronomers at the time, these ideas acquired sexist nicknames that would never be acceptable today:

1. **The "pickup" or "capture" hypothesis:** For decades, some scientists suggested that the moon was a foreign body from far outside Earth's orbit that was captured as it flew by Earth and was pulled into orbit by Earth's gravity. There were numerous problems with this model from the beginning. For one thing, the moon's orbit is in almost the same plane as Earth's orbit around the sun (just 5° out of alignment with our plane around the sun) and moves in the same direction that Earth rotates. This would be unlikely if an object coming at any other angle from outer space was captured. Such an orbit would most likely swing around Earth in any plane except the plane of the Earth-sun system. In addition, when gravitational capture of a large body occurs, the result is either collision or the object flies off into space with an altered orbit. To allow the moon to slowly be captured by Earth's gravity and stay in orbit without collision or escape, Earth's atmosphere would have to have been very thick and to have extended much farther out than it does now to produce the friction and drag necessary to slow the object down. No evidence supports the idea that Earth's atmosphere was once much thicker. Finally, if the moon were an exotic object captured by Earth's gravity, its composition would be radically different from Earth's composition. When the Apollo missions brought back samples, scientists studying those moon rocks were able to test this.

2. **The "daughter" or "fission" hypothesis:** First proposed by the astronomer George Darwin (son of Charles Darwin) in the late 1800s, this hypothesis suggests that the moon is made of original rapidly spinning undifferentiated earth matter. During this rapid spin, molten earth material flew off into space to form the moon. Some astronomers even suggested that the Pacific Ocean basin is the remnant scar of that event. This scenario seemed plausible for many years, although by the 1960s plate tectonics had shown that the Pacific basin is not an ancient scar but is floored by very young lavas, most less than 140 million years old. The "daughter" model also doesn't account for the angular momentum of the Earth-moon system. Once again, the crucial test would be the moon rocks brought back by the Apollo missions. If they were the same composition as the primordial Earth (before it separated into core and mantle and crust), then it would be plausible.

3. **The "sister" hypothesis:** Similar to the "daughter" hypothesis, this model suggests that the original Earth-moon system began as two separate blobs of matter that got locked into gravitational attraction with each other. Again,

there are problems with the angular momentum of the Earth-moon system in this model. But like the "daughter" hypothesis, it predicts that moon rocks would have a composition very similar to the primordial Earth before it differentiated into core and mantle layers.

These ideas and more were hanging in the balance in 1969 when *Apollo 11* and later moon missions brought lunar samples back to labs on Earth to study. To everyone's surprise, their composition did not support any of the earlier ideas. Instead, the samples proved a new model that no one had ever thought of before.

The lunar rocks brought back by *Apollo 11* through *Apollo 17* (figure 3.3) were not similar to early Earth in composition. Nor did they have some exotic composition that would confirm that they were from a body outside Earth that

Figure 3.3 ▲
A sample of lunar anorthosite gabbro collected on the Apollo missions. It is nearly identical in composition to Earth's mantle. (Courtesy of Wikimedia Commons)

was captured by gravity. Instead, the lunar rocks were made of a form of calcium-plagioclase-rich gabbro known as *anorthosite*, and its volcanic equivalent, the familiar black lava known as basalt. In other words, their composition was very much like that of the upper mantle, where the lavas that erupt basalt on the seafloor or in volcanoes such as Kilauea on Hawaii have their source.

This was a surprise. If the moon was made almost entirely of mantle material, with very little iron or nickel such as found in Earth's core, it must be a piece of Earth's mantle that formed *after* the primordial Earth had separated into a core of iron and nickel and the mantle made of silicate minerals. In other words, the moon was formed after Earth had cooled and coalesced and its layers had differentiated and separated.

Even more startling, the only way to get a lot of mantle material into space was to blast early Earth with a giant impact from another body (figure 3.4). Geologists now call this body Theia (the Greek name for the mother of Selene, the moon goddess), and they postulated that Theia was

Figure 3.4 ▲

Artist's conception of the planetary body Theia slamming into Earth to produce the debris from the mantle that would become the moon. (Courtesy of Wikimedia Commons)

a Mars-sized protoplanet that hit Earth obliquely with an impact that blew material sideways off Earth and into orbit. The energy of this collision would have been amazing! Trillions of tons of material would have been vaporized, and the temperature on Earth would have risen to 10,000°C (18,000°F). Once this debris began to orbit Earth (at one-tenth the distance that the moon is today), it would have gradually clumped together and coalesced over the course of about 1,000 years to form the moon, a ball of matter about 2,000 miles wide. It was only 14,000 miles away then, not the 250,000 miles away the moon is now. Earth was spinning so fast at that time that each day was only six hours long, with three hours of nighttime and three hours of daylight.

The heat from its own radioactive minerals would later have remelted parts of the moon completely, and most of the moon would have contained the same composition as Earth's mantle, with the melting of the anorthosites also causing huge eruptions of basaltic lava flows forming magma oceans that now make the dark "maria" or "seas" on the moon's surface. The moon has a tiny iron core, only 330 to 350 kilometers in diameter, which is thought to be a relict of the core of Theia left behind after the collision. Most of Theia's own iron-nickel core would have accreted to Earth's core. In contrast, if the "sister" or "daughter" models (favored before the Apollo missions) were correct, the moon would have a large core, roughly proportional to the size of Earth's core relative to its mantle.

When did this all occur? Once again, moon rocks give us the answer. Using the same uranium-lead and lead-lead dating methods discussed in chapter 2, many labs have dated moon rocks. Most are at least 4 billion years old, suggesting that the moon's surface formed early and has not changed much. After all, it has none of the forces that change Earth's surface—it has no atmosphere, no water, no weathering, and no plate tectonics. The only major modification of its surface are huge impacts leaving craters, and most of the crater debris has been dated as being older than 3.9 billion years. Most of the impacts occurred early, and not much has happened since then. However, the oldest pre-impact rock dates from the moon are currently 4.44 billion years old. This is much younger than the meteorites that date to the origin of the solar system, so the moon is definitely younger than the events that formed the solar system and Earth and the melting and differentiation episode that separated Earth's core from its mantle.

Since the initial proposal of the giant impact hypothesis, much evidence has come from analysis of moon rocks that supports the mantle as the source of the moon. Nearly all of the geochemical isotopes (oxygen, titanium, zinc, and many others) that have been studied in the 50 years since the moon rocks were collected have shown that the moon and Earth's mantle have identical chemical compositions. There are also many refinements to the impact model, with some versions having more than one impacting body or positing different size impactors or different impact mechanics. But no matter which version is currently favored by scientists, the Apollo samples inescapably point to the moon as being a chunk of Earth's mantle.

THE MOON AND CLIMATE

The impact of Theia that created the moon occurred long before any of our modern atmosphere had formed, and long before there was much evidence of climate or oceans on Earth. However, there are real and surprising ways the moon does affect Earth and its climate. Astronomers and physicists point to an interesting dynamic: the large size of our moon acts as a stabilizer, keeping Earth's rotation fairly steady so it doesn't spin on its side like Neptune does. The reason the rotational axis is tilted 23.5° from the plane of the orbit around the sun is due to the impact as well. When the impact occurred, it knocked Earth's rotational axis 23.5° off of vertical, and it now wobbles like a spinning top. The angle of 23.5° determines that the Tropic of Cancer and the Tropic of Capricorn are 23.5° north and south of the equator, respectively (forming the tropics), and the Arctic and Antarctic circles are 23.5° away from the north and south poles (66.5° from the equator, north and south), which has a huge effect on climate, especially in the polar regions where there is total darkness for months in the winter and "midnight sun" in the summer.

Just as amazing as the original impact is the story of how the Earth-moon system got to be the way it is now. Today the tidal pull of Earth has completely stopped the moon's rotation on its axis so it always show the same face to Earth (figure 3.5A). This is the only side of the moon any human saw until the Apollo 8 mission first flew around to the far side of the moon (figure 3.5B) and photographed it. (This is not the same as Pink Floyd's "Dark Side of the Moon," which is a misconception; there is no permanent "dark side" because all sides are dark and light depending on the position of the sun.) Meanwhile the moon's tidal pull on Earth has been gradually slowing down Earth's own rotation, losing about 1.5 milliseconds every century, and more than a minute in a few thousand years. At the beginning of each New Year, and especially after the millennium in 2000, the world's

Figure 3.5 ▲

(A) The near side of the moon facing Earth at full moon. (B) The far side of the moon as photographed by the *Apollo 8* astronauts. (Courtesy of the National Aeronautics and Space Administration [NASA])

most precise clocks needed to be adjusted to account for this, otherwise they would be out of synch with the atomic clocks.

A few milliseconds a year might not seem like much, but over millions and billions of years, it adds up. Physicists have done the calculations and found that Earth has slowed down so much that there were far more spins of Earth (Earth days) in the geologic past than there are now. Confirmation of this startling idea came from the paleontology of humble corals. In the early 1960s, the paleontologist John W. Wells of Cornell University was looking at fossil corals that had both daily growth lines and larger marks that showed the annual cycle of the seasons. He was able to slice the corals very thin, polish them, and count the growth rings under the microscope. Sure enough, in the Devonian Period (about 400 million years ago), Earth spun on its axis so much faster that it had 400 turns (400 days) in one revolution around the sun (one year). About 600 million years ago, the day was only 21 hours long, not 24 hours, and there were 430 days in a year. And only 150 million years ago, there were 380 days in a year.

What does this mean? Although it's extremely slow by human standards, Earth is gradually slowing down. About 5 billion years from now, Earth will also be tidally locked: only one side will face the moon, and the other side will never see the moon. The energy of the entire Earth-moon system is also decreasing, and both systems are slowly pulling apart. When the moon debris first was blasted out into space, the moon was only 10 percent of its current distance away from Earth, and it has been slowly receding since then. When the trilobites roamed the earth 600 million years ago, the moon was much closer and would have looked huge in the sky. Its tidal pull at that distance would have been so powerful that immense true tidal waves (not tsunamis formed by earthquakes, which have nothing to do with tides) would have swept across Earth as the tides rose and fell.

Eventually the two bodies will not only be tidally locked and much farther apart but their motion could also come to a halt. The side of Earth facing the sun would be baked to a crisp, and the opposite side that does not face the sun would be in frozen darkness. (This is what happened to the planet Mercury.) However, this is billions of years in the future, and the sun will probably explode before then and wipe out the inner planets, so tidal locking of Earth will never get a chance to occur.

It's pretty humbling when you think about it. Just a few pounds of rocks brought back in the Apollo spacecraft have revolutionized our understanding of the moon, Earth, and the solar system. The next time you read poems about the moon or hear romantic lyrics about the "moon in June," consider what we now know about the future of our only natural satellite. Will you ever think about the moon in the same old way again?

FOR FURTHER READING

Canup, Robin M., and Kevin Righter, eds. *Origin of the Earth and Moon*. Tucson: University of Arizona Press, 2000.

Chaikin, Andrew. *A Man on the Moon: The Voyages of the Apollo Astronauts*. New York: Penguin, 2007.

Chambers, John, and Jacqueline Mitton. *From Dust to Life: The Origin and Evolution of Our Solar System*. Princeton, NJ: Princeton University Press, 2013.

Gargaud, Muriel, Hervé Martin, Purificación López-Garcia, Thierry Montmerle, and Robert Pascal. *Young Sun, Early Earth, and the Origins of Life: Lessons for Astrobiology*. Berlin: Springer, 2013.

Hartmann, William K., and Ron Miller. *The History of Earth: An Illustrated Chronicle of An Evolving Planet*. New York: Workman, 1991.

Hazen, Robert. *The Story of Earth: The First 4.5 Billion Years from Stardust to Living Planet*. New York: Penguin, 2013.

McKenzie, Dana. *The Big Splat: Or How Our Moon Came to Be*. New York: Wiley, 2003.

04

FAINT YOUNG SUN

Here comes the sun, and I say

It's all right.

—THE BEATLES, "HERE COMES THE SUN" (1969)

THE HERTZSPRUNG-RUSSELL DIAGRAM

The women astronomers working at the Harvard College Observatory were known as the "Harvard Computers," and one of their first major achievements was the complete catalog of the stars made by Antonia Maury. She not only listed all the known stars in the sky that they could document in those days but also looked at the electromagnetic spectrum of the light the stars emitted (see chapter 2). Other astronomers were keenly aware of the work that Maury and the other Harvard Computers were doing, and as soon as their work was available for study, they eagerly began to comb through the data looking for patterns. In addition to cataloging all the stars in the sky that were then visible, Maury had categorized them by the width of their spectral lines, from thin and narrow lines to broader lines.

One of the astronomers who immediately noticed some patterns in the Maury catalog of stars was the Danish chemist and astronomer Ejnar Hertzsprung. As a chemist, he was especially interested in the spectra emitted by different elements as they burned, and he made a special study of the details of the spectral lines that were emitted by different elements—not only their wavelength on the spectrum but also their other characteristics. As an astronomer, he was keenly aware that the spectral lines of stars showed different patterns, and he wanted to understand what they meant. He started with Maury's data about line width and noticed that stars with narrower lines tended to appear to move less in the sky whereas those with broader bands moved more. In other words, the furthest stars had narrower bands, and the closer stars had broader bands.

By 1913, he had worked out a method to determine the distance to the nearby stars using stellar parallax, or the shift in the apparent position in the sky as we see it from one part of Earth's orbit to the opposite part of Earth's orbit six months later. This allowed him to calibrate the relationship, first established by Henrietta Leavitt of the Harvard Computers, between the periodic fluctuation of the Cepheid variables and their apparent brightness (luminosity). His discovery allowed the estimation of the distance to the Cepheid variables for the first time, giving us a reliable measure of the true distance of each star from us.

Hertzsprung also realized that spectral lines were a proxy for temperature. Using the complex relationship between distance, luminosity, and temperature, by 1911 he began to classify stars by those parameters (figure 4.1). One way to analyze these data was to plot each star on a graph of luminosity (a measure of the size of the star, correcting for distance) and temperature (as determined by the spectral lines). As he plotted more and more stars, he found a relationship: the stars were not randomly scattered on a plot of luminosity vs. temperature but formed a broad band running diagonally, from the upper left (hotter and brighter) to the lower right (cooler and dimmer). Over the next few years, more and more stars were analyzed, and the vast majority of them fell on this diagonal band that came to be known as the "main sequence." The enormous stars known as "red giants" and "supergiants" were especially intriguing, and they plot in the upper right part of the main sequence (relatively cooler and brighter). Eventually a number of stars that were dim and hot were found in the lower left part of the diagram, and they came to be known as "white dwarfs" (hotter and dimmer stars).

Hertzsprung was not the only astronomer to find patterns in Maury's catalog. The American astronomer Henry Norris Russell was on the same trail that Hertzsprung followed. Independently he too noticed that there was a relationship between spectral class (temperature) and absolute magnitude.

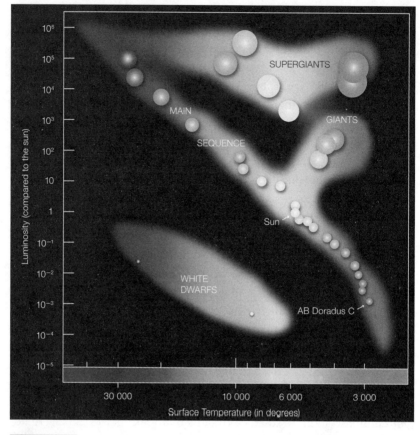

Figure 4.1 ▲

The Hertzsprung-Russell diagram. (Courtesy of Wikimedia Commons)

His own plot of this relationship was published in 1913, just two years after Hertzsprung's first mention of this trend.

This is not unusual in the history of science—when an idea is ripe for being discovered, multiple people may independently come across the same idea. In 1858, Alfred Russel Wallace came upon the idea of natural selection, forcing Darwin to stop procrastinating on his book about the idea (which he first wrote down in 1838) and publish it. Both scientists had their ideas read at the same meeting of the Linnaean Society, so they got equal credit. In 1900, three different genetics labs independently rediscovered Gregor Mendel's obscure and long-forgotten 1865 paper on heredity in plants like peas. In the early 1800s, both the British surveyor William Smith and the French paleontologists Georges Cuvier and Alexandre Brongniart noticed that fossil faunas change from one formation to the next. When the time is ripe to propose a scientific idea, many people often come up with the idea independently.

In this case, although Hertzsprung first published a preliminary plot of the main sequence in 1911, Russell's 1913 publication showed that they were working on the same idea at about the same time. Subsequent astronomers have given them both credit, and as a result the diagram shown in figure 4.1 is known as the Hertzsprung-Russell diagram (or HR diagram), honoring them both. It is now considered to be one of the most important discoveries in the history of astronomy.

A STAR IS BORN—AND DIES

The discovery of the relationship between temperature and luminosity in the Hertzsprung-Russell diagram was the first step in deciphering the way stars themselves change over their lifetimes. The process is too slow for us to watch in real time, so astronomers look at stars in various stages of evolution and calculate the physics of how one kind of star turns into another. From these discoveries, the life history of most stars is quite well known.

The story begins when collapsing clouds of dust and gas in giant molecular clouds begin to condense more and more until they form a massive body known as a protostar. Through its life span (which may be only 500,000 years if it is a low-mass star), a protostar continually draws in more matter that is gravitationally attracted to its center. Finally, as the infalling matter begins to diminish, the protostar becomes a pre-main-sequence star. This massive body continues to contract until the pressure is sufficient for hydrogen to fuse and form helium. Then the nuclear fusion process begins, fueling the star for most of its life span. Once the nuclear fusion process has taken over and the star reaches an equilibrium of contraction versus energy output, it is officially a main sequence star.

As stars burn through their hydrogen fuel, they continue to change. Objects with a very low initial mass never reach the main sequence and become brown dwarfs. But stars that have enough initial mass to become main sequence stars continue to change as well. They burn through their fuel at a known rate based on their original mass. Once the fusion reaction begins, the star changes yet again. Hydrogen is fused in a spherical outer shell around the core of the main sequence star, and this outer shell makes the star grow dramatically in size. Depending on the amount of fuel left to fuse, the star may grow into a subgiant and leave the main sequence. If the star is truly massive, it will have a series of concentric shells of hydrogen fusing to helium. This can expand into a red giant or even a supergiant, as the star begins to fuse the helium in its core as well as fusing hydrogen and helium to form heavier elements. The red giant phase ends when nearly all the fuel is exhausted, and the star collapses on itself. Most lower-mass stars collapse from a red giant down to a super dense white dwarf, often forming a planetary nebula around them. If a main sequence star is more than 10 times the mass of the sun, it turns into a supernova as the iron core inside collapses into a super dense neutron star or black hole.

What about our sun? Based on dating and the physical models that describe solar evolution (see chapter 2), our sun was born at the same time as the rest of the solar system. Chondritic meteorites (like the Allende meteorite) from the earliest solar system tell us that the sun formed before 4.567 billion years ago. This date backs up computerized models of stellar evolution known as nucleocosmochronology. Using this information, we can place the sun on the main sequence and compare its life history to other stars of the main sequence. Since it first formed, the sun's radius has expanded by 15 percent, and its surface temperature has increased from 5620 K to 5777 K, producing a 48 percent increase in luminosity (from 0.677 solar luminosities about 4.6 billion years ago to 1.0 solar luminosities today). Judging by other stars in the main sequence, it will be about 5 billion years before the sun has consuming all its fuel and hydrogen fusion ends, turning our sun into a red giant. So our sun is a middle-aged star, about halfway through its life span.

THE "SAGAN EFFECT"

Most people remember Carl Sagan from his numerous public appearances as one of the most successful science communicators in history. Not only did Sagan break new ground with his award-winning 1980 TV series *Cosmos*, but he was often in the public eye, appearing on news media and on the *Tonight Show with Johnny Carson*. In the years since his untimely death in 1996 at the relatively young age of 62, Carl's pioneering role in popularizing science for the public has become more and more appreciated. Today public figures such as Neil DeGrasse Tyson and Bill Nye the Science Guy explain science to the public, and they are highly respected and valued. But they would be the first to admit that no one could popularize science like Sagan did. (Both of them knew Carl Sagan well: Nye was a student of Sagan's at Cornell and Tyson almost became Sagan's grad student before choosing Harvard instead.)

Most people today would be surprised to learn that during his lifetime Sagan's popularization of science was frowned upon and discouraged in scientific circles, and many scientists shunned or dismissed him. If he was such a good science communicator, they thought he must be deficient as a research scientist. In the 1950s and 1960s, scientists focused on their narrow research programs and shunned public appearances as a waste of time. They could not imagine a scientist being both a good communicator with the public and also a good researcher.

This bias showed up early. After a top-notch education at the University of Chicago and a post-doc at the University of California Berkeley, Sagan so impressed the astronomers at Harvard that he was offered an assistant professor post. In 1961, he had already published a brilliant and highly speculative paper in *Science* about the atmosphere of Venus, which made him famous in the scientific community (when the first probes sampled the Venusian atmosphere, Sagan's idea was found to be accurate). After seven years at Harvard, in 1968 Sagan was denied tenure! No official reason was given, but researchers later found that his talent was dismissed because he published research across a number of areas. Most scientists were becoming renowned experts in a narrow specialty, and it was thought that you couldn't be any good if you worked in multiple areas of research. In addition, Sagan's high profile as a science advocate was perceived by some as borrowing the ideas of others for little more than self-promotion. Some jealousy was certainly involved because Sagan was already one of the most famous and popular of all scientists in the public sphere. Even one of his former mentors, the Nobel Prize-winning chemist Harold Urey of the University of Chicago, wrote a letter to Harvard recommending that they not give Sagan tenure! Fortunately, the astronomers at Cornell University appreciated his talents, hired him away from Harvard, and that is where he spent the rest of his career.

In addition to the denial of tenure at Harvard, Sagan was perpetually snubbed by the National Academy of Sciences, a body of the most famous and highly respected scientists in the United States. Once again it was felt that his work must be somehow inferior and not a good match with the productivity of scientists who were in the National Academy of Sciences. A later study showed that Sagan's scientific research was as good or better than most of the scientists elected in his time.[1] This snubbing of science communicators has been called the "Sagan effect" because Carl's treatment was one of the most egregious examples of academic snobbery, jealousy, and insularity. Christina Parker described the effect this way in 2017:

> According to the Sagan Effect, scientists' research productivity (quantity and quality) is perceived as inversely proportional to the amount of time spent on outreach efforts or popularity. In particular, scientists who spend a considerable amount of time and effort communicating science to the public are often viewed as less successful than their colleagues and their science is considered less rigorous. Some of their colleagues worry that these public scientists, or popularizers, are more focused on media presence than discovery.[2]

Recent studies show that scientists who do public outreach are actually *more* active and productive in research than most scientists who shun publicity.[3] Yet the Sagan effect persists in the scientific community. My good friend, the late Harvard paleontologist and evolutionary biologist Stephen Jay Gould, was also dismissed as a scientist by his rivals because he was an extremely popular science communicator, with many books, a monthly column in *Natural History*, and some groundbreaking papers that challenged the conventions in evolutionary biology.[4] As Parker writes:

> 5 percent of scientists. . . . . do not participate in outreach because "they do not see it as part of their role as a scientist; these scientists believe that it is not their job to interpret their work for a broader audience." Because it is not their responsibility, several scientists propose having non-scientists organize outreach efforts. From their perspective, scientists should focus on what they do best and share their work with intermediate people, who are very good at communicating science to the public.[5]

The times are gradually changing, but people like Bill Nye and Neil deGrasse Tyson are still occasionally attacked because they are popular science communicators. However, today these attacks are usually driven by right-wing science deniers who are trying to undermine theories proposed by scientists regarding evolution or climate change; they are not being attacked by the mainstream scientific research community.

Today the pressure is increasing for scientists to become better at outreach and communicating the results of their research, and most researchers must find a way to publicize their work and justify it to the public, especially if it is supported by public tax dollars. Since the 1990s, every proposal to the National Science Foundation has included a "broader impact" section, which the researcher must fill out to justify the societal importance of the research (even though much of what scientists do is pure research that is seeking answers for the sake of curiosity and their findings would not have an obvious immediate public application). The Sagan effect is not extinct, but public pressure is on for scientists to be better communicators as well as expert researchers.

FAINT YOUNG SUN?

One of Sagan's lesser-known but very important ideas was the "faint young sun hypothesis." In a groundbreaking paper written with George Mullen and published in *Science* in 1972, the two astronomers pointed out that knowing the history of the sun's evolution and temperature had important implications for Earth's history and climate as well. When the sun first formed 4.6 billion years ago, it would have been much less luminous, and Earth would have received only about 70 percent of the solar energy Earth receives today. Earth's surface temperature would have been well below freezing, making it just as inhospitable and lifeless as Mars. However, there is clear evidence that Earth had liquid water very early, even as early as 4.4 billion years ago (see chapter 5). If Earth was not frozen, it must have had some source of warming to compensate for the diminished solar output.

What explains the fact that Earth did not freeze over in its earliest history, even though the solar energy input was so weak? In 1972, Sagan and Mullen argued that ammonia (NH_3), a greenhouse gas, could have provided some of the warming needed. But later studies show that the photochemical reactions of ammonia under solar radiation breaks ammonia down to hydrogen gas and nitrogen gas (N_2). Sagan suggested that some of the ammonia was part of a photochemical haze (a kind of "smog") that prevented solar radiation from reaching most of the ammonia in the atmosphere, but this solution doesn't work. Furthermore, dense clouds of haze would block so much sunlight that Earth would freeze. So ammonia is out as a greenhouse solution for the problems with the "faint young sun."

What else might explain the unfrozen Earth during the early history of the sun? Since Sagan and Mullen's original paper, 52 years of research has gone into answering this question. Some have suggested that the lack of ice on Earth's surface produced a highly absorbent surface with low albedo (reflectivity). That is, most of the energy was trapped when it was absorbed by the dark ocean, and that helped warm the earth. But there would always be highly reflective clouds in the atmosphere (as there are now), so that solution doesn't work either. Others argued that the heat coming from Earth's interior was much greater and that this "great heat engine" kept the earth warm. Almost every calculation of the geothermal energy needed to make this work is implausible, so that is not a solution either.

That leaves greenhouse gases as the only likely possibility. Calculations of how much carbon dioxide, the main greenhouse gas today, would be needed to do the job suggests that it must have been 1,000 to 10,000 times as much CO_2 as exists today, which isn't plausible either. Consequently, a lot of scientists have turned to a more potent greenhouse gas, methane (CH_4, or "natural gas"), the simplest form of hydrocarbon. Methane could come from many sources: from comets, from the weathering of lavas on the deep seafloor, and eventually, as life evolved, from methane-producing organisms such as bacteria. An important thing to remember is that methane is a much more powerful greenhouse gas—about 30 times as effective as carbon dioxide—so a small amount of methane provides a *lot* of warming. Concentrations as low as 10 parts per million (ppm) to 100 ppm of methane in the atmosphere, combined with carbon dioxide, could easily achieve the amount of warming needed. Without methane, carbon dioxide would need to be about 60 times present levels to keep the earth from freezing. With methane at only 1/1000 the concentration of carbon dioxide, the levels of carbon dioxide would only need to be about 15 times present levels (which are currently 415 ppm) to make it work. It appears very likely that methane plus carbon dioxide were the main factors that overcame the effects of the faint young sun.

Startling confirmation of this idea occurred when measurements were made of Titan, one of the moons of Saturn. Sagan's own advisor at Chicago, Gerard Kuiper (who later described the Kuiper belt of comets outside the solar system), looked at the spectrum of Titan in 1944 and found that it had significant methane in its signature. In 1980, *Voyager I* flew by Saturn and Titan and sent back further proof that the atmosphere was about 94 percent nitrogen and 5.7 percent methane. Sagan was heavily involved in planning the *Voyager I* flight. He even helped design the message to any aliens that found the probe, which was attached to a gold-plated plaque on the space probe. Finally, on July 1, 2004, the Cassini-Huygens probe flew even closer to Titan and sent back excellent images and data. Not only does Titan have the amount of methane that had been predicted, but it has an orange color in the light spectrum due to the methane-rich smog on its surface (figure 4.2).

If you want to imagine early Earth as it was 4.5 billion years ago, don't think of a blue planet but an orange planet, with a thick layer of nitrogen, carbon dioxide, and methane smog dominating the skies!

Figure 4.2 ▲

Titan, a moon of Saturn, has an orange color because its atmosphere is loaded with methane along with nitrogen and ammonia. (Courtesy of the National Aeronautics and Space Administration [NASA])

NOTES

1. Susana Martinez-Conde, "Has Contemporary Academia Outgrown the Carl Sagan Effect?," *Journal of Neuroscience* 36, no. 7 (2016): 2077–2082, https://doi.org/10.1523/JNEUROSCI.0086-16.2016.
2. Christina Parker, "Science Outreach and the Sagan Effect," *The Pipette Pen*, February 13, 2017, http://www.thepipettepen.com/science-outreach-and-the-sagan-effect/.
3. Martinez-Conde, "Has Contemporary Academia Outgrown the Carl Sagan Effect?"
4. Michael Shermer, "Stephen Jay Gould as Historian of Science and Scientific Historian, Popular Scientist, and Scientific Popularizer," *Social Studies of Science* 32, no. 4 (2002): 489–524, https://michaelshermer.com/articles/stephen-jay-gould-historian-science-popularizer/.
5. Parker, "Science Outreach and the Sagan Effect"; see also Elaine Howard Ecklund, Sarah A. James, and Anne E. Lincoln, "How Academic Biologists and Physicists View Science Outreach," *PLoS One* 7, no. 5 (2012), https://journals.plos.org/plosone/article?id=10.1371/journal.pone.0036240.

FOR FURTHER READING

Ecklund, Elaine Howard, Sarah A. James, and Anne E. Lincoln. "How Academic Biologists and Physicists View Science Outreach." *PLoS ONE* 7, no. 5 (2012): e36240. https://journals.plos.org/plosone/article?id=10.1371/journal.pone.0036240.

Feulner, George. "The Faint Young Sun Problem." *Reviews of Geophysics* 50, no. 2 (2012): RG2006. https://doi.org/10.1029/2011RG000375.

Hart, Michael H. "The Evolution of the Atmosphere of the Earth." *Icarus* 33, no. 1 (1978): 23–39.

Hertzsprung, Ejnar. "Über die Sterne der Unterabteilung *c* und *ac* nach der Spektralklassifikation von Antonia C. Maury." *Astronomische Nachrichten* 179, no. 24 (1908): 373–380.

Kasting, James F., and Thomas P. Ackerman. "Climate Consequences of Very High Carbon Dioxide Levels in the Earth's Early Atmosphere." *Science* 234, no. 4782 (1986): 1383–1385.

Kasting, James. "Faint Young Sun Redux." *Nature* 464, no. 7289 (2010): 687–689.

Kuhn, W. R., and S. K. Atreya. "Ammonia Photolysis and the Greenhouse Effect in the Primordial Atmosphere of the Earth." *Icarus* 37, no. 1 (1979): 207–213.

Martinez-Conde, Susana. "Has Contemporary Academia Outgrown the Carl Sagan Effect?" *Journal of Neuroscience* 36 (2016): 2077–2082. https://www.jneurosci.org /content/36/7/2077.

Maury, Antonia C., and Edward C. Pickering. "Spectra of Bright Stars Photographed with the 11-Inch Draper Telescope as Part of the Henry Draper Memorial." *Annals of Harvard College Observatory* 28 (1897): 1–128.

Minton, David, and Renu Malhotra. "Assessing the Massive Young Sun Hypothesis to Solve the Warm Young Earth Puzzle." *Astrophysical Journal* 660, no. 2 (2007): 1700–1706.

Parker, Christina. "Science Outreach and the Sagan Effect." *The Pipette Pen.* February 13, 2017. http://www.thepipettepen.com/science-outreach-and-the-sagan-effect/.

Pavlov, Alexander A., James F. Kasting, Lisa L. Brown, Kathy A. Rages, and Richard Freedman. "Greenhouse Warming by CH_4 in the Atmosphere of Early Earth." *Journal of Geophysical Research* 105, no. E5 (2000): 11981–11990.

Prialnik, Dina. *An Introduction to the Theory of Stellar Structure and Evolution.* Cambridge: Cambridge University Press, 2000.

Roe, Henry G. "Titan's Methane Weather." *Annual Review of Earth and Planetary Sciences* 40, no. 1 (2012): 355–382.

Rosing, Minik, Dennis K. Bird, Norman Sleep, and Christian J. Bjerrum. "No Climate Paradox Under the Faint Early Sun." *Nature* 464, no. 7289 (2010): 744–747.

Russell, Henry Norris. "Relations Between the Spectra and Other Characteristics of the Stars." *Popular Astronomy* 22 (1914): 275–294.

Ryan, Sean G., and Andrew J. Norton. *Stellar Evolution and Nucleosynthesis.* Cambridge: Cambridge University Press, 2010.

Sagan, Carl, and George Mullen. "Earth and Mars: Evolution of Atmospheres and Surface Temperatures." *Science* 177, no. 4043 (1972): 52–56.

Sagan, Carl, and Christopher Chyba. "The Early Faint Sun Paradox: Organic Shielding of Ultraviolet-Labile Greenhouse Gases." *Science* 276, no. 5316 (1997): 1217–1221.

Shermer, Michael B. "Stephen Jay Gould as Historian of Science and Science Historian, Popular Scientist, and Scientific Popularizer." *Social Studies of Science* 32, no. 4 (2002): 489–524. https://michaelshermer.com/articles/stephen-jay-gould -historian-science-popularizer/.

Sleep, Norman H., and Kevin Zahnle. "Carbon Dioxide Cycling and Implications for Climate on Ancient Earth." *Journal of Geophysical Research: Planets* 106, no. E1 (2001): 1373–1399.

Walker, James C. G. "Carbon Dioxide on the Early Earth." *Origins of Life and Evolution of the Biosphere* 16, no. 2 (1985): 117–127.

05

THE OCEANS FORM

In a great number of the cosmogonic myths the world is said to have developed from a great water, which was the prime matter. In many cases, as for instance in an Indian myth, this prime matter is indicated as a solution, out of which the solid earth crystallized out.

—SVANTE ARRHENIUS, *THEORIES OF SOLUTIONS*

With every drop of water you drink, every breath you take, you're connected to the sea. No matter where on Earth you live. Most of the oxygen in the atmosphere is generated by the sea.

—SYLVIA EARLE

THE PRECAMBRIAN OR CRYPTOZOIC

The time before abundant fossils were found in Earth's rock record has long been known as the Precambrian because it is based on rocks found beneath Cambrian beds with abundant fossils such as trilobites. This immense duration of time includes almost 4.0 billion of Earth's 4.567-billion-year history, or about 88 percent of Earth's history (figure 5.1). Although the old-fashioned term "Precambrian" is still widely used by geologists, the entire time interval has also been called the Cryptozoic (in Greek *crypto* means "hidden" and *zoic* means "life") because fossils are scarce and mostly microscopic. More recently, geological societies have recommended formally dividing this time interval into three eons: the Hadean Eon (4.567 to 4.0 Ga (billion years ago), the Archean Eon (4.0 to 2.5 Ga), and the Proterozoic Eon

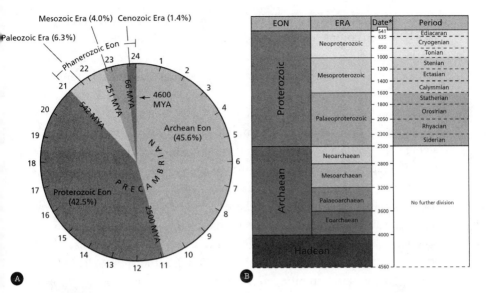

A

Figure 5.1 ▲

Most of geologic time is part of the Precambrian. (A) Pie chart showing the portion of Earth history that is Precambrian; (B) time scale of the Precambrian. (Redrawn from several sources)

(2.5 Ga to 542 million years ago (Ma). Each eon is further subdivided into eras. The Archean consists of the Eoarchean, Paleoarchean, Mesoarchean, and Neoarchean eras; and the Proterozoic is broken down into the Paleoproterozoic, Mesoproterozoic, and Neoproterozoic eras. Finally, many of the eras are subdivided into periods; for example, the Neoproterozoic Era consists of the Tonian, Cryogenian, and Ediacaran periods. Most geologists don't need to know the time scale at this level of detail unless they become specialists in Precambrian geology. However, it is important to know the three big subdivisions of this immense span of time—Hadean, Archean, and Proterozoic—because each represents distinct geological events and the kinds of rocks that were produced during each time interval.

THE HADEAN (4.567 TO 4.0 GA): HELL ON EARTH

The first half-billion years of Earth's history is known as the Hadean Eon, after Hades, the Greek mythological version of Hell or the Underworld.

And indeed it was truly a hellish time to be on the earth. The earth's surface had been completely destroyed and remelted after the impact of Theia blasted off a chunk of the mantle to form the moon (see chapter 3). Studies show that even this huge event could not have completely remelted Earth; it takes incredible amounts of energy to melt this much rock. A lot of the material that was blasted off into orbit eventually formed the moon through gravitational condensation, but not all of the material reached a stable orbit and coalesced to form the moon. Some material only went a few kilometers above the earth's surface and would have rained down as a rock vapor atmosphere for at least 2,000 years after the moon-forming impact while the enormous scar where the mantle had been ruptured was healing. Think about it! The earliest atmosphere of Earth after the moon was blasted into space would have been vaporized rocks for thousands of years. As the rock vapor finally settled out and the atmosphere cleared, the atmosphere was rich in nitrogen gas, carbon dioxide, methane, hydrogen gas, and water vapor.

In the beginning Earth was very hot, and its surface was still molten. If any crust cooled above the molten surface, it didn't last very long or get very big. No crustal rocks from the earliest Earth survive any more. The oldest dates derived from crustal rocks are currently about 4.32 Ga and are from the Nuvvuagittuq greenstones on the east shore of Hudson Bay (first dated in 2008). The next oldest crustal rocks known are the 4.03 Ga Acasta Gneisses of northwestern Canada, near Great Slave Lake (first dated in 1999). Geologists are constantly looking for even older rocks and are dating more and more samples, so there's a good chance they will find a rock with an older date by the time you are reading this. Indeed, in 2018, they found mantle rocks brought up by volcanoes that dated from the beginning of Earth, about 4.5 Ga.

Despite the lack of a rock record for most of the Hadean, we can infer many things about the conditions on Earth then. The crust was hellishly hot and had a molten surface, and it was continually bombarded by meteorites as the debris from the early solar system was pulled in by Earth's gravity. This debris must have pulverized Earth's surface many times, disrupting formation of a permanent solid surface. However, the bombardment may have sped up the formation of the continental crust. Each time a rock from space hit, the churning and differentiation of lighter crustal rocks from the original magmas increased. It is analogous to shaking a bottle of a carbonated beverage, which causes the tiny light bubbles of carbon dioxide to combine until they become visible bubbles—and your beverage becomes foamy. This heavy bombardment apparently went on continuously from 4.5 to about 3.9 Ga, and then it slowed down tremendously. How do we know this? The Apollo moon samples of impact craters and their melt rocks provide dates in this age range but almost no dates that are younger and none that are older.

Geochemical data suggest that the Hadean crust was extremely thin and simple, without the geological complexity we see in our modern crust. Various lines of evidence suggest that the crust was only about 20 kilometers (12 miles) thick, whereas our modern continental crust is 30 to 50 kilometers (20 to 30 miles) thick. The crust was made of material from the mantle that had cooled into lavas near the surface, with the deeper crust undergoing high-temperature, high-pressure metamorphism. These small, thin, hot proto-continental blocks did not yet experience true plate tectonics; they simply bumped against each other as they moved around on the highly mobile mantle. We know that there was no plate tectonics in the modern sense because there is no geochemical evidence of volcanoes being produced by subduction zones.

The moon had just been blasted away from Earth's mantle and was still condensing and was much closer to Earth than it is now (see chapter 3). The moon would have looked enormous in the sky. A much closer moon would have immense tidal pull on Earth's surface and could cause the earth's crust to rise and fall as the pull of the moon's gravity changed every hour. The moon's effects were especially strong in Earth's early oceans. The tidal range must have been incredible, with periods of complete exposure and drying when the tide went out, followed by huge walls of water roaring across the surface of the early ocean with the incoming tide—a true "tidal wave." (The immensely destructive waves produced by earthquakes are not "tidal waves" but are properly known as tsunamis, or seismic sea waves. They are caused by earthquakes and have nothing to do with tides.)

Earth was also spinning much faster then. A complete rotation on its axis occurred in about 10 hours in the Hadean, so the day was 10 hours long, not 24 hours long, and each day and night cycle was about 5 hours long in the spring or fall. There were almost 600 days in a year, not the 365.25 days we know now, and that rate continues to slow down thanks to the tidal friction caused by the moon.

So what kind of atmosphere would early Earth have held? We know from looking at other planetary bodies such as Jupiter and Saturn as well as our sun that the most abundant elements in the solar system were hydrogen and helium. They are the most common elements in those bodies, but Earth is not massive enough to have had the gravitational pull necessary to hold in the original hydrogen and helium. These elements must have escaped into space (as they have with the other small inner planets such as Mercury, Venus, and Mars). Mars and Venus have an atmosphere that is mostly carbon dioxide, and that was probably true for early Earth, along with inert nitrogen gas, which makes up 78 percent of our atmosphere today and was probably equally abundant in the early Precambrian (figure 5.2). We know that abundant methane formed a warming blanket that kept Earth from freezing, and a compound of nitrogen and ammonia is found in many smaller planetary bodies (see chapter 4). These elements are essential for the formation of life. After the hydrogen and helium escaped to space very early, so did the ammonia and methane. What remained was mostly nitrogen with some carbon dioxide. As water began to escape from the early earth and eventually formed oceans, water vapor became a third important atmospheric gas.

There was no free oxygen in the atmosphere until about 2.4 Ga, and only a tiny percentage of free oxygen was available until about 500 million years ago. By the late Precambrian, the high volume of carbon dioxide in the early atmosphere had been drawn down into crustal rocks and into the shells of the first multicellular organisms. Although carbon dioxide may have been as high as 25 to 30 percent of the total atmosphere in the early Precambrian, by the late Precambrian it was less than 1 percent, and currently it is only 0.4 percent (415 ppm).

AN EARLY COOL EARTH?

In the Hadean eon, Earth had a hot core and mantle, the surface of the magma ocean was molten, and there was no crust. It is hard to know much about the earliest crust of the earth because none of it survives. Weathering, erosion, and plate tectonics have destroyed nearly all of it. The oldest crustal rocks we have found have a very thin crust that erupted directly from the mantle.

An even bigger question is when did the surface of the earth cool down below the boiling point of water so liquid water could condense out of it? In other words, when did water collect on Earth's surface to form the first oceans? Even though the boiling point of water on Earth's surface today is 100°C (212°F), it didn't need to be that cold in the Hadean. A dense carbon dioxide-nitrogen atmosphere could have created a surface pressure

Composition of earth's atmosphere

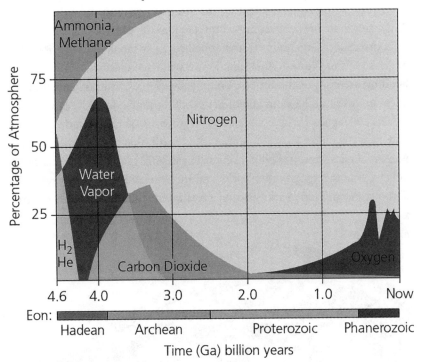

Figure 5.2 ▲

A history of the early atmosphere. Hydrogen and helium made up most of the solar system and were once abundant in Earth's atmosphere, but these elements escaped into space because Earth's mass and gravitational attraction were not strong enough to hold them. Ammonia and methane were also once abundant (based on other planetary bodies) and were important to the emergence of life, but these elements eventually vanished or were oxidized and converted into carbon dioxide and nitrogen, which are the main gases of the early atmosphere (nitrogen still makes up 78 percent of our atmosphere). Water vapor became important as Earth cooled and eventually formed oceans. But free oxygen was a very late arrival, first appearing 2.4 billion years ago but not becoming abundant until 500 million years ago. (Redrawn from several sources)

27 times stronger than the atmospheric pressure today. A lot more heat is needed to boil water and vaporize it under that much air pressure. Even if the surface temperature was as high as 230°C (446°F), it's estimated that liquid water could have remained on the earth's surface and oceans could form without boiling away.

For decades, most geologists thought Earth's crust could not have cooled more than about 4 Ga and that 500 million to 1 billion years would be required before Earth's crust cooled down below the boiling point of water. A few years ago, however, scientists made a startling discovery when examining a few grains of sand from Australia. A handful of zircon (zirconium silicate, or $ZrSiO_4$) sand grains from a much younger sandstone were found in the Jack Hills of Western Australia (figure 5.3). Each individual grain can be dated by uranium-lead methods, so they provide a scatter of ages. But the oldest grains of all have an age of 4.374 billion years, dated by John Valley and his colleagues at the University of Wisconsin. Thus the current record holder for the oldest crustal material from Earth (that is, not a meteorite or moon rock) is almost 4.4 billion years. These sand grain dates put us closer and closer to the age of moon rocks and meteorites, but we still have a gap of about 160 million years between 4.374 billion and 4.56 billion. In 2018,

Figure 5.3 ▲

The zircon from Jack Hills, Australia, is dated at 3.4 billion years ago and contains bubbles of the original atmosphere, suggesting that Earth was cool enough to have liquid water at that time. (Courtesy of J. W. Valley)

scientists announced measurements of chemical isotopes from several samples of mantle rocks that has been brought up by volcanoes. These had chemical signatures suggesting that they were part of the original mantle 4.5 Ga.

But those tiny zircon sand grains held even more surprises. Not only did they provide the oldest known dates, but when scientists analyzed the ratio of the two isotopes of oxygen (oxygen-16 and oxygen-18) within them, they found evidence of early surface water. These zircons had oxygen isotopes suggesting that Earth had liquid water on its surface as early as 4.374 Ga!

Prior to this discovery, geologists had assumed that it took a long time for Earth to cool from its molten state 4.56 Ga. The Jack Hills zircons turn that assumption inside out. If they truly indicate the presence of liquid water on Earth 4.374 Ga, Earth cooled from its molten state to a condition below the boiling point of water in about 180 million years. A duration of 180 million years seems short when you consider the billion year timescales of the early Earth. For comparison, remember that 180 million years before today Earth was in the Middle Jurassic heyday of the Age of Dinosaurs. That was a long time in the context of the history of life on Earth. This evidence also suggests that there may not have been as many meteoritic impacts during this time interval, or the oceans would have been vaporized over and over again. Taken together, these data suggest what is now called the "Cool Early Earth hypothesis."

Where did this early earth water come from? Traditionally, geologists thought that it was water trapped inside Earth's mantle when it cooled, which gradually escaped through volcanoes in a process called degassing. But recent chemical analyses of extraterrestrial objects match the chemistry of Earth's oceans (especially carbonaceous chondrite meteorites). This suggests that there was a lot of water trapped in the debris of the early solar system (of which the chondrites are remnants). The same is true of moon rocks, which do not have much water in them today but apparently were wetter when the solar system formed. If this is so, Earth was born with water already present as it cooled and condensed. The surface temperature only had to drop below 100°C for that water to form the first oceans.

One source we can rule out is comets. Although comets are often called "dirty snowballs" because they are made mostly of water ice and dust, the chemical analyses of four comets now shows that their geochemistry is very different from Earth's water. So the popular idea that comets impacted the early earth and melted to form its oceans can be dismissed.

FOR FURTHER READING

Cloud, Preston. "A Working Model of the Primitive Earth." *American Journal of Science* 272, no. 6 (1972): 537–548.

Drake, Michael J. "Origin of Water in the Terrestrial Planets." *Meteoritics & Planetary Science* 40, no. 4 (2005): 515–565.

Hartmann, William K., and Ron Miller. *The History of Earth: An Illustrated Chronicle of an Evolving Planet.* New York: Workman, 1991.

Hazen, Robert. *The Story of Earth: The First 4.5 Billion Years from Stardust to Living Planet.* New York: Penguin, 2013.

Hopkins, Michelle, T. Mark Harrison, and Craig E. Manning. "Low Heat Flow Inferred from >4 Gyr Zircons Suggests Hadean Plate Boundary Interactions." *Nature* 456, no. 7221 (2008): 493–496.

Marchi, S., W. F. Bottke, L. T. Elkins-Tanton, M. Bierhaus, K. Wuennemann, A. Morbidelli, and D. A. Kring. "Widespread Mixing and Burial of Earth's Hadean Crust by Asteroid Impacts." *Nature* 511 (2014): 578–582.

Shaw, D. M. "Early History of the Earth." *Proceedings of the NATO Advanced Study Institute* (1975): 33–53.

Shaw, George H. *Earth's Early Atmosphere and Oceans, and the Origin of Life.* Berlin: Springer, 2015.

Sleep, N. H., K. Zahnle, and P. S. Neuhoff. "Initiation of Clement Surface Conditions on the Earliest Earth." *Proceedings of the National Academy of Sciences* 98, no. 7 (2001): 3666–3672.

Valley, John W. "A Cool Early Earth?" *Scientific American*, October 2005, 58–65.

Valley, John W., W. H. Peck, E. M. King, and Simon A. Wilde. "A Cool Early Earth." *Geology* 30 (2002): 351–354.

Valley, John W, J. S. Lackey, A. J. Cavosie, C. C. Clechenko, M. J. Spicuzza, M. A. S. Basei, I. N. Bindeman, et al. "4.4 Billion Years of Crustal Maturation: Oxygen Isotopes in Magmatic Zircon." *Contributions to Mineralogy and Petrology* 150, no. 6 (2005): 561–580.

Ward, Peter, and Joseph Kirschvink. *A New History of Life: The Radical New Discoveries About the Origin and Evolution of Life on Earth.* New York: Bloomsbury, 2015.

Wilde, S. A., J. W. Valley, W. H. Peck, and C. M. Graham. "Evidence from Detrital Zircons for the Existence of Continental Crust and Oceans on the Earth 4.4 Gyr Ago." *Nature* 409 (2001): 175–178.

Windley, Brian F. *The Evolving Continents.* New York: Wiley, 1984.

GASP

An adult human can last 40 days without food, a week without any sleep, three days without water, but only five minutes without air. Yet nothing is more taken for granted than the air we breathe. However, not just any air will do—it must be exquisitely designed to meet our needs. Too little oxygen in the atmosphere will kill us, as will too much.

—HUGH ROSS

MOUNTAINS OF IRON

In 1844, the pioneering scientist and inventor William Austin Burt was exploring and mapping on the Upper Peninsula of Michigan and first reported huge deposits of raw iron. Once the word was out, other miners and prospectors began to explore other areas, and rich deposits of iron were eventually found elsewhere in the area, especially in Michigan and in northern Minnesota. Iron ore was found in the Menominee Range in 1867, the Gogebic Range in 1884, the Vermilion Range in 1885, the Mesabi Range in 1890, and the Cuyuna Range in 1903.

In the 1870s, Charles R. Van Hise and his successor Charles Kenneth Leith—among the first geologists at the University of Wisconsin— began formally exploring and mapping northern Michigan, Wisconsin, and Minnesota. They mapped lots of mountains largely made of iron, which had been mined without much understanding for decades. They called these strange rocks "banded iron formations" or "BIFs" for short. As their name suggests, these rocks have red bands of iron a few millimeters to a centimeter thick, alternating with bands of pure silica in the form of chert or jasper (figure 6.1). In some places thousands of alternating bands in a row extend over huge areas of outcrop. When these were first discovered in the mid-1800s, their meaning was a mystery. As Van Hise and Leith noted, the rock is made of pure iron plus chert with little or no mud or sand, which you might normally expect to find washing out into the ancient seas when the iron was being deposited. How such a rock could form was a great puzzle. No one knew that these rocks would eventually be the key to unlocking the mystery of Earth's earliest atmosphere.

Some of the early mines in the Iron Ranges were simple shaft mines, but soon heavy equipment for strip-mining entire mountains of iron away

Figure 6.1 ▲ ▶

Banded iron formations (BIFs) consist of alternating centimeter-scale bands of red iron and black chert, often extending over immense distances and enormous thicknesses. (A) Contorted BIF from the Mesabi Range, Soudan, Minnesota; (B) BIF from Fortescue Fall, Karijini National Park, Western Australia; and (C) a thick sequence of BIFs from the Hamersley Range, Australia, the largest deposit of iron in the world. (Courtesy of Wikimedia Commons)

Figure 6.1 ▲

(*continued*)

at relatively low cost was developed. If you visit the Mesabi Iron Range, or other major ranges such as the Vermilion and Cuyuna Ranges in Minnesota, the Gunflint Range stretching into Canada, or the Marquette and Gogebic and other ranges in the Upper Peninsula of Michigan, you will see amazing sights. Entire mountains made mostly of iron have been removed, leaving immense open-pit mines that today are flooded with water. The Hull-Rust-Mahoning iron mine near Hibbing, Minnesota, is one of the largest mines in the world (figure 6.2). Its pit alone spans 1.5 by 3.5 miles and is more than 600 feet deep. When you stand on the rim, the pit below looks like a small ocean basin. As you look across the water flooding the bottom, you might see modern operations still at work. The bucket of the dragline to remove overburden is larger than a house, and the mining equipment is on a heroic scale. The gigantic excavators and dump trucks have tires that are more than 12 feet in diameter.

Opened in 1895, the Hull-Rust-Mahoning iron mine has produced more than 700 million tons of ore, and more than 500 tons of waste rock lies piled

Figure 6.2 ▲

Panorama of the huge open pit of the Hull-Rust-Mahoning mine. (Courtesy of Wikimedia Commons)

up all over the barren, desolate landscape. The town of Hibbing (birthplace of Bob Dylan) even had to be moved because the mine expanded and swallowed the original location of the town. Almost 25 percent of all the iron produced in the United States came from this hole in the ground, which once was a mountain of iron. Much of the iron that supplied the Industrial Revolution in the late 1800s and early 1900s came from this mine, especially during the huge demand for steel for ships, tanks, and aircraft during World War I and World War II.

The Hull-Rust-Mahoning iron mine is not unique. The Rouchleau mine near Virginia, Minnesota, is 3 miles long, half a mile wide, and 450 feet deep. Opened in 1893, it has produced 300 million tons of iron ore and is still expanding, forcing the Mineview in the Sky mining museum to move to a new location. U.S. Route 53 had to be rerouted away from the expanding pit, and a bridge now spans part of the chasm. Other huge mines are near Soudan, Minnesota, the town waggishly named by miners daydreaming of the hot desert temperatures of the Sudan in Africa while suffered through long, bitter, freezing Minnesota winters.

These mines and their wealth have had a huge influence on U.S. history. The iron of the Lake Superior region meant that American steel could be used to build our great buildings and ships, along with millions of automobiles and other steel machines. Iron from the Iron Ranges was crushed into pellets of iron oxide ore called taconite, then shipped in trains to ports on Lake Superior, especially to Duluth, Minnesota. The iron boats took their cargo across Lake Superior, down Lake Huron, and into Lake Erie and Cleveland, where it was shipped to the steel mills in eastern Ohio and western Pennsylvania. Rivers (such as the Allegheny, Monongahela, and Ohio that surround Pittsburgh) brought barges full of coal from the Appalachian mines nearby to power the furnaces for the smelters that turned raw taconite iron into high-grade steel. Stories of the iron mines and the iron boats even inspired the 1976 hit song "The Wreck of the Edmund Fitzgerald," by the late Canadian folk singer Gordon Lightfoot, which described a 1975 tragedy that sank a famous iron boat.

The resources and culture of the iron mines left a huge imprint on the United States, but by the 1970s and 1980s most of the mines had closed. Cheaper iron ore was coming from many other places in the world, including from the immense Hamersley iron range in the Pilbara region of western Australia. The Hamersley open pit mines are so large that they can be seen from space, and Australia is now the largest iron producer in the world. In 2014, Australia produced 430 million tons of ore, most of it from the mines in the Hamersley Range. Some geologists estimate that Australia has 24,000 million tons of ore remaining in its iron ranges. In contrast, the United States only produced 58 million tons in 2014. However, the huge demand for steel from China in recent years has exceeded what the Australian mines can produce, and some of the U.S. iron mines have reopened.

BIFs, GIFs, AND LIPs

How did so much iron become concentrated in limited places such as the Iron Ranges of Minnesota and the Hamersley Range in Australia? Most of these deposits are BIFs, and geologists have been puzzling about these rocks for many years. Hundreds of studies have been done, and a 1968 paper by the pioneering Precambrian geologist Preston Cloud first suggested the version of the model that is widely held today: These concentrations of iron could only have formed in a world with very low or no free oxygen.

How could sediments consisting of dissolved iron and silica be deposited and settle out on the seafloor without being mixed with sand and mud? The first thing to know is that iron cannot stay dissolved in modern seawater today because it is rapidly oxidized into various forms of iron oxide ("rust") and clings to other minerals or settles out. The only way to transport and concentrate huge amounts of iron in seawater is if the oceanic oxygen content is so low that iron cannot rust. When the iron formations were deposited, the ancient seas must have been completely anoxic. Most geologists think the atmosphere must have been very low in oxygen as well.

Next, the seafloor needs to be far enough from land so almost no sand or mud from the land can mix in the deep ocean basin with the chemical deposits of iron and silica. Perhaps the iron basins were in the center of ancient seas, and the sands and muds were trapped in basins on the edge of the ancient continents. However, the Hamersley deposits seem to have formed on a shallow marine shelf, so this model is not true of all BIFs.

Finally, it would be a lot easier to deposit huge concentrations of iron if an abundant source of dissolved iron was entering the ocean. Most geologists think the iron came largely from the weathering of the basaltic lavas (which are iron rich) in the ancient midocean ridges and possibly from dissolved iron from the weathering of land rocks (which would only be possible if the rivers were completely anoxic as well). Analyses of rare trace elements such as europium suggest that the older deposits (Archean "Algoma-type" deposits in the Great Lakes region) derived their iron from the dissolution of iron from deep-sea volcanic eruptions along ancient midocean ridges. The younger late Proterozoic ("Superior-type") deposits have geochemical signatures that suggest they came from weathering of the land. Geologists working on BIFs recently have noticed that some of the largest deposits occurred when Earth experienced gigantic eruptions of flood basalt, known as "large igneous provinces" (LIPs). This excessive eruption of lava would have weathered to produce a lot of iron as long as the atmosphere and ocean were low enough in oxygen for iron to remain in solution and not rust.

Banded iron formations are found in some of the oldest rocks on Earth, including the 3.7-billion-year-old rocks of Greenland, the Isua Supracrustals mentioned in chapter 7. Most of the world's BIFs were produced during the Archean, 4.0 to about 2.5 Ga (figure 6.3), when Earth not only had an anoxic atmosphere but also was covered by small protocontinents bashing around in proto-oceans made of a weird lava called komatiite. By 2.6 to 2.4 Ga, the largest volume of BIFs had been deposited, especially the huge

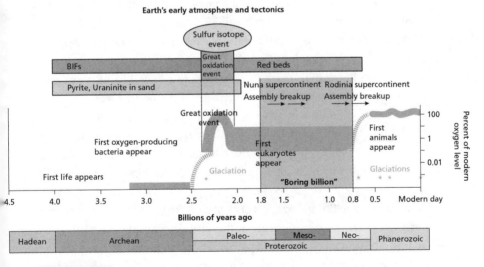

Figure 6.3 ▲

Timescale of the appearance of banded iron formations, stromatolites, and oxygen concentrations in the Precambrian. (Redrawn from several sources)

mountains of iron in the Hamersley Range of Australia, the Iron Ranges around Lake Superior, and similar deposits in Brazil, Russia, Ukraine, China, and South Africa. This time window was also when huge eruptions of LIPs was at its peak.

THE OXYGEN HOLOCAUST

Around 2.4 to 2.3 Ga something happened. The BIFs began to disappear, although large deposits of iron in granular form rather than banded form remained, known as granular iron formations (GIFs). By 1.9 Ga, BIFs and GIFs had vanished completely, with the exception of a few late occurrences during a "snowball earth" episode 750 to 580 Ma (see chapter 8). Most geologists regard the disappearance of iron formations as the time when oxygen finally began to reach significant levels in Earth's atmosphere, and possibly in the ocean as well. Geologists refer to this as the "Great Oxidation Event" (GOE). Oxygen was still nowhere near the 21 percent now found in Earth's atmosphere, but it went from almost nothing before 2.4 Ga to about 1 percent of present levels in the oceans, which is enough to begin rusting the dissolved iron in the oceans. By 1.9 Ga, geologists believe oxygen levels in the oceans were high enough for oxygen to escape into the atmosphere, possibly weathering rocks on land, although oxygen was still not abundant in the atmosphere. Only in the last 500 million years did oxygen reach today's levels (about 21%), so they saturated the oceans and the atmosphere so they are now completely oxygenated.

How do we know that oxygen levels were this low? The best evidence comes from the BIFs, which could only form if the oceans were so low in oxygen that iron could stay dissolved rather than rusting and being left behind. There are other geochemical indicators as well. Before 1.9 to 1.8 Ga, sand grains and pebbles in ancient river deposits were often made of pyrite or "fool's gold," which is iron sulfide (FeS_2). Today pyrite only forms in places with very low oxygen levels, such as at the bottom of stagnant bodies of water or in deep hot springs and crustal rocks that are far from the atmosphere. Once pyrite grains weather at the surface, they quickly break down into lumps of iron oxide. I have collected specimens of iron oxide (hematite) that still had the crystal form of pyrite, even though the mineralogy had changed. As the pyrite breaks down, some iron is released and the sulfur is oxidized into sulfate becoming minerals such as gypsum (calcium sulfate, or $CaSO_4$). Few significant gypsum deposits are more than 1.8 billion years old, and few pyrite pebbles or sand grains have been found beyond that same time frame. Sand grains made of uranium oxide (uraninite, UO_2) are known before 1.7 Ga, but they are never found after that time. Like pyrite sand and dissolved iron, these minerals are unstable in an oxygen-rich atmosphere.

If we look at the record of isotopes of carbon through time, we no longer see really low values associated with low oxygen after about 2.2 Ga. Sulfur isotope values in Archean rocks are highly variable, fluctuating all over the place. But after 2.4 Ga, the sulfur isotopes are highly stable: they are no longer floating free in minerals like pyrite but are stabilized in gypsum and other minerals common in an oxygen-rich world.

The world went through a dramatic transformation when oxygen became available. The Great Oxidation Event has also been nicknamed the Oxygen Holocaust because life on the planet that was used to anoxic conditions found the appearance of such a reactive molecule as O_2 poisonous (see chapter 7). Today bacteria and other microbes adapted to low-oxygen conditions must live in oxygen-starved places like the bottoms of stagnant lakes and marine basins like the Black Sea. Before 2.3 Ga, however, they ruled the planet. Once the atmosphere became too rich in oxygen, it truly was a holocaust for them, and they were replaced by microbes that could survive oxygen-rich conditions.

The burning question then arises: Where did Earth's atmosphere get its free oxygen? The answer is clear: photosynthesis, first from blue-green bacteria or cyanobacteria, and around 1.8 Ga when eukaryotic "true" algae evolved from plants as well. The big puzzle is that we have cyanobacterial fossils from 3.5 and possibly even 3.8 Ga, but the GOE didn't begin until about 2.3 to 1.9 Ga. Was their oxygen production so meager that they didn't make a dent on the planet? Did they produce lots of oxygen, but most of it was locked into oxidized crustal rocks (such as BIFs)? Was so much oxygen finally produced that the crustal reservoirs were saturated and free oxygen was left over? Perhaps true eukaryotic algae didn't evolve until 2.3 Ga, with their much larger cells and bigger oxygen production. Maybe only true algae could produce enough oxygen to overwhelm Earth's oxygen sinks.

The argument remains controversial and speculative, and there is no consensus on the answer. But it is clear is that 1.7 Ga there were true eukaryotic algae everywhere, and there was an atmosphere with about 1 percent or maybe even more oxygen, forever changing Earth's oxygen balance. Without some free oxygen, multicellular animals could not have evolved—and we would not be discussing this issue because humans could never have evolved either. In fact, the evolution of all life as we know it depends on an oxygen-rich planet, which isn't possible without the evolution of photosynthetic microbes and plants.

This places a severe restriction on the speculative ideas about extraterrestrials and alien life on other planets as well. It's true that astronomers have found lots of other planets with Earth-like properties, including the right size, the right temperatures, and possibly even liquid water oceans on their surface. But so far not one has evidence of free oxygen in its atmosphere. Without it, there is no multicellular animal life and no alien being similar to those appearing in science fiction movies (and in the entire culture of people who believe in aliens and UFOs). It's possible that there are anoxic microbes in the deep crustal rocks of other planets, but without abundant free oxygen the aliens of other planets and in our imagination don't exist.

FOR FURTHER READING

Canfield, Donald E. *Oxygen: A Four Billion Year History*. Princeton, NJ: Princeton University Press, 2014.

Hazen, Robert M. *The Story of the Earth: The First 4.5 Billion Years from Stardust to Living Planet*. New York: Penguin, 2013.

Knoll, Andrew H. *Life on a Young Planet: The First Three Billion Years of Evolution on Earth*. Princeton, NJ: Princeton University Press, 2003.

Lane, Nick. *Oxygen: The Molecule That Made the World*. Oxford: Oxford University Press, 2003.

Schopf, J. William. *Cradle of Life: The Discovery of Earth's Earliest Fossils*. Princeton, NJ: Princeton University Press, 1999.

Shaw, George H. *Earth's Early Atmosphere and Oceans, and the Origin of Life*. Berlin: Springer, 2015.

Ward, Peter, and Joe Kirschvink. *A New History of Life: The Radical New Discoveries About the Origin and Evolution of Life on Earth*. New York: Bloomsbury, 2015.

07

PLANET OF THE SCUM

If the theory [of evolution] be true, it is indisputable that before the lowest Cambrian stratum was deposited, long periods elapsed . . . and the world swarmed with living creatures. [Yet] to the question why we do not find rich fossiliferous deposits belonging to these earliest periods . . . I can give no satisfactory answer.

—CHARLES DARWIN, *ON THE ORIGIN OF SPECIES* (1859)

DARWIN'S DILEMMA

When Charles Darwin first published *On the Origin of Species* in 1859, he knew that one of the weakest links in his argument was the poor fossil record of early life. At that time, the lowest fossiliferous beds of the Cambrian (full of trilobites and other archaic fossils) produced the oldest fossils known, and there were no apparent megascopic fossils in the beds below. Darwin confessed the problem and admitted he had no answer, although he did spend two chapters in his book bemoaning the poor quality of the fossil record. As the fossil record got better, Darwin suggested that we would find these missing fossils (which indeed we have).

Finding such early fossils is no easy task. First of all, just a handful of places have rocks from the first half-billion years of Earth's history, and most of those are highly metamorphosed and all evidence of fossils has been destroyed. The older the rock, the greater the chance it was eroded away long ago, is still buried under younger rocks, or is highly metamorphosed so much of the key evidence is missing. The oldest rocks on Earth are just over 4.32 billion years old (from the Quebec shore of eastern Hudson Bay), and there are structures and biochemicals in these rocks that are suggestive of life. Some of the Australian Jack Hills zircons have organic carbon in them, and they are 4.1 billion years old (chapter 5). Lower-grade metasedimentary rocks in Greenland (the Isua supracrustals) have yielded organic molecules suggestive of life 3.9 Ga, and they have the oldest structures that might be megascopic fossils. Whether these structures are true fossils or pseudofossils and whether the carbon in these rocks is truly ancient or is a later contaminant is still being discussed, but it seems likely that some form of life was established at least 4.1 Ga.

In addition, the dates from meteorite impacts on the moon show that the early solar system was heavily bombarded by junk left over from the coalescence of the planets, with large impacts pounding both Earth and the moon from 4.6 to 3.9 Ga. This period of "impact frustration" prior to 3.9 Ga is thought to have vaporized the planet's surface waters over and over again, preventing life from becoming established in the world's shallow oceans or lakes (but not affecting life that arose in deep sea volcanic vents, as many think).

The solution to Darwin's dilemma about the lack of apparent fossils in Precambrian rocks is that we were looking for them in the wrong way. The fossils were there all along, but they are nearly always microscopic. It wasn't until the 1940s and 1950s that Stanley Tyler and Elso Barghoorn found cherts and flints, such as the 2-billion-year-old Gunflint Chert in Canada, that preserved these delicate microfossils and made it possible for us to study them.

Finally, scientists found several localities in Australia and South Africa dated from 3.4 to 3.5 Ga that yield specimens that are undoubtedly fossils of the most ancient forms of life (figure 7.1). The most famous of these are found in a rock unit called the Apex Chert, from the Warrawoona Group in Western Australia, and are dated at 3.5 Ga. They are on an isolated ridge in mountains deep in the dry outback, a place so hot and inhospitable and remote from civilization that Australians jokingly called it the "North Pole." Slightly younger fossils are known from the Fig Tree Group in South Africa, dated at 3.4 Ga.

These fossils are of blue-green bacteria, or cyanobacteria—prokaryotes with the ability to photosynthesize and produce their own food. These were

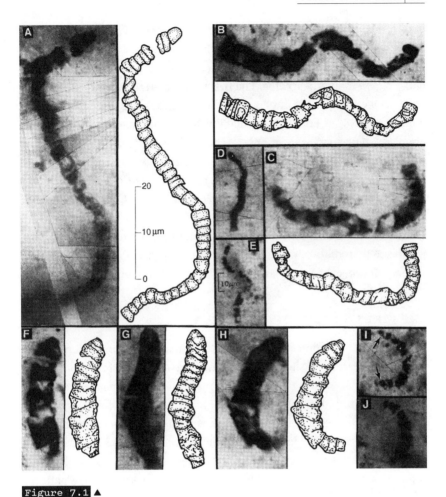

Figure 7.1 ▲

Bacterial microfossils from the 3.5-billion-year-old Apex Chert, Warrawoona Group, Australia. (Courtesy of J. W. Schopf)

called "blue-green algae" in the past, but they are not true algae. Algae are eukaryotic members of the Plant Kingdom, and cyanobacteria are prokaryotes. These fossils show that long strings of filamentous cyanobacteria were well established in the shallow oceans all over the world.

As it was in the beginning, so it was for almost another 2 billion years. There are hundreds of Precambrian microfossil localities around the world in rocks dated between 3.5 and 1.75 Ga, and they yield plenty of good examples of prokaryotes. These fossils show an extraordinarily slow rate of evolution;

in fact, it appears that cyanobacteria evolve slower than anything we know. They show almost no visible change in 3.5 billion years. Everywhere we look in rocks between 3.5 and about 1.8 billion years old, we see nothing more complicated than prokaryotes. The first fossil cells that are large enough to have been eukaryotes (complex cells with a nucleus and organelles) do not appear until 1.8 Ga, and multicellular life does not appear until 600 Ma. For almost 2 billion years, or about 60 percent of life's history, there was nothing on the planet more complicated than a bacterium or a microbial mat, and for almost 3 billion years, or 85 percent of Earth's history, there was nothing more complicated than single-celled organisms. As J. William Schopf of UCLA put it, Earth was truly the "planet of the scum." If aliens existed and had visited the planet long ago, odds are there was nothing more interesting to see than mats of cyanobacteria—and they probably would have left immediately because this planet was so primitive and uninteresting!

THE FIRST MEGASCOPIC FOSSILS

The search for microfossils took a long time to develop, and when megascopic Precambrian fossils were discovered, no one knew for sure whether they were fossils. During the 1800s, numerous false alarms were recorded of structures found in the rocks and claimed to be early fossils that were debunked at a later date. In retrospect, it is easy to see why people were fooled. Most geologists learn early in their careers that the world is full of pseudofossils, objects that look like possible fossils until you look closer. Almost every amateur rock hound is fooled by the very plantlike patterns of pyrolusite dendrites, a mineral structure of manganese oxide that looks just like a branching fern. The most common pseudofossils are concretions, which are grains of sand or mud cemented together to form a variety of shapes. Most are shaped like spheres or balls or odd shapeless blobs, but many have bizarre shapes that untrained amateurs visualize as a "fossil brain" or "fossil phallus" or many other shapes that fool our tendency to see a "pattern" where there is none.

Like seeing "patterns" in clouds, or "patterns" of stars in a constellation, humans are hardwired to infer meaning and pattern in nearly any collection of random images, a phenomenon known as *pareidolia* or "patternicity." Veteran geologists learn to be skeptical of interpreting any odd-shaped rock as a fossil, and it takes years of experience to tell one from the other—especially in the early days of geology when most sedimentary structures, and structures formed by burrowing, had not yet been defined and distinguished from true body fossils.

One of these controversial structures was formations known as stromatolites ("layered rocks" in Greek). In cross section, they looked like layers of a cabbage (figure 7.2A), and in top view they showed layers wrapped around the central core (figure 7.2B). Some were short and formed low mounds, and others seemed to form tall cylindrical pillars. For the first half of the twentieth century, the geological and paleontological community was deeply divided

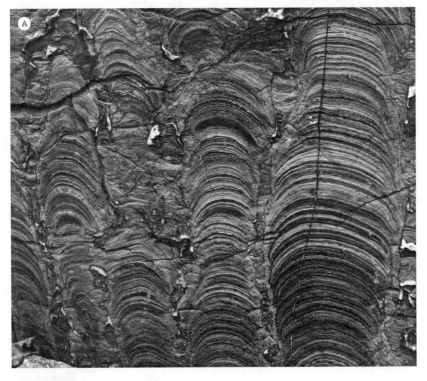

Figure 7.2 ▲ ▶

Stromatolites are sedimentary structures formed by domed algal and cyanobacterial mats that trap sediment as the filaments grow up toward the light. (A) In vertical cross section, they resemble layers of a cabbage. (B) A top view of stromatolites showing layers around a central core. This specimen was sliced open by glaciers and is from the Hoyt Limestone, Lester Park, Saratoga, New York. ([A] Courtesy of Wikimedia Commons; [B] photo by author)

Figure 7.2 ▲
(*continued*)

about what stromatolites were. Study after study had produced no signs of organic material or preserved cells in layers, so the case seemed weak. As long as there were no modern examples of these structures living and growing today, there was no convincing evidence to silence the doubters.

In 1956, the geologist Brian W. Logan of the University of Western Australia in Perth and some other geologists were exploring the northern coast of Western Australia. Logan and his colleagues came across a lagoon known as Shark Bay, about 800 kilometers (500 miles) north of

Figure 7.3 ▲
In Shark Bay, Australia, modern stromatolites grow as tall pillar-like shapes because the lagoon at Hamelin Pool is too salty for grazers to eat the cyanobacterial mats. (Courtesy of Wikimedia Commons)

Perth. When the tide went out in Hamelin Pool on the southern shore of the bay, they saw a landscape that had not been seen on Earth in 500 million years! The bottom of the bay was covered by 1 to 2 meter tall cylindrical towers with domed tops (figure 7.3). They were dead ringers for many of the Precambrian stromatolites—but they were still alive and growing! Closer inspection showed that these pillars and towers were made of millimeter-scale finely layered sediment, just like ancient stromatolites. On top were the organisms that produced these mysterious structures: sticky mats of blue-green bacteria, or cyanobacteria. Cyanobacteria are not only among the most primitive and simple forms of life on Earth but were probably the first photosynthetic life on Earth. Most important to the subject of this book, photosynthesis by cyanobacteria produced the free oxygen that transformed the atmosphere and caused the Great Oxidation Event and the Oxygen Holocaust (see chapter 6).

Studies of the Shark Bay stromatolites revealed how they produce their finely layered structure. These slimy mats of blue-green bacteria grow very rapidly toward the sun when the tide comes in and immerses them in the daytime. The freshly growing mats have a sticky surface that traps more sediment, especially when it is dark or the tide is going out and the cyanobacteria stop growing for a few hours. When the tide is high and the sun is up again, the bacteria grow new filaments reaching up to the sun, and they completely engulf the layer of sediment that accumulated the previous night. This goes on, day after day, year after year. In an area with favorable conditions, hundreds of individual growth layers of sediment are trapped by this daily mat growth. The organic material of the bacteria gradually decays away, leaving just the layered sediments with no organic structures or chemical traces of their existence.

If this process is so easy, why don't we have stromatolites growing everywhere on Earth, just as they did in the Precambrian? Shark Bay suggested an answer to that as well. In the shallow waters of Hamelin Pool, a bar of sand across the mouth of the bay restricts the flow of water in and out, so the waters are extremely salty. In addition, the subtropical desert-belt location is extremely hot and sunny, and as the water evaporates, the shallow bay sediments become saltier and saltier. They have twice the salinity of the ocean (over 7 percent salt compared to 3.5 percent in the oceans), and only the cyanobacteria can tolerate these conditions. Grazing snails (limpets and periwinkles and abalones in our modern tide pools) would normally eat these bacterial mats, but they cannot live in such salty waters so the mats just keep growing, uncropped. This is very much like the world of the Precambrian, when advanced marine grazers like snails had yet to evolve. For 3 billion years, the most complex forms of life were microbial mats, and eventually algal mats, and nothing hindered their growth.

Since the discovery in Shark Bay in 1956 (first published in 1961), living stromatolites have been found in many places on Earth. Most of them have one key feature in common: they grow in places where the conditions are too hostile for more advanced forms of life (such as grazing snails that eat these mats) to survive. I've seen them close-up, growing in salty lagoons along the Pacific Coast of Baja California. They grow in the salty waters of the west coast of the Persian Gulf, and huge dome-topped pillars like those at Shark Bay also grow in the salty lagoons of Lagoa Salgada ("salty lagoon" in Portuguese) in Brazil. A few grow in normal marine salinity in Exuma Cays in the Bahamas, but the water currents are so strong there that even snails like limpets and periwinkles cannot hang on.

Since the discovery at Shark Bay, more and more fossil stromatolites have been found, including some as old as life itself. These include probable stromatolites from the previously mentioned Warrawoona Group in Western Australia (only a few hundred kilometers east of Shark Bay) that are 3.5 billion years old, along with the oldest microscopic evidence of cells of cyanobacteria. There are well-established stromatolites from the 3.4-billion-year old Fig Tree Group in South Africa. By 1.25 Ga, stromatolites were at the peak of their diversity in shapes and size and abundance and were the only visible evidence of life on the planet. They began a slow decline over the next 500,000 years, and by the Cambrian stromatolites were reduced to 20 percent of their original abundance—this was probably due to the huge number of new grazing creatures that cropped them anywhere they grew in normal marine waters. (The Lester Park stromatolites in figure 7.2B are in the Middle Cambrian Hoyt Limestone and are among the few exceptions of stromatolites that survived into the Cambrian.) By the time of the huge radiation of invertebrate life in the Ordovician (about 500 million years ago), stromatolites had nearly vanished from the planet.

However rare they have been for the last 500 million years, microbial mats seem ready to spring back and flourish any time their predators are suppressed. After three of Earth's great mass extinctions (the end-Ordovician, the late Devonian, and the biggest mass extinction of all at the end of the Permian Period), stromatolites returned in abundance during the post-mass-extinction "aftermath." In each case, stromatolites grew like weeds, taking advantage of the wide-open landscape and flourishing whenever the creatures that ate them were wiped out.

FOR FURTHER READING

Bell, Elizabeth A., Patrick Boehnke, T. Mark Harrison, and Wendy L. Mao. "Potentially Biogenic Carbon Preserved in a 4.1 Billion-Year-Old Zircon." *Proceedings of the National Academy of Sciences* 112, no. 47 (2015): 14518–14521.

Grotzinger, John P., and Andrew H. Knoll. "Stromatolites in Precambrian Carbonates: Evolutionary Mileposts or Environmental Dipsticks?" *Annual Review of Earth and Planetary Sciences* 27 (1999): 313–358.

Knoll, Andrew H. *Life on a Young Planet: The First Three Billion Years of Evolution on Earth.* Princeton, NJ: Princeton University Press, 2003.

Lepot, K., K. H. Williford, T. Ushikubo, K. Sugitani, K. Mimura, M. J. Spicuzza, and J. W. Valley. "Biogenicity of 3.4 Gyr Old Carbon Indicated by Texture-Specific Isotopic Compositions." *Geochimica et Cosmochimica Acta* 112 (2013): 66–86.

Morag, N., K. H. Williford, K. Kitajima, P. Philippot, M. J. Van Kranendonk, K. Lepot, C. Thomazo, and J. W. Valley. "Microstructure-Specific Carbon Isotopic Signatures of Organic Matter from ~3.5 Ga Cherts of the Pilbara Craton Support a Biologic Origin." *Precambrian Research* 275 (2016): 429–449.

Schopf, J. William. *Cradle of Life: The Discovery of Earth's Earliest Fossils.* Princeton, NJ: Princeton University Press, 1999.

——, ed. *Earth's Earliest Biosphere: Its Origin and Development.* Princeton, NJ: Princeton University Press, 1983.

08

SNOWBALL EARTH

I think about the cosmic snowball theory. A few million years from now
the sun will burn out and lose its gravitational pull. The earth will turn
into a giant snowball and be hurled through space. When that happens, it
won't matter if I get this guy out.

—BILL LEE, BASEBALL PITCHER

THE "SNOWBALL EARTH"

As early as the 1930s, Australian geologists such as the legendary polar
explorer Sir Douglas Mawson noticed some peculiar rock formations from
the Flinders Ranges of South Australia. In the middle of upper Precambrian
sequences of marine limestones (formed in shallow warm tropical seas),
there were layers of angular boulders and cobbles mixed with sand and
mud that suggested a glacial deposit called glacial till. Till is an unsorted,
unstratified mass of boulders, gravels, sand, and mud that are randomly
dumped by the snout of the glacier as it melts in one place. These sediments
are distinctive and almost no other geological process produces anything
like them, so ancient glaciation can be recognized in the rock record. However, many geologists prefer the term "diamictite" (meaning "thoroughly
mixed" in Greek) or sometimes "tilloid" to describe rock with this texture
without directly implying that is it glacial.

Following work by Oskar Kulling in 1934 and by Walter Howchin, Mawson was eventually convinced that this was evidence of a global glacial
event in the late Proterozoic because the deposits in Australia were not

far from the equator today. By the late 1950s and early 1960s, geologists began to dismiss Mawson's argument because plate tectonics had shown that Australia and other continents had moved long distances over time. Conceivably, the Australian Precambrian glacial beds could be from a time when the continent was closer to the South Pole. Ironically, we now know that Australia was right on the equator at that time, even more tropical than Mawson thought, so his evidence was stronger than anyone realized.

In 1964, Cambridge geologist W. Brian Harland published a famous paper showing that tropical upper Precambrian glacial deposits were not restricted to Australia (figure 8.1). Like Mawson, Harland was familiar with

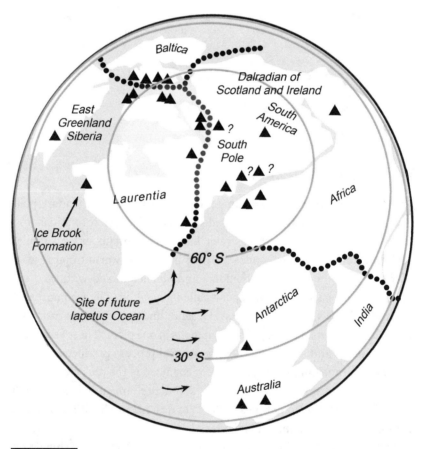

Figure 8.1 ▲

Map showing the location of major late Proterozoic glacial deposits. (Redrawn by K. Marriott from several sources)

glaciers and ice sheets firsthand because over his lifetime he spent a lot of time in the Arctic. He established the Cambridge Arctic Shelf Program, and he was part of an amazing 43 polar field seasons (he led 29 of them) from 1938 to the 1960s, mapping the geology of the archipelago of Svalbard (= Spitzbergen), a group of islands north of Norway and Greenland. In addition to deposits of recently melted glaciers, Harland saw upper Precambrian glacial deposits in abundance—not only on Svalbard but also in Greenland and Norway.

Harland pointed out that many of the upper Precambrian ice deposits were sandwiched between layers of limestone (figure 8.2). This was surprising because today limestones are only formed in warm tropical or subtropical shallow marine settings such as in the Bahamas, Florida, Yucatan, the Persian Gulf, and the South Pacific. If this ice bed–limestone sandwich had been formed by modern processes, the glacial deposit surrounded by limestones had to form in the tropics and at sea level. Tropical glaciers are known in a few places today—at the top of Mt. Kilimanjaro in Tanzania, on Mt. Kenya in Kenya, and also in the Peruvian Andes—but these places are in the high mountains. It seemed impossible to imagine sea level tropical glaciers, but Harland's evidence was inescapable. If the tropics were glaciated, so were the poles, and so was the whole planet in the late Precambrian.

Harland backed up his conclusions with a new line of evidence: paleomagnetism. He was one of the first to measure the magnetic directions frozen in the ancient rocks, which can tell us the latitude when a given rock unit was formed. The paleomagnetic directions for all of these Svalbard, Greenland, and Norwegian rocks were tropical or subtropical in the Precambrian, so the limestone-till-limestone sandwich was not some fluke. The paleomagnetic data from Australia was not very good then, but later analyses showed that Mawson's Precambrian ice bed–limestone sandwiches were right on the equator. Clearly something weird was going on if there was indisputable evidence of tropical sea level ice at that time.

How do you freeze over a planet once the ice sheets start to grow? The answer to this came from a surprising direction: climate modeling. In 1969, the Russian geophysicist Mikhail Budyko of the Leningrad Geophysical Observatory published a paper that showed how easy it is to freeze over a planet once the ice sheets start to grow; he described this as the albedo feedback loop.[1] Albedo is just a fancy word describing the reflectivity of a surface. If you've ever spent time skiing or snowboarding, you know that

Figure 8.2 ▲

The Elatina sequence in the Flinders Ranges of Australia were at the equator in the late Proterozoic and show the classic sequence of limestone–glacial till–limestone that could only
be explained by sea level glaciers in the tropics. (*A*) The Elatina glacial deposits capped by
limestones; (*B*) a close-up of the Elatina till, with huge boulders dumped into finer-grained
sediments. (Courtesy of Wikimedia Commons)

snow or an ice sheet has a high albedo, reflecting most of the sunlight that hits it. Skiers and snowboarders wear dark goggles with polarized lenses that reduce glare and tinted lenses so they don't go "snowblind." In contrast, dark surfaces (such as forests or the open ocean) absorb a lot more sunlight and reflect very little of it.

The albedo feedback system is sensitive to small changes, which can produce rapid changes from frozen to ice-free and back to frozen quickly (figure 8.3). For example, let's say Earth's surface is covered by ice, and it

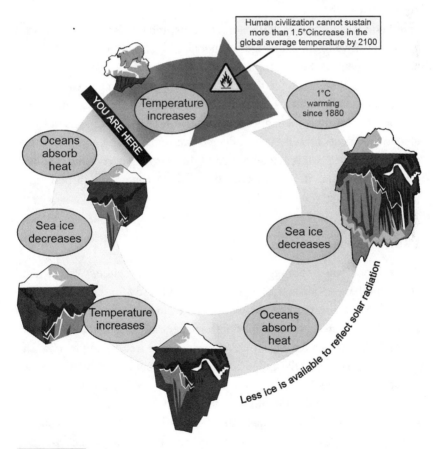

Figure 8.3 ▲

The albedo feedback loop, starting with an ice-covered surface with high albedo, which gradually melts, exposing dark rocks or oceans that amplify the warming until the ice is completely melted. (Drawing by K. Marriott)

has a high albedo and reflects most of the sun's energy back into space. If the planet begins to warm slightly, that ice sheet melts back a bit, exposing dark land and water. This absorbs more sunlight and generates heat, which melts even more ice. Back and forth these two processes go in a feedback loop that eventually melts the ice in a very short time. Now let's image that this dark land and ocean surface has a few really cold winters and cool summers and the reflective snow and ice layer lasts a bit longer. The increased ice cover reflects more energy into space, and the land gets colder; even more ice sticks around the next few winters, and the ice sheet expands. Before you know it, the entire system has switched back into a complete ice age.

Albedo is a key feature of the polar regions, and that is why these regions are so sensitive to small changes in global temperature. If you had even a small ice sheet in the subtropical or tropical latitudes, climatic modeling predicts that the albedo feedback loop would kick into high gear, and the entire planet could freeze over rapidly. The only dilemma with this model was how to thaw the planet once it is completely frozen and has such a high albedo that most of its energy is reflected back into space. A completely frozen reflective ice ball is a dead end, and the warming part of the feedback loop cannot rescue it. So how does the planet escape being turned into a permanently frozen snowball?

ESCAPING THE ICE PLANET

The solution was first suggested in a 1981 paper by James Walker, Paul Hays, and James Kasting.[2] They were focused mostly on the way in which weathering of silicate minerals in soils in the landscape can absorb carbon dioxide, but in the last paragraphs of the paper, they talked about Budyko's models and how an ice cap would shut down the weathering mechanism and lead to what Budyko called an "ice catastrophe." In a brief sentence in the last paragraphs, they suggested another possible mechanism: volcanoes. Earth is unlike any other frozen planet in space because it has an active crust with plate tectonics that powers lots of volcanoes. Volcanic eruptions release lots of gases, especially greenhouse gases such as carbon dioxide, water, methane, and sulfur dioxide. If Earth were indeed completely frozen, the volcanic gases would eventually build up and warm the planet through the greenhouse effect, and the ice would finally begin to melt. Once enough dark surface had been exposed, the albedo feedback loop could kick into high gear and quickly melt a frozen planet into an ice-free subtropical planet capable of forming limestones in the tropics.

The idea was in print and discussed by the few people working on the late Precambrian ice deposits but was not widely known. This all changed with a legendary paper by my good friend Joe Kirschvink, now the Niko and Marilyn Van Wingen Professor of Geobiology at Caltech. Kirschvink is one of the most brilliant people I have ever met, and he has more great ideas in a single month than most people have in a lifetime. He is one of the world's best paleomagnetists. He also does research on all sorts of problems on the boundary between geology and biology, from magnetic bacteria and butterflies and human biomagnetism, magnetofossils, and biomineralization, to innovative ideas about the Cambrian explosion, to climate change and geochemical modeling, to true polar wander and reconstructing ancient continental positions. In addition, Joe designs, builds, and maintains his own lab equipment, and he even writes his own computer programs. He is an outstanding, provocative, mind-expanding teacher who challenges his brilliant students at Caltech to push the boundaries. He has won the Feynman Prize for teaching at Caltech, the William Gilbert Award in Paleomagnetism from the American Geophysical Union, and even has an asteroid named after him.

In 1989, Kirschvink put his mind to the problem of how the snowball earth escapes being totally frozen over, and he revived the solution proposed by Walker, Hays, and Kasting in 1981. They were all part of the PPRG (Precambrian Paleobiology Research Group), organized by my good friend Bill Schopf of UCLA. At a PPRG meeting in 1989, Kirschvink not only revived the volcanic solution to a frozen-over earth but pointed to evidence from Mawson's Elatina Formation in Australia that it had actually occurred. Most important, he coined the term "snowball earth," which was catchy and memorable and shifted the focus from soil weathering to volcanoes and a frozen planet.

Kirschvink has one of the world's best paleomagnetic labs at his disposal, so he did new analyses of the ancient latitudes of these Proterozoic ice bed–limestone sandwiches, showing that many of them (especially Mawson's Elatina sequence in Australia) were tropical or subtropical. Then he wrote up these ideas in a short paper tucked away in a huge expensive symposium volume from the 1989 PPRG meeting about Precambrian life.

After many delays, it was published in 1992 (it was so expensive that few people could own it), and Kirschvink moved on to other problems. Most people would have published such a groundbreaking paper in *Nature* or *Science*, but Kirschvink doesn't need the glory. He has so many great ideas all the time that he doesn't spend time promoting each one for long. The idea of a "snowball earth" was formally named and proposed, with a clear mechanism for how to make it work, and other geologists soon caught on.

Further study of some of the limestone-glacial deposit-limestone sequences yielded interesting clues. One of the unusual features is that the limestones on top of the glacial till are particularly thick and well developed, and they have some peculiar geochemical and mineralogical characteristics. These "cap carbonates" sitting on top of the glacial tills may be products of direct precipitation of limestones once the ocean geochemistry, saturated with dissolved carbonate, has been released from the grip of the ice (figure 8.4). They are clearly not the normal kind of

Figure 8.4 ▲

These glacial deposits in Namibia full of boulders and poorly sorted till are covered by cap carbonates. Dan Schrag on the left and Paul Hoffman on the right for scale. (Courtesy of P. Hoffman)

limestones formed today, which are precipitated from shells of sea creatures such as corals, mollusks, and other marine creatures, as well as calcareous algae.

Another suggestive piece of evidence is the brief return of banded iron formations (BIFs) during the peak of the late Proterozoic snowball conditions (see chapter 6). This would make sense if Earth was frozen because freezing would shut down the oceans. They would be anoxic and saturated with dissolved carbonate and become highly acidic (similar to what we are doing to our oceans now thanks to our greenhouse gases). With runoff from the sediment flowing down rivers completely frozen, the sulfate input to the oceans would be shut off, resulting in abundant dissolved iron in these acidic, low-oxygen, low-sulfur oceans. Under these conditions, iron could accumulate on the bottom as it did between 3.7 and 1.7 Ga.

The main "snowball earth" model is shown in figure 8.5. Something (such as the uplift of Rodinia and increased weathering) causes the planet to begin to cool down dramatically, and large ice sheets begin to form. Without abundant and complex life (such as we have today) to regulate the carbon cycle and keep pumping carbon dioxide into the atmosphere, Earth would begin to experience a runaway albedo feedback loop and eventually freeze over to the equator. Once it was a frozen snowball, it would be stuck in that state for millions of years, just like Mars is completely frozen now (although it once had liquid surface water with oceans and rivers). Oceanic circulation would shut down, BIFs would accumulate on the anoxic seafloor, and lots of carbon would be frozen into little cages of ice (known as methane hydrates) in the seafloor sediments. If nothing else happened, Earth would have stayed frozen, and we would not be here.

Unlike Mars or any other planet, however, Earth has plate tectonics and volcanoes, and over a long enough time greenhouse gases erupt and finally begin to warm the planet. Once that occurs, another runaway albedo feedback loop kicks in, and the ice melts rapidly until it is almost all gone. The carbon caged in ice in methane hydrates on the seafloor releases huge amounts of methane, further accelerating global warming. The ocean geochemistry is now so rich in dissolved carbonate that huge deposits of calcite precipitate directly out of seawater to form cap carbonates. Finally, Earth is stable again with warm tropics and cooler poles.

How Snowball Earth Happens

Extreme greenhouse and ice house states are cyclical.

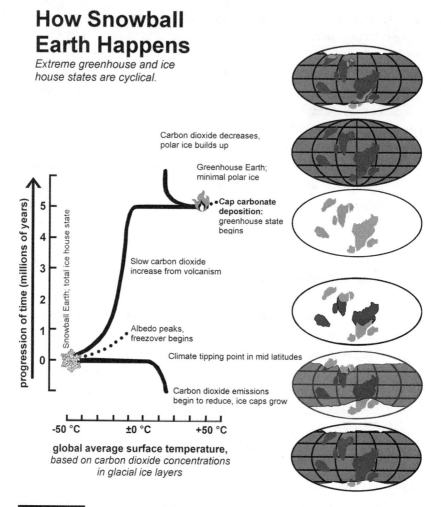

Carbon dioxide decreases, polar ice builds up

Greenhouse Earth; minimal polar ice

• Cap carbonate **deposition:** greenhouse state begins

Slow carbon dioxide increase from volcanism

Albedo peaks, freezover begins

Climate tipping point in mid latitudes

Carbon dioxide emissions begin to reduce, ice caps grow

progression of time (millions of years)

Snowball Earth; total ice house state

-50 °C   ±0 °C   +50 °C

global average surface temperature, *based on carbon dioxide concentrations in glacial ice layers*

Figure 8.5 ▲

Paul Hoffman's model for the "snowball earth" cycle. (Redrawn by K. Marriott from several sources)

Further research has revealed at least two or three separate events like this in the late Proterozoic, and one in the early Proterozoic (about 2 Ga) known as the Huronian (figure 8.6). It was based on the well-known Gowganda tillite on the shores of Lake Huron, and it showed that snowball earth conditions are not unique but can happen multiple times if the conditions are right.

Figure 8.6 ▲

The early Proterozoic Huronian glaciation was the first "snowball earth" event, around 2.1 billion years ago. (*A*) Finely bedded glacial deposits of the Gowganda tillite, with small drop-stones from icebergs that once floated above the fine-grained bottom muds; (*B*) larger boulders from a glacial till in the Gowganda Formation near Lake Huron, Ontario. (Courtesy of Wikimedia Commons)

SLUSHBALL OR SNOWBALL?

Like all scientists, geologists are naturally skeptical of new ideas, especially those that seem beyond the norm. The snowball earth model had piled up an increasing volume of data in the past 30 years, so most of the geological community had no choice but to accept the obvious conclusions that something like a snowball earth must have happened at least three or four times.

Still, there are dissenters. A number of geologists accept the hard evidence that there were equatorial sea level glaciers in the late Proterozoic— but they do not believe that the entire tropical region had become a frozen snowball. They prefer a slightly less extreme idea, nicknamed the "slushball." In this model, there was some glaciation on the equator (the data demand it), but much of the tropical region was cold but ice free. They point to geological evidence of sediments that could only be formed in water, not ice. However, even the original "snowball" model allowed for some ice-free regions in the tropics, so this is not really a meaningful criticism. Also, many geologists see evidence that the snowball earth episodes had rapid fluctuations of glacial-interglacial cycles as seen in the most recent ice ages, which allows for both glacial sediments and sediments formed in running water and unfrozen oceans. Most important, the dating of the separate snowball events in the latest Proterozoic shows that they were globally synchronous and occurred from pole to equator at the same time. This favors a more extreme snowball earth because in slushball models the ice lines retreat when the carbon dioxide levels go up—and that is not what we see in the late Proterozoic snowball model.

NOTES

1. M. I. Budyko, "The Effect of Solar Radiation Variations on the Climate of the Earth," *Tellus* 21, no. 5 (1969): 611–619
2. James C. G. Walker, P. B. Hays, J. F. Kasting, "A Negative Feedback Mechanism for the Long-term Stabilization of Earth's Surface Temperature," *Journal of Geophysical Research: Oceans* 86 (C10, 1981): 9776–9782.

FOR FURTHER READING

Harland, W. Brian. "Critical Evidence for the Great Infra-Cambrian Glaciation." *Geologische Rundschau* 54 (1964): 45–61.

Hazen, Robert M. *The Story of the Earth: The First 4.5 Billion Years from Stardust to Living Planet*. New York: Penguin, 2013.

Kirschvink, Joseph L. "Late Proterozoic Low-Latitude Global Glaciation: The Snowball Earth." In *The Proterozoic Biosphere: A Multidisciplinary Study*, ed. J. William Schopf and C. Klein, 51–52. New York: Cambridge University Press, 1992.

Macdougall, Doug. *Frozen Earth: The Once and Future Story of Ice Ages*. Berkeley: University of California Press, 2013.

Schopf, J. William. *Cradle of Life: The Discovery of Earth's Earliest Fossils*. Princeton, NJ: Princeton University Press, 1999.

Schopf, J. William, and C. Klein. *The Proterozoic Biosphere*. Cambridge: Cambridge University Press, 1992.

Shaw, George H. *Earth's Early Atmosphere and Oceans, and the Origin of Life*. Berlin: Springer, 2015.

Ward, Peter, and Joe Kirschvink. *A New History of Life: The Radical New Discoveries About the Origin and Evolution of Life on Earth*. New York: Bloomsbury, 2015.

Walker, Gabrielle. *Snowball Earth: The Story of a Maverick Scientist and His Theory of Global Catastrophe That Spawned Life as We Know It*. New York: Broadway, 2004.

09

PLANKTON POWER

The ocean is teeming with organisms so small you can't see them, populations of microorganisms called phytoplankton. Tiny they may be, but over
recent decades these microscopic plant-like organisms have been shown to
help drive the global carbon cycle.

—PAUL FALKOWSKI, *THE POWER OF PLANKTON* (2012)

THE POWER OF PLANKTON

The popular Nickelodeon cartoon *SpongeBob SquarePants* featured a villain-
ous character called Plankton. Thousands of times bigger than the actual
microscopic plankton he represents (he looks like a common planktonic
crustacean called a copepod, which is only a few millimeters long in real
life), he nonetheless is much smaller than the other characters and suf-
fers from a Napoleonic complex due to his size. He also is driven by meg-
alomania and an urge to rule the world, or at least to steal the recipe for
Krabby Patties from Mr. Krabs, who operates a successful restaurant across
the street from Plankton's failing eatery, the Chum Bucket. Of course, his
schemes are defeated at the end of each episode.

The characters are, of course, silly cartoons and are not biologically real-
istic in any sense. The late Stephen Hillenburg, who created the series, was
a trained marine biologist, and he knew a lot about sea creatures. He began
his career trying to write children's books about marine biology before
his experience in animation led him to the wildly successful *SpongeBob*

SquarePants. (Sadly, he died in 2018 of ALS, also known as Lou Gehrig's disease, at the young age of 57.)

The irony is that in real life plankton actually *do* rule the world in some very important ways. Planktonic plants (phytoplankton) live in immense numbers in all the world's oceans, and they are the base of the food chain for the entire ocean, from tiny microorganisms to great whales. More important, you can thank them for every breath you take. Phytoplankton produce about 85 percent of all the oxygen in the atmosphere, far more than rain forests (whose animals and soils quickly consume most of the oxygen their plants produce) or any other source of oxygen. They also pull more carbon dioxide out of the atmosphere than any other phenomenon and are more effective than any artificial means of withdrawing carbon dioxide we have created to solve our problem with greenhouse gases. In a very real sense, phytoplankton are one of the principal regulators of Earth's climate. They have a complex relationship with the oceans and their supply of nutrients, the atmosphere, and the rest of the earth, so much so that they can be thought of as controlling how Earth works.

DISCOVERY OF THE PLANKTONIC WORLD

Early oceanographers and marine biologists pulled fine-mesh nets along the surface of the ocean and discovered lots of tiny organisms in the plankton. As the quality of microscopes improved in the 1800s, they began to identify and classify the different microorganisms they had found and determine how each kind lived. These observations were mostly local and published in a scattershot fashion; there was no systematic understanding of the plankton that lived in the world's oceans.

The first great scientific expedition to explore the oceans was the voyage of the HMS *Challenger*, from 1872 to 1876 (figure 9.1). The expedition was dreamed up and arranged by Sir Charles Wyville Thomson, a Scottish marine biologist and a natural historian, who used his clout in the scientific community along with the urging of the Royal Society to launch the expedition. He argued that the Royal Navy ruled the seas and should use some of their resources to better understanding it. (The same spirit of exploration and discovery led to the voyage of the HMS *Beagle* from 1831 to 1836, whose main job was to survey the coast of Argentina and Brazil; this ship also carried a passenger named Charles Darwin.) In the 1870s, many

Figure 9.1 ▲

The HMS *Challenger* conducted the first truly scientific voyage around the world's oceans and laid the foundation for modern oceanography. (*A*) Contemporary lithograph of the HMS *Challenger*; (*B*) some key personnel on the voyage. (Courtesy of Wikimedia Commons)

scientists thought the unknown and unexplored deep ocean still contained trilobites that had hidden from human view, and the British natural history establishment was anxious to explore the wonders of the oceans that were controlled by their fleets.

Thomson convinced the Royal Navy to refit the HMS *Challenger*, a 225-foot-long Royal Navy *Pearl*-class corvette, for research purposes. She was propelled by two sails and had a steam engine on board, mainly to operate equipment. She was stripped down and modified from a warship (in service since 1858) to become a research vessel. All but two of her cannons were removed, her spars reduced, and extra cabins were built on board for the scientists. Equipped with numerous microscopes, specimen jars, and alcohol for preserving specimens; trawls and dredges and 291 kilometers (181 miles) of rope for deep ocean depth sounding; plus thermometers, water-sampling bottles, and other important equipment, the ship left Portsmouth on December 21, 1872. The *Challenger* sailed more than 127,000 kilometers (68,000 nautical miles), traveling from the Atlantic to the Indian Ocean to the Southern Ocean, then spending over a year exploring the Pacific before returning to the Atlantic and reaching Britain more than four years later. The voyage began with a handful of scientists, 21 officers, and 216 crew; by the end of the voyage, they were down to 144 due to deaths during the long voyage. This loss of life was typical in those days. The ships were dangerous to work on, and more often than not the crews contracted malaria and other tropical diseases when they docked in equatorial regions. (The same thing happened to Captain Cook's voyage around the world. They lost so many sailors to malaria that they had to recruit a whole new crew in what is now Indonesia to sail the ship home.)

The *Challenger* returned with gigantic amounts of oceanographic data: they made 492 deep sea soundings, 133 bottom dredges, 151 open water trawls, and 263 water temperature measurements. They were the first to find the deepest part of the ocean in the Marianas Trench in the Pacific. Their sounding lines went down 8,184 meters (26,850 feet) to reach the most extreme ocean depths ever documented; today it is called the Challenger Deep. Altogether about 4,700 new species of marine life were discovered. *The Reports of the Scientific Findings of the Voyage of the HMS Challenger, 1873–1876*, was published a few years after they returned, and it documented all of their discoveries. In it were the first detailed maps of the ocean depths, what kinds of sediments were found where, water temperatures, ocean currents, and many other studies. John Murray, one of the chief scientists on the expedition (who also described the basic pattern of oceanic sedimentation that we still recognize today), wrote that the voyage was "the greatest advance in the knowledge of our planet since the celebrated discoveries of the fifteenth and sixteenth centuries." It was the beginning of modern oceanography.

Among the many scientists who worked on the scientific data from the voyage was the famous German biologist Ernst Haeckel. He is best known today for being Darwin's biggest supporter in Germany from the 1860s until his death in 1919. His work on embryology pioneered its importance as evidence for evolution. However, he was a versatile scientist, and he worked on microorganisms as well. He was sent most of the samples of the tiny shelled amoebas known as radiolarians (figure 9.2), which he described in a lavishly illustrated volume in 1887 for the *Challenger* reports, as well as separate volumes on the sea jellies and some other groups. His experience with the simplest organisms, as well as his work on embryology, gave him insights that few scientists could match in his time. With work like his and other scientists around the turn of the century, the plankton of the world's oceans began to be much better known.

THE BIRTH OF MODERN OCEANOGRAPHY

After the *Challenger* expeditions, a few isolated oceanographic cruises were mounted by different countries in the early twentieth century, but it was difficult to acquire funding for such expensive research especially with World War I, the Great Depression, and World War II changing societies around the world. Oceans make up almost 71 percent of Earth's surface, but before World War II we knew almost nothing about the deep ocean. The importance of submarine warfare in World War II made it apparent to the world's navies that we needed to understand the deep ocean. After the war, the United States and a few other countries began pouring funds into oceanographic institutions to correct this long-term blind spot in human knowledge. Funding was provided, and war-surplus navy vessels were repurposed as oceanographic vessels for research instead of being cut up for scrap.

By the late 1940s and early 1950s, the major American oceanographic institutions (Scripps in San Diego, Woods Hole in Massachusetts,

I – 5. CROMYOMMA. 6 7 CROMYODRYMUS. 8 9 CROMYOSPHAERA.

Figure 9.2 ▲

One of many beautiful lithographic plates produced for Haeckel's pioneering study of the radiolarians of the *Challenger* voyage. (Courtesy of Wikimedia Commons)

Lamont-Doherty in New York, and a few others) all had vessels sailing around the world all year, collecting data on the temperature, salinity, density, and chemistry of ocean water; on the depth to the seafloor and the nature of the rocks and sediments beneath the seafloor; dropping long tubes called piston corers off the side of the boat to take 33-foot-long cores of sediment recording millions of years of oceanic history; and towing torpedo-like proton precession magnetometers (once used to hunt submarines) to measure the magnetism of ocean floor rocks.

Deep sea sediment cores recovered the shells of plankton that recorded how the oceans had changed through time and how climate had changed around the world, even unlocking the causes of the ice ages (see chapter 22). Soon the most important scientists on board were the micropaleontologists. When the sediment cores first came up from the deep waters, micropaleontologists could look at the cores and date them based on their preserved planktonic microfossils. They could also estimate the water depth when the core sediments were deposited. From their interpretations, the lead scientists of each deep sea mission could decide whether to keep coring deeper or go somewhere else to take the next deep sea core.

These cores provided the first long record back in time through the ocean sediments, something that ordinary coring by dropping steel tubes off the ship could not do. The scientists then realized that drilling sediments around the world could provide an even more detailed history of all the oceans and their changes through many millions of years. Unlike the sedimentary record on land, which is incomplete and episodic, the muds and plankton that rain down from the ocean surface to the seafloor provide a nearly continuous record of Earth's history over millions of years with few missing intervals of time.

In June 1966, Scripps and a consortium of oil companies formed a new project cosponsored by the National Science Foundation and private oil companies; it was named the Deep Sea Drilling Project (DSDP). By October 1967, they had begun building a newer, more advanced ship called the *Glomar Challenger* (figure 9.3). Its name came from the shipbuilder Global Marine Inc. ("Glomar" in the trade) and in honor of the HMS *Challenger*, the British sailing ship that traveled the world's oceans on the first true oceanographic expedition. Launched on March 23, 1968, the *Glomar Challenger* was 120 meters (400 feet) long, 20 meters (65 feet) wide, and could sail at 22 kilometers per hour (14 miles per hour) for up to three months. Topped by a 60 meter (200 foot) drilling derrick, it drilled in water depths of 6,100 meters (20,000 feet), and eventually it could send down a drill string that could drill through 800 meters (2,500 feet) of sediments on the seafloor.

Figure 9.3 ▲

The *Glomar Challenger* was the first modern oceanographic drilling vessel; it launched a new era in marine geology and oceanography. (Courtesy of Wikimedia Commons)

The first two legs of the ship's history were shakedown cruises in the Gulf of Mexico to make sure everything worked. The next outing marked the first true scientific project, and naturally they set out to test the hottest idea in geology in 1968—whether or not seafloor spreading was real. They sailed to the South Atlantic and drilled a series of cores on each side of the Mid-Atlantic Ridge. Sure enough, the sediments at the bottom of each core were older the farther they were from the ridge, proof that the seafloor was indeed spreading.

By 1983, *Glomar Challenger* had been in almost continuous operation for 15 years, making 96 separate expeditions. It had logged 695,670 kilometers (375,632 miles) of sailing, drilled 624 holes in the seafloor, and recovered 19,119 cores. In the process, it had obtained an amazing record of the history of the world's oceans, which solved all sorts of mysteries, from the causes of the ice ages to the extinction that killed the dinosaurs (see chapter 18), as well as how the changing oceanic currents had affected climate for the last 150 million years. Many people regard this work as one of the most important projects in the history of science, and certainly in marine geology and oceanography. All of this knowledge came from studying plankton fossils and how the chemistry of their shells record ocean temperature and climate changes.

PLANKTONIC PLAYERS

Let's take a closer look at these organisms at the base of the ocean food pyramid that influence the world's oceans and climate so much. Phytoplankton are single-celled algae, and they are organized into several groups. Cyanobacteria float near the surface of the ocean along with many types of single-celled algae from different families. These creatures are common in plankton samples grabbed from the sea surface, but they have no internal shells and don't leave fossils.

Three groups of planktonic algae do have internal shells and leave abundant fossils when they die and their shells sink to the seafloor. They can be recovered from deep sea cores and studied. The tiniest planktonic organisms are known as nannoplankton (*nano* is Greek for "dwarf"; it is spelled "nanno" in Latin), and they secrete calcareous plates known as nannofossils. The majority of the nannofossils are made of the skeletal plates of tiny golden-brown algae known as coccolithophores. These algae form spherical cells about 15 to 100 microns in diameter and are enclosed in a ball of calcareous plates called coccoliths, which are about 2 to 25 microns in diameter (figure 9.4A). Most coccoliths are so tiny that they can fit into the pores of larger plankton, such as the foraminiferans. Because of their extremely small size, nannofossils were rarely studied by micropaleontologists until about 35 years ago. Since then, their study has grown exponentially because they have proven to be one of the best tools for examining Mesozoic and Cenozoic biostratigraphy. They are extremely abundant in nearly all pelagic calcareous marine sediments deposited since the Jurassic. In fact, the great Cretaceous chalks are primarily made of the clay-sized residue of coccoliths, with a few foraminiferans. Even shallow marine sandy and muddy sediments, which may have few other plankton, can contain enough coccoliths to be biostratigraphically dated.

Diatoms (figure 9.4B) are the other major group of phytoplankton in the fossil record. Other than their skeletal chemistry and similar size (typically

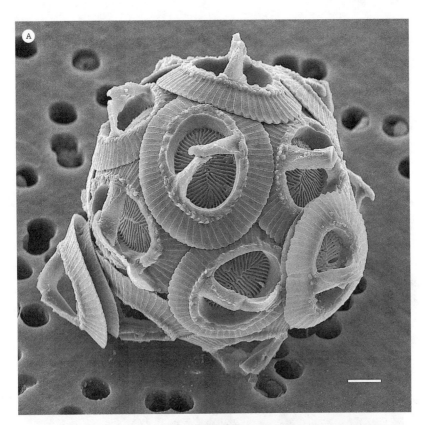

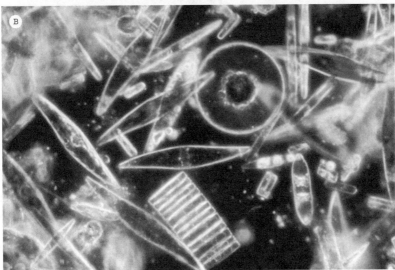

Figure 9.4 ▲ ▶

The major groups of shell-secreting phytoplankton. (*A*) Coccolithophorids are algae that secrete button-shaped coccoliths around the cell in the center and are the major producers of calcareous ooze and chalks. (*B*) Diatoms are algae that secrete siliceous plates around them and nest one inside the other like a petri dish. (*C*) Dinoflagellates are algae that secrete an organic-walled cyst around themselves. (Courtesy of Wikimedia Commons)

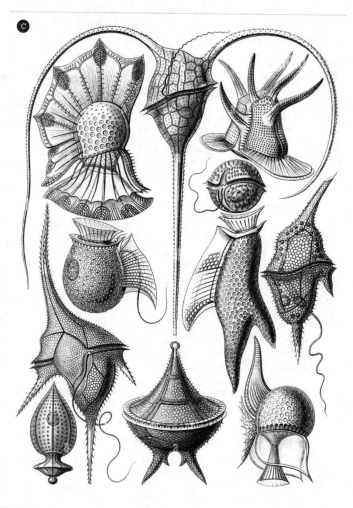

Figure 9.4 ▲

(*continued*)

20 to 200 microns), however, diatoms have little in common with other siliceous plankton, such as the amoeba-like protists known as radiolarians. For
one thing, diatoms are photosynthetic protists more like plants (members
of the Phylum Chrysophyta, the golden-brown algae), so they require light
for photosynthesis and are therefore restricted to the shallow surface waters
(less than 100 meters deep). Along with coccolithophores, they are important parts of the food chain for much of the marine realm. But they also occur

in polar latitudes that are too cold for other phytoplankton. In these regions, they are the most important organisms at the base of the food chain.

Diatoms are not restricted to the marine plankton and benthos; they are found just about anywhere there is light and moisture. They are far more diverse and abundant in freshwaters, where they are the most common organism and make up the base of many freshwater food pyramids. They also occur in soils, ice, and attached to rocks in the splash zone of the surf. Diatoms are among the first algae to colonize any available surface in the photic zone, including rocks, shells, sand, sea grass, and the bottoms and tops of polar ice. They can even cause a yellowish coating on the skin of whales. Of the more than 600 living and fossil diatom genera, 70 percent are marine, 17 percent are freshwater, and the remaining 13 percent are predominantly either marine or freshwater with only a few species occurring in both habitats. In offshore waters, hundreds to thousands of cells occur per liter of seawater; in nearshore waters and estuaries, their concentrations may reach millions per liter. A square meter of shallow seafloor in the photic zone may contain over a million cells.

Coccolithophores make their shells out of calcite, diatoms use silica, and the third major group of shelled planktonic algae, the dinoflagellates, make their shells from organic material (figure 9.4C). Their shells are typically shaped like spheres or oblong capsules, often with horns or spikes sticking out in different directions. They get their name from the whip-like flagellae on the outside of their shells that help propel them through the water column (*dinos* means "whirling" in Greek, and the second part of their name refers to their flagellae). Like diatoms, dinoflagellates live in both marine and fresh waters, although they tend not to be as abundant as diatoms in most settings. More than 2,000 species of marine dinoflagellates are known, and another 220 live in freshwater. Dinoflagellates are most familiar to people as the creatures that cause "red tides." When they encounter a nutrient-rich patch of water (such as when nitrogen from fertilizers run from rivers into the oceans), they bloom in tremendous numbers, often in concentrations of millions of cells per milliliter. Red tides are dangerous for marine life. Once dinoflagellates bloom and absorb all of the nutrients, they quickly die, and their rotting tissues make the water toxic for larger marine life. Some kinds of dinoflagellates are bioluminescent and glow in the dark when they are disturbed in the ocean, making the waves glow with an eerie blue light.

These major groups—cyanobacteria, green algae, coccolithophores, diatoms, and dinoflagellates—make up the base of the entire marine food chain. They are fed upon by larger creatures, such as the shelled amoebas known as foraminiferans (which have a spiral of bubble-shaped chambers made of calcite) and radiolarians (whose shells are delicate "Christmas ornaments" made of silica as seen in figure 9.2). All of these microorganisms are smaller than 1 to 2 millimeters, some only a few tens of microns. Feeding on them are slightly larger predators, especially a variety of tiny crustaceans, such as copepods, krill, tiny shrimp, and many other groups of marine crustaceans. Their predators include filter-feeding fish and even baleen whales, and small fish that are then eaten by bigger fish, and so on up the food chain. Without phytoplankton, none of the other members of the marine food web would survive.

NUTRIENT CYCLES AND THE PLANKTONIC PUMP

The past 70 years of research in oceanography and marine geochemistry have revealed important connections between phytoplankton and the way ocean currents and nutrients move through the oceans. The first thing to remember is that phytoplanktons are plants and require light, so they live only in the uppermost surface waters of the ocean and disappear when the waters become too dark and murky for them to photosynthesize. Most planktonic groups only flourish in the shallow surface waters, forcing most other marine life to live there as well. Lots of creatures live in the world of darkness below the photic zone, but many of them make a living harvesting the rain of organic matter from the crowded surface waters (known as "marine snow"). Others are predators such as larger fish and squid and whales that prey on these filter-feeders. As these surface-dwelling plankton die, they begin sinking to the bottom of the ocean at a rate of about a meter a day, taking their nutrients (organic matter, calcite, silica, nitrogen, phosphorus, etc.) to the deep ocean as they sink. Ultimately, it all ends up on the sea bottom, which is rich in nutrients.

This "biological pump" is constantly moving nutrients from the shallow waters to the deep ocean (figure 9.5), where it could stay forever, trapped in the deep ocean, stagnant and undisturbed. The surface waters would end up depleted of nutrients, with concentrations no richer than those found in the freshwater rivers that flow into them. It is estimated that this pump

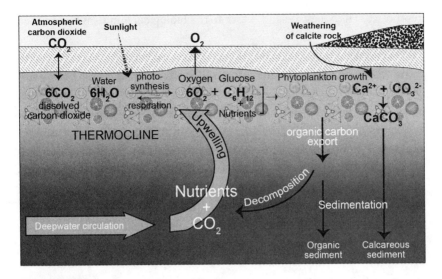

Figure 9.5 ▲

The "biological pump" in the oceans, where plankton grow and flourish at the surface, then die and their nutrients sink to the deep ocean. The system then completes its circulation and redistribution of nutrients trapped in the deep water through upwelling. (Redrawn by E. Prothero from several sources)

transfers about 11 gigatonnes of carbon from the shallow ocean to the deep seafloor every year. This is a major part of the global carbon cycle, pulling out all this carbon in the form of decaying organic matter as well as calcium carbonate locked in the shells of coccolithophores and foraminiferans. If that carbon stayed in deep water, the oceans would soon become stagnant. Without nutrients in the surface waters, most of the shallow plankton that sustain life would die off.

How do these nutrients make it back up to the surface waters again? The answer involves ocean currents, especially upwelling currents that bring water from the deep ocean back to the surface of the photic zone. Upwelling currents occur for a number of reasons. If winds are blowing the surface waters offshore, upwelling from below brings water to the surface to replace it. If an obstacle such as an island blocks the flow, deep water is diverted upward to the surface. If ocean currents collide with one another in a certain way, part of that current flow will rise to the surface, again bringing nutrients to the photic zone. When exposed to sunlight, the plankton bloom and explode in numbers, using up all the nutrients as fast as they can reproduce. Then they die and return to the biological pump on their way to the seafloor.

We can see how voraciously phytoplankton consume nutrients by examining their activity in surface waters. Silica is normally depleted in surface waters because diatoms and radiolarians bloom so fast when silica is available that they consume all of it and deplete it again. The shallow waters around Antarctica are a huge zone of upwelling, and they support enormous blooms of diatoms, which are the base of a complex food web through krill and fish that is the major feeding ground of whales. There is also a midocean upwelling zone along the equator that causes tremendous blooms of siliceous plankton, although radiolarians are more common in the tropics than are diatoms. Likewise, along shorelines with upwelling, there are always huge blooms of siliceous plankton. Even rivers, like the enormous flow of the Amazon River, bring huge amounts of silica to the ocean near Brazil, resulting in huge plankton blooms.

When upwelling shuts down, the surface waters are soon depleted of key nutrients, the plankton don't bloom, and the ocean becomes barren. The most famous example is the El Niño phenomenon in the eastern Pacific Ocean off the coast of Peru and Chile. It occurs when the ocean currents reverse, and the strong westbound trade winds of the southern equatorial zone and the corresponding equatorial current below it shut off abruptly. When El Niño strikes, the current dragging the surface waters away from Peru and Chile weakens and the forces bringing up nutrient-rich deeper waters from below also shut off or flow east, back to South America. The upwelling turns to downwelling, and the eastern Pacific becomes a marine dead zone. In fact, El Niño was first noticed by Peruvian fisherman; in certain years, their nets were empty, and they also starved. Because it tended to start around Christmas time, it was called "El Niño" ("the boy child" in Spanish) in reference to the Christ child born at Christmas time.

Silica is the limiting nutrient for siliceous plankton such as diatoms and radiolarians, but a different substance might be crucial for other organisms. Dinoflagellates use organic matter in their shells and bloom in huge numbers when they get a lot of a scarce nutrient such as nitrogen. When this nutrient is gone, they die off, decay, and make the water stagnant and toxic to larger organisms, causing a deadly "red tide." In other parts of the ocean, phosphorus might be a scarce limiting nutrient, and pollution from detergents can lead to huge amounts of phosphates in the water, causing blooms of plankton that consume it.

How efficient and important is this rapid consumption of nutrients? In 1934 the American oceanographer Alfred Redfield measured the chemical composition of seawater during his voyage on the research vessel *Atlantis*. He found that the ratio of carbon to nitrogen to phosphorus was 106:16:1, not only in the seawater samples but also in the tissues of the plankton living in that same water. For some time, Redfield and other scientists debated what might be causing this similarity of the ratios of seawater and plankton. Some argued that the plankton were passively absorbing whatever nutrients were available to them, making them chemically similar to their surrounding seawater. But after a few years it became clear that the opposite was true: plankton consume nutrients so aggressively that they change the chemistry of the seawater to match their own tissues and control the chemistry of seawater. The proof came when they conducted experiments on this seawater, artificially adding an excess of one of the nutrients. The plankton would quickly bloom and multiply until all the excess nutrients were consumed, and soon the seawater again matched the Redfield ratio of nutrients found in the tissues of the phytoplankton. Plankton power is so great that the chemistry of the ocean is completely controlled by their activities.

STRANGELOVE OCEANS AND GLOBAL WARMING

We have seen how plankton control climate and rule the marine realm. They are at the base of their food chain, and if something disrupts the plankton pump or changes the abundance of key nutrients, everything in the oceans also changes. But when did the plankton pump begin? As soon as the oceans developed planktonic cyanobacteria and eventually algae, most scientists think that some sort of planktonic pump was in effect. The tiny organic-walled cyst fossils known as acritarchs are thought to be the resting spores of algae, and their abundance in the late Precambrian suggests that green algae were abundantly represented in the plankton.

Some have suggested that acceleration of the plankton pump might have drawn down the carbon dioxide in the atmosphere and triggered the global cooling events that caused the snowball earth events (see chapter 8). But by the Early Cambrian, we begin to see the first fossils of the major planktonic groups, showing that they are part of the oceanographic system. The oldest known radiolarian fossils come from the latest Precambrian and the earliest Cambrian, at a time when multicellular life is also evolving rapidly. The earliest dinoflagellates occur in the Silurian. Coccoliths, on the other hand, appear in the Late Triassic, and diatoms do not appear in the fossil record until the Late Jurassic, along with the first members of the modern planktonic foraminiferans known as the globigerinids.

A number of geoscientists, such as the paleontologist Steve Stanley and the sedimentologist Laurie Hardy, have suggested that the abundance of these calcareous plankton in the late Mesozoic suggests that they were the major producer of limestones in the Mesozoic, changing the chemistry of the oceans in the process. Their incredible productivity can be seen in the thick chalk beds of the Cretaceous, from the White Cliffs of Dover to the Cretaceous chalk deposits of Alabama and Texas and Kansas (see chapter 16).

Others have suggested additional ideas about how the plankton pump has evolved over time. In 1985, the Chinese-Swiss geologist Kenneth Hsü (legendary for his many discoveries, from the Mediterranean salinity event to the explanation of mélange in the subduction zone) proposed the idea that an asteroid impact after the Cretaceous caused the "nuclear winter" and darkened the skies with dust long enough to completely shut down the phytoplankton for years. This would also shut down the plankton pump, reducing the upwelling of nutrients that had once sunk to the deep ocean. Most ocean life that had survived the Cretaceous catastrophe would starve, contributing to a mass extinction in the oceans. The evidence for this idea was in the unusual concentrations of carbon isotopes in deep sea sediments in the earliest Paleocene, suggesting that planktonic productivity had shut down. Hsü named this the "Strangelove ocean" because the devastation would be like a nuclear winter and a sort of "doomsday machine," first suggested in the Stanley Kubrick satire, *Dr. Strangelove or: How I Learned to Stop Worrying and Love the Bomb*. A masterpiece of Cold War filmmaking, Kubrick's dark comedy revolves around the madness of MAD (mutually assured destruction) policy during the Cold War. In the movie, an accidental launch of U.S. nuclear missiles by a deranged American officer (General Jack Ripper) leads to a nuclear attack with a Soviet response, which triggers their doomsday machine that would destroy the entire planet. Ironically, the idea of a doomsday machine was a fictional device in that movie until the 1980s, when scientists discovered that the exchange of just a handful of missiles would trigger a "nuclear winter" that would indeed lead to a real doomsday machine and end life on this planet.

Since Hsü's original proposal, most scientists have argued that the plankton are so resilient and can bloom so quickly whenever enough light and nutrients are present that the ocean surface waters would be blooming with plankton as soon as the darkness of the clouds ended. If there were really a "Strangelove ocean" at the beginning of the Paleocene, it was short lived and quickly replaced by a normal ocean once the plankton pump was restored. In addition, the Cretaceous extinctions would not stop upwelling currents, so as soon as light was sufficient, the phytoplankton would have access to plenty of nutrients and would explode in abundance once again.

Finally, many people have suggested that the plankton pump might be a good way to solve the problem of excess carbon dioxide and global warming. Calcareous plankton such as coccolithophores and foraminiferans are by far the most efficient organisms for absorbing carbon from the atmosphere; they are much more effective than replanting trees on land. In addition, they can bloom and multiply in a matter of hours and days, whereas growing a new forest and trapping carbon in trees takes decades. The suggestion is that we fertilize certain parts of the ocean with a rare nutrient (such as powdered iron that would dissolve quickly in seawater) and allow a huge bloom of plankton to pull carbon dioxide out of the seawater and trap it in the deep ocean, thus removing it from the atmosphere.

Of course, the key limitation on this idea is that it must be delivered to a part of the ocean that has only downwelling. Where there is upwelling, the nutrients and carbon would return to the surface waters again and release their carbon dioxide back into the atmosphere. One such place would be the centers of the giant oceanic gyres in the center of the North Pacific, South Pacific, North Atlantic, South Atlantic, and southern Indian Ocean. These gyres result from complex oceanic currents that pile up the seawater into a "hill" of water about 2 meters (6 feet) high near the center of the gyre. Because the water is piled up and can't go any higher, most of it sinks in that region of downwelling. Normally the centers of gyres are barren zones in the ocean because they lack nutrients. If it were done in the right way, the centers of gyres could be "fertilized" with iron, plankton would bloom and withdraw huge amounts of carbon dioxide, then they would die and sink to the bottom, trapping enormous amounts of carbon on the ocean floor.

Of course, where this fertilization takes place, the seafloor should be shallow enough that it is not below the carbonate compensation depth (about 4 to 5 kilometers deep). Below this depth, carbonate dissolves because the seawater is undersaturated in carbonate.

There are places like the midocean ridges in certain gyres where downwelling occurs and the ocean is shallow enough that carbonate sediment would still be trapped in the deep ocean. This idea has not yet moved past the discussion stage because it's enormously expensive and may be too risky. But as global climate change gets worse and the cost of doing nothing about our excess carbon dioxide becomes intolerable, it may be implemented at some point.

FOR FURTHER READING

Chisholm, Sallie W. "The Iron Hypothesis: Basic Research Meets Environmental Policy." *Reviews of Geophysics* 33, no. S2 (1995): 1277-1286.

De la Rocha, Christina, and Uta Passow. "The Biological Pump." In *Treatise on Geochemistry*. Vol. 6, *The Oceans and Marine Geochemistry*, ed. H. Elderfield, 93-122. Oxford: Pergamon Press. 2006.

Falkowski, Paul. "The Power of Plankton." *Nature* 483 (2012): 517-520.

Falkowski, Paul, R. J. Scholes, E. Boyle, J. Canadell, D. Canfield, J. Elser, N. Gruber, K. Hibbard, and P. Högberg. "The Global Carbon Cycle: A Test of Our Knowledge of Earth as a System." *Science* 290, no. 5490 (2000): 291-296.

Hain, M. P., D. M. Sigman, and G. H. Haug. "The Biological Pump in the Past." In *Treatise on Geochemistry*, 2nd ed. Vol. 8, *The Oceans and Marine Geochemistry*, 482-517. Amsterdam: Elsevier, 2014.

Hsu, Kenneth., and Judith A. McKenzie. "A 'Strangelove Ocean' in the Earliest Tertiary." In *The Carbon Cycle and Atmosphere CO$_2$: Natural Variations Archean to Present*, ed. E. T. Sundquist and W. S. Broecker, 487-492. Washington, DC: American Geophysical Union, 1985.

Kump, Lee R. "Interpreting Carbon-Isotope Excursions: Strangelove Oceans." *Geology* 19, no. 4 (1991): 299-302.

Lam, Man Kee, Keat Teong Lee, and Abdul Rahman Mohamed. "Current Status and Challenges on Microalgae-Based Carbon Capture." *International Journal of Greenhouse Gas Control* 10 (2012): 456-469.

Ridgwell, A. "The Evolution of the Ocean's 'Biological Pump.'" *Proceedings of the National Academy of Sciences* 108, no. 40 (2011): 16485-16486.

Robinson, Josie, E. E. Popova, A. Yool, M. A. Srokosz, R. S. Lampitt, and J. R. Blundell. "How Deep Is Deep Enough? Ocean Iron Fertilization and Carbon Sequestration in the Southern Ocean." *Geophysical Research Letters* 41, no. 7 (2014): 2489–2495.

Sigman, Daniel M., and Gerald H. Haug. "The Biological Pump in the Past." In *Treatise on Geochemistry*. Vol. 6, *The Oceans and Marine Geochemistry*, ed. H. Elderfield, 491–528. Oxford: Pergamon Press, 2006.

Steinberg, Deborah K., and Michael R. Landry. "Zooplankton and the Ocean Carbon Cycle." *Annual Review of Marine Science* 9, no. 1 (2017): 413–444.

Volk, Tyler, and Martin I. Hoffert. "Ocean Carbon Pumps: Analysis of the Relative Strengths and Efficiencies of Ocean-Driven Atmospheric CO_2 Changes." In *The Carbon Cycle and Atmosphere CO_2: Natural Variations Archean to Present*, ed. E. T. Sundquist and W. S. Broecker, 99–110. Washington, DC: American Geophysical Union, 1985.

10

GREENHOUSE OF THE TRILOBITES

For everything there is a season, and a time for every purpose under heaven: a time to be born, and a time to die; a time to plant, and a time to pluck up that which is planted; a time to kill, and a time to heal; a time to break down, and a time to build up; a time to weep, and a time to laugh; a time to mourn, and a time to dance; a time to cast away stones, and a time to gather stones together; a time to embrace, and a time to refrain from embracing; a time to seek, and a time to lose; a time to keep, and a time to cast away; a time to rend, and a time to sew; a time to keep silence, and a time to speak; a time to love, and a time to hate; a time for war, and a time for peace.

ECCLESIASTES 3:1-8

There is grandeur in this view of life, with its several powers, having been originally breathed by the Creator into a few forms or into one; and that, whilst this planet has gone cycling on according to the fixed law of gravity, from so simple a beginning endless forms most beautiful and most wonderful have been and are being evolved.

—CHARLES DARWIN, *ON THE ORIGIN OF SPECIES* (1859)

"WHILST THIS PLANET HAS GONE CYCLING ON"

As geologists began to piece together the sequence of the rock record in the 1820s and 1830s, they began to notice some patterns. By the late 1830s, they realized that Earth's history was characterized by three long periods of consistently stable fossil groups and rock types that were transformed by mass extinctions. In 1838, the pioneering British geologist Adam Sedgwick (who named the Cambrian and Devonian and was Darwin's geology professor at Cambridge) coined the word Palaeozoic (Greek for "ancient life") for the rocks that yielded ancient trilobite and brachiopod fossils. In 1840, the British geologist John Phillips named the Mesozoic ("middle life" in Greek) and Cainozoic ("recent life" in Greek) for the eras of the Age of Dinosaurs and the Age of Mammals. Those three eras have come to be universally accepted among geologists because their typical animals are so fundamentally different. This is largely because mass extinctions at the end of the Paleozoic and at the end of the Mesozoic rearranged and reshuffled the dominant groups in the oceans and on land.

In addition to noticing the changes in fossil assemblages through time, geologists began to identify other patterns. At certain times in Earth's history, such as the early-middle Paleozoic or the late Jurassic-Cretaceous, the seas must have covered most of the continental landmasses. Fossiliferous limestones (which form in warm shallow water) and abundant deep water shales were deposited in the center of today's continents. If you examine Paleozoic roadcuts in Wisconsin, for example, far from the edge of the North American continent, you will find lots of marine shales and limestones full of fossils of sea creatures that once lived in the center of the continent. Trilobites were swimming in Cambrian seas where Minnesota and Wisconsin are right now, and outcrops in Iowa are chock full of long conical shells of primitive relatives of the nautilus and squid. During the Late Cambrian, almost all of North America (and much of the rest of the world) was covered by warm tropical seas like those that surround the Bahamas today. Coral reefs have been found in Chicago and Michigan in the Silurian and the Devonian, and the roadcuts around Cincinnati, Ohio, are legendary for the marine shells of brachiopods and bryozoans ("moss animals") and corals and crinoids ("sea lilies") and trilobites that can be found there. This evidence has long been explained as a "greenhouse" world, in which there were no polar ice caps, the planet was much warmer, and oceans covered most of the continents.

The end of the early Paleozoic greenhouse cycle occurred in the Early Carboniferous (Mississippian in the United States). For example, in the Mississippi Valley between Iowa and Illinois is the Lower Mississippian Burlington Limestone, a typical unit. Its volume is about 30×10^{10} cubic meters, and it is estimated to contain the remains of 28×10^{16} individual crinoids! That's just one formation among many. Everywhere Mississippian rocks crop out there are thick deposits of crinoidal limestone, with other kinds of limestone and occasionally minor shales represented as well. They have different names in different parts of the world—Mountain Limestone in the United Kingdom, many different names in the Appalachians, Pahasapa Limestone in South Dakota, Madison Limestone in the northern Rockies, Redwall Limestone in the Grand Canyon, Monte Cristo Limestone in southern Nevada—but the pattern is always the same. Abruptly above it, however, the Upper Carboniferous (known as the Pennsylvanian in the United States) rocks are very different, with coal beds, lots of sandstones and shales, and almost no clean mud-free limestones anywhere. These are the cyclothems, and they are a combination of the cyclicity of glaciers in Gondwana plus the abundant sands and muds eroding down from the Appalachian, Ouachita, and Ancestral Rocky Mountains. The seas were too muddy for clean limestones, and the only muddy limestones you might find would be in the middle of the seaway in Kansas.

If we traveled in a time machine to scuba dive in the world of the Mississippian, we would see a shallow carbonate shoal setting similar to that in the Bahamas today, but on an immense scale. These shallow waters would have been packed with extensive meadows of crinoids (and other invertebrates that lived among them), with their umbrella-like filtering arms and heads swaying back and forth with the currents. Earth was rotating much faster then than it is today (400 days in a year rather than the 365.25 days we now have); and the moon was much closer than it is today, so the tidal currents sweeping across these shoals would have been very powerful. This type of environmental setting extended all the way from Nevada, Arizona, and Montana to the Midwest (especially Michigan, Illinois, Indiana, Kentucky, and the South). In fact, the only place where limestones were not being deposited in the Mississippian was in the foothills of the vanishing Acadian Mountains in what is now the Appalachian belt, where red sandstones and shales formed in rivers and deltas eroded off the mountains. These include units such as the brick-red Mauch Chunk Formation, the Big Stone Gap Shale, the Grainger Sandstone, and the Pocono Sandstone in Pennsylvania, Virginia, and West Virginia, and the New Albany Shale in Indiana and Illinois.

In contrast to times when the planet was covered by shallow seas, during the late Paleozoic and the Triassic conditions were much drier and harsher. The marine rocks had disappeared; in their place were dune deposits with giant cross-bedded sandstones or brick-red shales that represented arid floodplains or dry lake deposits. It is clear that Earth alternated between episodes when high sea levels covered the continents and episodes when the oceans retreated and harsh desert climates with sand dunes, dry floodplains, and dry lakes prevailed over many of the continents.

SMALL-SCALE CYCLES

In the early twentieth century, geologists began to notice finer-scale patterns of cyclicity as well. Glacial lake beds are famous for their regular pattern of fine-scale bedding, known as glacial varves (figure 10.1A). Sometimes these alternate between dark and light bands, which are thought to represent periods when the lake was frozen and organic material accumulated on the bottom (dark bands) and periods when the lake was thawed and well-oxygenated (light-colored bands). In addition to what appear to be annual fluctuations of lake sedimentation, larger-scale patterns or rhythms of finer-scale beds often alternate with thicker or coarser beds, suggesting cyclic patterns running from decades to centuries to thousands of years.

Similar fine-scale layering has been seen in many other ancient lake deposits, including the oil shales of the Eocene Green River Formation (figure 10.1B) in Wyoming, Utah, and Colorado, which alternate between light oxygenated layers (suggesting the lake waters were circulating and oxygenating the bottom) and dark organic-rich layers (suggesting periods of lake stagnation when dark organic material accumulated at the bottom without oxidizing).

Fine-scale rhythms have even been found in water bodies (both lakes and shallow seas) that were full of evaporating minerals, such as salt and gypsum. The most famous examples are the Permian Castile evaporites of western Texas and southern New Mexico (figure 10.1C), which have couplets of varves each a few centimeters thick, with more than 260,000 varve couplets spanning millions of years of the Permian. Some of these couplets can be traced over 113 kilometers (70 miles). Apparently, these water bodies fluctuated between extreme drying (precipitating halite, or rock salt) and slightly less concentrated brines, producing gypsum (hydrous calcium

Figure 10.1 ▲

Varved sediments consist of many long continuous layers of uniform thickness in the milli-meter to centimeter bed thickness. (A) Glacial varves, deposited in a proglacial lake by the seasonal alternation of thawing (creating light colored layers from the oxygenation of bottom sediments) and freezing (dark layers, formed when organic matter is trapped at the bottom and does not get consumed). (B) The fine-scale seasonal banding of the Eocene Green River Formation lake deposits. (C) The fine-scale lamination of the Permian Castile evapo-rites, which later were tectonically deformed. (Courtesy of Wikimedia Commons)

sulfate). These rocks also show signs of bundles of couplets with regular patterns on a decadal or century scale, suggesting that external climatic forces caused these rhythmic patterns.

CYCLES, CYCLES, EVERYWHERE

One of the most influential geologists of the second half of the twentieth century was Alfred Fischer (figure 10.2). Born December 10, 1920, in Rothenburg, Germany, he was the son of a traveling shoe salesman who was born in the United States but had German roots. George had made a fortune selling shoes in the Old West and moved to Germany to raise a family. They lived in a chalet in the Alps for 14 years, where young Alfred got to know the rocks of the Alps well. In 1935, Hitler and the Nazis came to power, and the Fischer family fled Germany. They moved to Watertown, Wisconsin, where Alfred finished high school and then enrolled at the University of Wisconsin, where he switched from majoring in forestry when he got hooked on geology. He earned his bachelor's degree and master's

Figure 10.2 ▲
Alfred Fischer at his microscope in his office at USC. (Courtesy of D. J. Bottjer).

degree, but had to interrupt his doctoral studies to take a teaching job at Virginia Polytechnic Institute in order to support his growing family.

After Pearl Harbor and the U.S. entry into World War II, his poor eyesight and the fact that he was a young father exempted him from military service, so he worked for Stannolind Oil, finding energy resources for the war effort. In 1946, Fischer completed a doctoral degree at Columbia University and then took a series of teaching jobs at the University of Rochester and the University of Kansas. He collaborated with the legendary paleontologists Raymond C. Moore and Cecil Lalicker to write a classic textbook on paleontology (Moore, Lalicker, and Fischer, *Invertebrate Fossils*, 1952) that is still in print today. He eventually ended up at Princeton in 1956, where he stayed until he reached retirement age at 65. But Fischer didn't like mandatory retirement and moved on to the University of Southern California where he finally retired in 1991 at the age of 71. He passed away in 2017 at age 96.

I remember Al Fischer well, and I saw him frequently at professional meetings. He was always warm and friendly and interested in any idea, no matter how wild or outlandish it might seem at first.

> Fischer was also a generous and devoted teacher. His elder son, Joseph Fred "Fritz" Fischer, recounted how after giving a guest lecture at Yale University in the late 1970s, his father turned down an offer to tour the university's geology department in favor of talking to its graduate students.
>
> "The moment the students heard about it, they dashed off to quickly read up on his literature," Fritz Fischer said. However, rather than talking about his own work, Al Fischer concentrated on the students' efforts, sitting down with each student individually and asking about their thesis or dissertation, then making suggestions.
>
> "The students were utterly amazed," Fritz Fischer said. "Here's this great man, he's made such a name for himself and he wasn't there to talk about himself. He was there to find out what they were doing. They all walked out of there saying, 'Oh my God, I want to be like that man.'"[1]

Fischer's greatest claim to fame was his lifelong obsession with discovering and interpreting cyclical patterns in Earth's history, a topic that had been neglected for decades. He was interested in limestones and how they formed, and in 1964 he published landmark studies of Triassic limestones in the Alps, not far from where he grew up. He was especially interested in the rhythmic bedding patterns and cycles that showed up in the Dachstein limestones of the Upper Triassic Keuper beds of Austria, Germany, and Switzerland. By the 1970s, his interest in the Triassic cyclicity of sedimentary beds had expanded to looking at all sorts of different rhythms and cycles in the rock record.

In addition to cycles that lasted decades to thousands of years to a few million years, Fischer was interested in larger-scale "supercycles." In the late 1970s and 1980s, the idea of greenhouse conditions with high carbon dioxide levels in the atmosphere and warm global climates was becoming more accepted, and scientists were also realizing that Earth was now undergoing a "greenhouse effect" of global warming. Fischer argued that Earth went through at least two complete greenhouse-icehouse "supercycles" that lasted about 250 million years each during the last 700 million years (figures 10.3, 10.4). During planetary greenhouse conditions, high carbon dioxide and other greenhouse gases warm the planet, which melts all the polar ice and causes sea level to rise across continents. When carbon dioxide levels drop, the planet cools down and ice caps form and grow by pulling water out of the oceans. This causes sea levels to drop and continents to be exposed to erosion and drier and harsher conditions (especially far from the coasts, where harsh continental climates generate deserts and other extreme habitats).

Fischer and his student Mike Arthur pointed to the last 700 million years of Earth's history, for which the record is more complete and well dated. About 700 to 600 million years ago, Earth was in the grip of the last of the "Snowball Earth" episodes, a peak of icehouse climates (see chapter 8). When the late Proterozoic snowball conditions melted, Earth rapidly warmed; and by the Cambrian, Earth was in full greenhouse mode with the seas flooding the Appalachians and the Grand Canyon in the Early Cambrian and reaching all the way to Wisconsin by the end of the Cambrian (figures 10.3, 10.4). These shallow marine epicontinental seas were where the trilobites roamed. Warm tropical seas flooded most of the continents, except where local tectonic collisions formed mountain belts, such as the Taconic orogeny in the later Ordovician in the Appalachians, the Acadian orogeny in the eastern United States in the Devonian, the Antler orogeny in Nevada in the Devonian, or the Appalachian orogeny in the late Carboniferous, which created the modern Appalachian Mountains as well as the Ouachita Mountains in Arkansas and Oklahoma and the Ancestral Rocky Mountains across the future area of the Rockies.

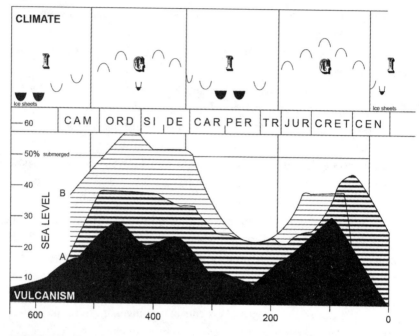

Figure 10.3 ▲

Fischer's depiction of the large-scale, 250-million-year-long "greenhouse" (G) and "icehouse" (I) cycles. At the bottom are Fischer's plots of the trends in vulcanism (black shape), and two other versions of global sea level history: (A) from Peter Vail and others, and (B) from various Russian authors. (Modified and redrawn from Alfred Fischer, "Long-Term Climatic Oscillations Recorded in Stratigraphy," in *Climate in Earth History: Studies in Geophysic* [Washington, DC: National Academy of Sciences Press, 1982], 97–104)

An enormous diversity of marine life could be found in these shallow seas. Just imagine what a trilobite might have seen in the Cambrian! The moon was much closer then, and the tides sweeping across these shallow marine shoals would have been huge (see chapter 3). Water would have flowed across some areas as huge tidal waves coming in, then drained out until only a few areas were immersed (somewhat like the giant tidal waves that cycle in and out of the Bay of Fundy in Canada every 12 hours). In many places (such as the Upper Cambrian rocks in Wisconsin), enormous storm deposits dumped huge boulders into the sea; the storms crossing this shallow seaway could be fierce.

This early and middle Paleozoic greenhouse effect began to show signs of changing by the Devonian, especially during the transition from the

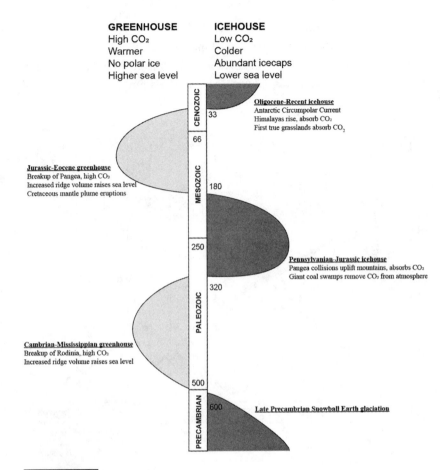

GREENHOUSE
High CO₂
Warmer
No polar ice
Higher sea level

ICEHOUSE
Low CO₂
Colder
Abundant icecaps
Lower sea level

CENOZOIC — 33, 66

Oligocene-Recent icehouse
Antarctic Circumpolar Current
Himalayas rise, absorb CO₂
First true grasslands absorb CO₂

Jurassic-Eocene greenhouse
Breakup of Pangea, high CO₂
Increased ridge volume raises sea level
Cretaceous mantle plume eruptions

MESOZOIC — 180, 250

Pennsylvanian-Jurassic icehouse
Pangea collisions uplift mountains, absorbs CO₂
Giant coal swamps remove CO₂ from atmosphere

PALEOZOIC — 320, 500

Cambrian-Mississippian greenhouse
Breakup of Rodinia, high CO₂
Increased ridge volume raises sea level

PRECAMBRIAN — 600

Late Precambrian Snowball Earth glaciation

Figure 10.4 ▲

The large-scale pattern of "greenhouse" and "icehouse" cycles over the past 600 million years. Greenhouse conditions have high atmospheric carbon dioxide, raising temperatures and melting icecaps, so sea level rises and drowns continents. Icehouse conditions occur when carbon dioxide drops, lowering temperatures and building up polar ice caps, which traps water in the icecaps and brings down sea level. In general, greenhouse events seem to be triggered by breakup of supercontinents, which have high rates of seafloor spreading. This seafloor volcanism not only releases a lot of greenhouse gases, but rapid spreading creates thicker mid-ocean ridges which push water out of the ocean basin and onto the continents. Icehouse conditions tend to be triggered by large-scale mountain building, which absorbs carbon dioxide during increased weathering. In addition, land events like the huge coal swamps of the Carboniferous, or the expansion of grasslands during the Oligocene-Miocene, also serve to trap carbon dioxide in the earth's crust.

Early to the Late Carboniferous (figures 10.3, 10.4, 10.5). Lower Carboniferous rocks are nearly all shallow marine limestones, full of millions of pieces of animals called crinoids, or "sea lilies," which lived in enormous numbers on the seafloor. These Lower Carboniferous limestones are capped

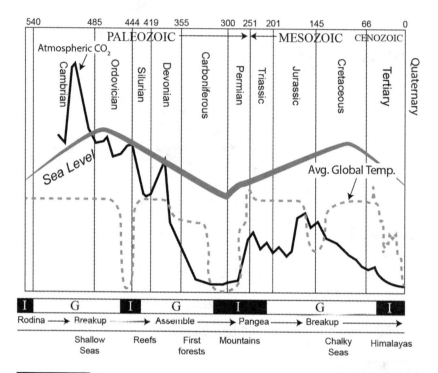

Figure 10.5 ▲

A more detailed version of the pattern of sea level (thick gray line), atmospheric carbon dioxide (thin black line), and average global temperature (dashed line) over the Phanerozoic, during the "icehouse" ("I") and "greenhouse" ("G") phases of Earth's climate. Notice there is a greenhouse cycle in the early Paleozoic (Cambrian to Early Carboniferous) with high sea level, high carbon dioxide, and mostly high temperatures (except during the Late Ordovician glaciation). This ends with the beginning of an "icehouse" cycle that runs from the Late Carboniferous through the Triassic, with low sea level, low carbon dioxide, and low temperatures as glaciers formed on Gondwana and on the northern parts of Pangea. It also coincides with the formation of Pangea and uplift of many mountain ranges (absorbing carbon dioxide through weathering), but also the growth of the first forests. The next "greenhouse" cycle begins in the Jurassic and runs through the Oligocene, with rising sea level, rising temperatures, and rising carbon dioxide. This corresponds with the breakup of Pangea and other tectonic events (see chapter 17). Finally, since the Oligocene, the earth has entered an "icehouse" mode which it is still in (although modern release of carbon dioxide is changing this climate fast). (Redrawn from several sources)

by Upper Carboniferous coals and sandstones and shales, with very little limestone—a legacy of the beginning of the late Paleozoic icehouse and huge tectonic events (figures 10.4, 10.5).

With the advent of full Gondwana glaciation (and some Arctic glaciers on Pangea as well), the planet was in full icehouse mode from the Late Carboniferous through the Permian and well into the Triassic. The next transition happened in the middle of the Jurassic, when greenhouse conditions and flooding of the continents began again (figure 10.3).

The Jurassic-Cretaceous "greenhouse of the dinosaurs" persisted well into the early Eocene, about 54 Ma, with the last gasp of a "greenhouse planet" that allowed alligators and primates to live above the Arctic Circle and in the Antarctic as well. That final greenhouse ended during the Eocene-Oligocene transition, when icehouse conditions returned with glaciation on Antarctica 33 Ma, and glaciation in the Arctic appeared 2.6 Ma (figure 10.4).

THE PULSE OF THE EARTH

What could be causing these long supercycles? Many suggestions have been made. For more than a century, geologists have been speculating that there were cyclical events in the history of mountain building, and they argued over whether there were "tectonic cycles." One famous advocate for these ideas was Johannes Herman Frederik Umbgrove, whose 1947 book *The Pulse of the Earth* laid out his ideas. He recognized a 250-million-year periodicity in sea level, mountain building events, subsidence of sedimentary basins, climate, and volcanic activity. At that time, good radiometric dates were not available to confirm that all of the ages on the igneous rocks were similar, and there was no real mechanism to explain why "pulses" might exist.

That all changed when potassium-argon (K-Ar) dating was developed in the 1950s and the theory of plate tectonics became popular in the early 1960s and 1970s. By the early 1970s, geologists began to recognize that the collision and separation and then another collision of continents together operated on a repeated basis. It took about 250 million years for supercontinents to split up or for many continents to reunite into a supercontinent. The Canadian geologist J. Tuzo Wilson first realized this when he discovered evidence of a proto-Atlantic Ocean during the Cambrian that had vanished completely when the African and South American portions of Gondwana collided with the Americas in the Late Carboniferous—and by the Jurassic, the modern Atlantic was formed in almost the same place. This pattern of collision between continents and then separation followed by yet another collision as the oceans open and close and then reopen is now known as a "Wilson cycle" in his honor. By the mid-1980s, Tom Worsley and others were postulating that there were multiple supercontinents and Wilson cycles going back at least 2 billion years, and that each of these Wilson cycles drove other cycles of climate, sea level change, and changes in life on Earth.

Is there really a link? A lot of geologists are a bit skeptical of the 250-million-year "supercycle." After all, there have been only three icehouse and two greenhouse episodes in the last 700 million years, not really enough events to conclude that something is truly cyclical rather than just random coincidence (figure 10.4). But assuming that these cycles are real, how might they work?

The first apparent link is the association of the breakup of supercontinents with the beginning of greenhouse episodes, and the coalescence of supercontinents with the development of icehouse episodes. When supercontinents break up, lots of things could trigger greenhouse conditions. The most obvious of these is that as continents drift apart there is rapid seafloor spreading as new oceans form between continents as they separate. Not only do the huge eruptions of mantle lava on midocean ridges contribute gigantic volumes of greenhouse gases to the atmosphere, they also contribute to the rapid melting of any ice caps from the previous icehouse world, raising global sea level.

In addition, the shape and depth of the ocean floor also changes. On modern midocean ridges with a slow rate of spreading, their cross section and volume are relatively small (figure 10.6). A good example is the modern Mid-Atlantic Ridge, which spreads only 6 centimeters per year, about as fast as your fingernails grow. Its steep, relatively small-volume profile develops because the hot rocks that erupted from the center of the ridge have plenty of time to cool and contract as they slowly move away from the ridge axis. But a much faster spreading ridge (such as the modern East Pacific Rise) spreads apart about 3 times as fast (18 centimeters per year). Its cross-sectional profile is much thicker and more gently sloped because the hot expanded rocks are carried a long way from their origin in the rift valley

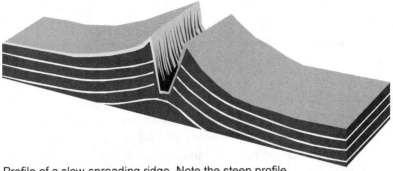

Profile of a slow-spreading ridge. Note the steep profile, which is the result of a low eruptive volume.

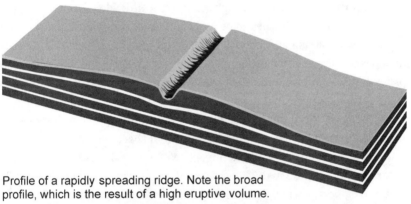

Profile of a rapidly spreading ridge. Note the broad profile, which is the result of a high eruptive volume.

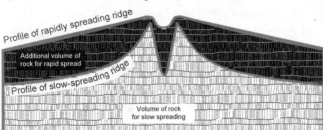

Profile of rapidly spreading ridge

Additional volume of rock for rapid spread

Profile of slow-spreading ridge

Volume of rock for slow spreading

Figure 10.6 ▲

One contributor to sea level change is the volume of midocean ridges. When they are spreading slowly, the erupted lava in the crust has a long time to cool, so the profile contracts and the volume of the ridge is less. When they are spreading rapidly, the hot expanded oceanic crust moves away from the rift valley in the center more quickly, and a thicker crust pushes outward from the central rift, displacing water from the ocean basins and onto land. (Drawing by K. Marriott)

where they erupted before they have much chance to cool and shrink down. Thus the profile of a fast-spreading ridge is much broader and has a much greater volume than a slow-spreading ridge. If you increase the volume of rocks at the bottom of the seafloor, the water in the ocean basins has only one place to go—up and out of the ocean. Water spills out of the oceans and begins to flood the continents.

If we look at the breakup of Rodinia, the Precambrian supercontinent, all of these factors seem to be in play and can explain at least part of the early Paleozoic greenhouse and the global flooding of the continents with shallow seas. The breakup of the supercontinent Pangea during the Triassic and Jurassic seems to fit this model as well: high spreading rates during the Cretaceous correlate well with the high sea levels worldwide (see chapter 17).

But how might the coalescence and collision of smaller supercontinents contribute to making an icehouse planet? The most widely accepted explanation is that the collision of continents generates huge mountains where the plates smash into each other. This has two consequences. First, huge collisions uplift the entire continental plate, and as they rise, the epicontinental seas drain away. Second, as these mountains rise, their rocks are exposed to much more rain and other atmospheric effects and become deeply weathered. The chemical reactions of weathering, known as the calcite-silicate reaction, absorb carbon dioxide in the breakdown of silicate rocks and produces soils that trap carbon. This might be an explanation for the cooling conditions of the latest Paleozoic, when the rise of many mountain belts in the Late Carboniferous and Permian seemed to correlate with the cooling that accompanied the growth of Gondwana glaciers. But there is another important factor in the Carboniferous: the rise of the first forests, with huge trees that absorbed carbon, then died and turned into coal and locked that carbon in the earth's crust—before coal mining released that carbon again (see chapter 13).

However, it's not nearly so clear that this mechanism works to produce the icehouse world of the later Cenozoic. For one thing, the pieces of Pangea are not yet united into a new supercontinent, and probably won't be until some 250 million years in the future. Some mountain-building events, such as the uplift of the Rockies in the latest Cretaceous, did indeed drain the Western Interior Seaway from the midcontinent for a short time, but there were epicontinental seas in the Dakotas and Montana in the Paleocene, so the effect was not permanent.

Even though the fragments of Pangea have not yet combined into a new supercontinent, some pieces have gotten together again, such as India leaving Gondwana in the Late Cretaceous and slamming into the belly of Asia by the early Eocene. Maureen Raymo, William Ruddiman, and some others have argued that the collision of India with the belly of Asia, which led to uplift of the huge Alpine-Himalayan belt, created enormous mountains and deep weathering that lowered carbon dioxide and ended the early Cenozoic greenhouse world. However, most of that Himalayan uplift took place in the Miocene, and the first Antarctic glaciers didn't show up until the early Oligocene. Arguments about the causes of the Cenozoic icehouse are discussed in chapter 20.

NOTE

1. Susan Bell, "In Memoriam: Alfred Fischer, 96, Prominent Geologist," *USC News* (University of Southern California), July 24, 2017, https://news.usc.edu /125246/in-memoriam-alfred-fischer-96-prominent-geologist/.

FOR FURTHER READING

Einsele, Gerhard, Werner Ricken, and Adolph Seilacher. *Cycles and Events in Stratigraphy.* New York: Springer-Verlag, 1991.

Fischer, Alfred G. "The Lofer Cyclothems of the Alpine Triassic." *Kansas State Geological Survey Bulletin* 169 (1964): 107–150.

——. "Gilbert-Bedding Rhythms and Geochronology." In *The Scientific Ideas of G. K. Gilbert,* ed. Ellis Yochelson. *Geological Society of America Special Papers* (1980): 93–104.

——. "Climatic Oscillations in the Biosphere." In *Biotic Crises in Ecological and Evolutionary Time,* ed. M. Nitecki, 103–131. New York: Academic Press, 1981.

——. "Climatic Rhythms Recorded in Strata." *Annual Reviews of Earth and Planetary Science* 14 (1981): 351–367.

——. "Long-Term Climatic Oscillations Recorded in Stratigraphy." In *Climate in Earth History: Studies in Geophysics,* 97–104. Washington, DC: National Academy of Sciences Press, 1982.

Fischer, Alfred G., and D. J. Bottjer. "Orbital Forcing and Sedimentary Sequences." *Journal of Sedimentary Petrology* 61 (1991): 1063–1069.

Fischer, Alfred G., and M. A. Arthur. "Secular Variations in the Pelagic Realm." In "Deep Water Carbonate Environments," ed. Harry E. Cook and Paul Enos. *Society of Economic Paleontologists and Mineralogists* 25 (1977): 18–50.

Heckel, Philip H. "Diagenetic Models for Carbonate Rocks in Midcontinent Pennsylvanian Eustatic Cyclothems." *Journal of Sedimentary Petrology* 53 (1983): 733–759.

——. "Sea Level Curve for Pennsylvanian Eustatic Marine Transgressive-Regressive Depositional Cycles Along Midcontinent Outcrop Belt, North America." *Geology* 14 (1986): 330–334.

Moore, Raymond C., Cecil G. Lalicker, and Alfred G. Fischer. *Invertebrate Fossils.* New York: McGraw-Hill, 1952.

Nance, R. Damian, and J. Brendan Murphy. "Origins of the Supercontinent Cycle." *Geoscience Frontiers* 4 (2013): 439–448.

——. "Supercontinents and the Case for Pannotia." *Geological Society of London Special Publication* 470 (2018): 1–10.

Umbgrove, Johannes H. F. "Periodicity in Terrestrial Processes." *American Journal of Science* 238 (1940): 573–576.

——. *The Pulse of the Earth.* The Hague, Netherlands: Martinus Nijhoff, 1947.

Wanless, Harold R. "Pennsylvanian Cycles in Western Illinois." *Illinois Geological Survey Bulletin* 60 (1931): 179–193.

——. "Local and Regional Factors in Pennsylvanian Cyclic Sedimentation." In "Symposium on Cyclic Sedimentation," ed. D. F. Merriam. *Kansas Geological Survey Bulletin* 169 (1964): 593–605.

Wanless, Harold R., John B. Tube Jr., Donald E. Gednetz, and John L. Weiner. "Mapping Sedimentary Environments of Pennsylvanian Cycles." *Geological Society of America Bulletin* 74 (1963): 437–486.

Weller, J. Marvin. "Cyclothems and Larger Sedimentary Cycles of the Pennsylvanian." *Journal of Geology* 66, no. 2 (1958): 195–207.

Wilson, J. Tuzo. "Did the Atlantic Ocean Close and Then Reopen?" *Nature* 211 (1966): 676–681.

Worsley, Thomas R., and Judith B. Moody. "Tectonic Cycles and the History of the Earth's Biogeochemical and Paleoceanographic Record." *Paleoceanography* 1 (1986): 233–263.

Worsley, Thomas R., R. Damian Nance, and Judith B. Moody. "Global Tectonics and Eustacy for the Past 2 Billion Years." *Marine Geology* 58 (1984): 373–400.

MASS EXTINCTIONS

Mass extinction is box office, a darling of the popular press, the subject of cover stories and television documentaries, many books, even a rock song. . . . At the end of 1989, the Associated Press designated mass extinction as one of the "Top 10 Scientific Advances of the Decade." Everybody has weighed in, from *The Economist* to *National Geographic*.

—DAVID M. RAUP, *EXTINCTION: BAD GENES OR BAD LUCK?* (1991)

All species that have ever lived are, to a first approximation, dead.

—DAVID M. RAUP, *THE NEMESIS AFFAIR* (1986)

THE CAUSES OF EXTINCTION

If 5 to 50 billion species have lived on this planet, but only about 5 to 15 million are alive today, then 99.9 percent of all species that have ever lived are now extinct. As the statistician might say, *to a first approximation, all species are extinct!* Extinction is a hard fact of nature, not a mark of obsolescence. We call something a "dinosaur" when it is obsolete because their extinction seems to prove their inadequacy, but dinosaurs dominated this planet for more than 150 million years. Our species has been on this planet for only 100,000 to 300,000 years, so we are in no position to laugh (especially when nuclear weapons, climate change, and our own destructive consumption habits make our probability of long-term survival much less likely). Most species in the fossil record survived for a few million years and then became extinct. We will be lucky to survive as long as the average species, and it is unlikely that we can match the record of life-forms today that have been around for tens to hundreds of millions of years.

Despite its importance to life's history, surprisingly little research has been done on extinction. Biologists have spent decades and written hundreds of books on how species evolve—but not on how they disappear. Until 1980 (when the asteroid extinction hypothesis for the end-Cretaceous event was first published) paleontologists had relatively little interest in extinction either, even though they were confronted with it constantly. Most paleontologists took extinction for granted, but there was little impetus to study the phenomenon itself. Although this has changed since 1980, extinction theory is still not a major part of evolutionary biology. As David Raup put it in 1991, extinction is still a "cottage industry" without the trappings or funding of Big Science, such as the Supercollider or Human Genome Project or Hubble Space Telescope. Raup points out that this is comparable "to a demographer trying to study population growth without considering death rates. Or an accountant interested in credits but not debits." Raup suggests that many discussions of extinction have little explanatory content and amount to saying "organisms become extinct because they become extinct":

> Textbooks of evolutionary biology contain little about extinction beyond a few platitudes and tautologies like "species go extinct when they are unable to cope with change" or "extinction is likely when population size approaches zero." The *Encyclopaedia Brittanica* (1987) says, "extinction occurs when a species can no longer reproduce at replacement levels." These statements are almost free of content.

When extinction is discussed at all, it is usually in strict Darwinian terms: organisms become extinct because they fail to adapt to new circumstances. Although this is probably true for the majority of extinctions, it is a difficult hypothesis to test in the wild or in the fossil record except to say that the organism is extinct. Rarely can we find geologic evidence (independent of the last appearance of the organism) that establishes a cause for that extinction. As Raup wrote, "Sadly, the only evidence we have for the inferiority of victims of extinction is the fact of their extinction—a circular argument. . . . The disturbing reality is that for none of the thousands of well-documented extinctions in the geologic past do we have a solid explanation of why extinction occurred."

MASS EXTINCTION

As the fossil record became better known and the idea of extinction first gained acceptance in the 1800s, scientists began to notice that lots of extinctions were concentrated at certain periods in Earth's history. When the pioneering geologist Adam Sedgwick named the Palaeozoic Era in 1838, it was in part based on the striking differences in the typical Paleozoic assemblage of fossils (such as brachiopods, bryozoans, and crinoids) from the fossils of the Triassic (mostly clams and snails and sea urchins, with few of the typical Paleozoic animals). The end-Paleozoic extinction, or Permian-Triassic extinction, radically rearranged life on the seafloor and pushed nearly all of the typical Paleozoic marine organisms to extinction or near-extinction (see chapter 15). Likewise, in 1840 the English geologist John Phillips coined the terms Mesozoic Era and Cainozoic Era because of the striking difference in Mesozoic marine fossils (mainly ammonites) and Cenozoic marine organisms (no ammonites, but lots of snails and clams and sea urchin fossils). Thus the original subdivision of Earth's history into three eras was largely based on how mass extinction events at the end of the Permian and at the end of the Cretaceous dramatically changed the fossils that were found in those rocks.

Yet another century passed before anyone made much of the fact that extraordinary events in the history of life also occurred among marine faunas. In 1952, the legendary paleontologist Norman D. Newell of the American Museum of Natural History was the first to publish an article summarizing how life had gone through multiple episodes of mass extinction and explosive evolution. He further developed his ideas about major mass extinction events in a paper published in 1967. (I knew Norman pretty well because he was still active at the American Museum when I started there as a student in 1976, even though he had formally retired. His influence was felt even more because he was the mentor and PhD advisor of such legendary paleontologists as Stephen Jay Gould, Niles Eldredge, Alan Cheetham, and Don Boyd. Since I was also a student of Eldredge, I am his academic grandson.)

It was another 28 years before the idea of mass extinctions caught on, largely because little direct evidence could be found about the causes of mass extinctions. In 1980, the Alvarez group published a famous paper arguing that the impact of an asteroid caused the end-Cretaceous extinction, and suddenly extinctions became the hottest topic in paleontology for the next 20 years. That debate rages on some 43 years later (see chapter 18). Interest in the end-Cretaceous extinction spilled over into research on other mass extinctions, until the dinosaurs and the asteroid made the end-Cretaceous extinction sexy enough to appear on the cover of *Time* magazine and grabbed 99 percent of the attention.

Ironically, the next phase of the study of mass extinctions was founded on a research project that was entirely unrelated to that topic. As a graduate student of Stephen J. Gould at Harvard, Jack Sepkoski began a huge research project compiling the time span of all the major groups of marine invertebrate fossils through the entire geological record. This work consumed more than a decade as he pored over journals in the library (he finished his PhD in 1977). At that time, information this detailed was not found in any of the major reference works. Sepkoski single-handedly collected data on least 27,000 fossil species, many thousands of genera, and hundreds of families of marine fossils.

By the mid-1970s, Sepkoski had some results that showed interesting patterns in the history of life. He did a complex statistical analysis of his database of marine invertebrate families through time and formally established that there had been three great "evolutionary faunas" in the history of life (figure 11.1). These "evolutionary faunas" were assemblages of fossil groups that expanded rapidly and dominated for some time interval and then, in some cases, faded and were replaced by new evolutionary fauna. In 1981, Sepkoski published his ideas in the journal *Paleobiology*, which focuses on theoretical paleontology. The "Cambrian Fauna" was dominated by trilobites, with minor components of other archaic groups, such as primitive brachiopods, snails, clams, and echinoderms, as well as the strange sponge-like reef builders known as archaeocyathans. These were gradually replaced and phased out as many new groups arose in the Ordovician, including advanced brachiopods, bryozoans, horn corals, tabulate corals, crinoids, and other advanced echinoderms, which were ruled over by large predators such as huge straight-shelled nautiloids. Sepkoski called this the "Paleozoic Fauna" because nearly all post-Cambrian fossiliferous deposits of the Paleozoic are dominated by shells of these animals. The Paleozoic fauna is nearly wiped out in the great Permian extinction event (see chapter 15), and was replaced by Sepkoski's "Modern Fauna" in the Triassic, which is dominated by the same shell fossils that make up most seashell

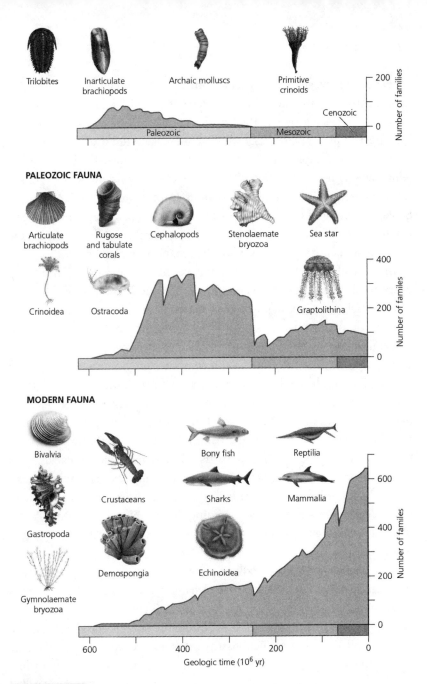

Figure 11.1 ▲

Sepkoski's "three evolutionary faunas," which were recognized by performing a multivariate analysis on a huge database of individual fossil ranges through time. The "Cambrian fauna" consists mostly of trilobites and archaic molluscs, echinoderms, and brachiopods, and fades after the Early Ordovician. The "Paleozoic fauna" ruled the seafloor from the Ordovician through Permian, and is dominated by brachiopods, bryozoans, tabulate and rugose corals, crinoids, graptolites, and primitive cephalopods, and was nearly wiped out in the great Permian extinction. The "modern fauna" consists mostly of clams and snails and sea urchins, plus bony fish and other marine vertebrates. They were present in the background through most of the Paleozoic, but took over the oceans in the Triassic when the Permian crisis and the Mesozoic marine revolution decimated the Paleozoic fauna. (Redrawn from several sources)

collections today: snails, clams, and sometimes sea urchins, along with the modern groups of marine swimmers such as fish and marine mammals.

Sepkoski was teaching at the University of Rochester in 1974 (before he had finished his Harvard PhD) when he met David Raup, another brilliant theoretical paleontologist. In 1978, both Raup and Sepkoski left Rochester, which was then home to a lot of famous paleontologists and was one of the most progressive paleontology programs in the country. Both Raup and Sepkoski were hired at the University of Chicago, and they remained colleagues and often collaborated on projects that broke new ground or looked at the fossil record in novel ways that had many implications.

In particular, Sepkoski's huge database of the time ranges of almost every fossil family and genus had lots of potential implications in addition to teasing out the major "evolutionary faunas." By replotting his data to focus on the last occurrences of each genus or family, Sepkoski could determine how many extinctions had occurred in each interval of time. Using Raup's statistical prowess and computer programming skills, Raup and Sepkoski published a famous paper in 1982 showing the pattern of extinctions that occurred throughout each time interval in the geologic past (figure 11.2). In this study, they showed that the percentage of extinction over time fluctuated wildly, from low rates in some time intervals to huge percentages in other intervals. They found a steady background extinction rate averaging between 2 and 4.6 families per million years, declining slightly over the duration of the Phanerozoic. The rate of background extinctions seems to be declining slightly, but whether this is a real phenomenon or an artifact of the way the data was collected and plotted is still unclear.

Five distinct intervals stood out against this steady background, with between 10 and 20 families dying out each million years. Each of these events contrasted with the steady background of extinction in other time intervals, and statistically they were distinct from the background. These are the so-called Big Five mass extinctions: the Late Ordovician extinction, the Late Devonian extinction, the end-Permian extinction, the end-Triassic extinction, and the end-Cretaceous extinction. (More recently authors have called the ongoing mass extinction caused by humans and their effects on the environment the "Sixth Extinction."). All of these events are discussed in later chapters, but here I address the first of these extinctions: the Late Ordovician extinction.

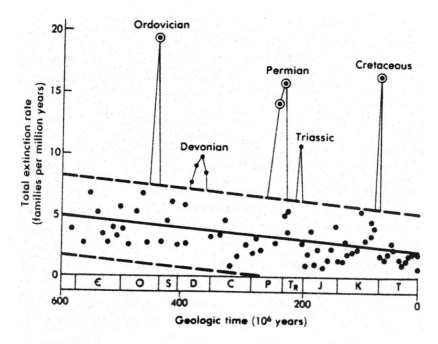

Figure 11.2 ▲

Raup and Sepkoski's plot of extinction rates through the Phanerozoic, showing the Big Five mass extinction peaks standing above the background rate of regular extinctions. (Modified from David M. Raup and J. John Sepkoski Jr., "Mass Extinctions in the Marine Fossil Record," *Science* 215, no. 4539 [1982]: 1501–1503.)

THE LATE ORDOVICIAN EXTINCTION

When Raup and Sepkoski first teased out their data and plotted it in the early 1980s, the Late Ordovician crisis around 445 to 443 Ma was one of the most severe of all the extinction events, possibly more severe than the event that wiped out the dinosaurs and second only to the great Permian extinction event. The Late Ordovician crisis was estimated to have wiped out 50 to 60 percent of marine genera and as much as 85 percent of marine species (only the Permian event, which wiped out 95 percent of marine species, was worse). Despite its severity, most people have never heard of the Late Ordovician crisis, and only a few paleontologists know much about it. Most of the organisms that were affected (specific families and genera of brachiopods, bryozoans, trilobites, and graptolites) were obscure except to the paleontologists who specialize in these groups. Although species loss was severe, this did not radically rearrange the marine ecosystems. There was almost nothing on land but lichens, nonvascular plants, and possibly some simple mosses and some millipedes, so there were no corresponding effects on the terrestrial realm. This contrasts with the Permian extinction and especially with the end-Cretaceous extinction, in which diverse plants and animals on land underwent significant changes during the mass extinctions.

So who were the victims of the second-largest mass extinction of all time, wiping out more than 100 marine families and hundreds of genera and species of marine organisms? To the layperson, their names don't mean much, but to a paleontologist they are very familiar organisms. Most people have heard a bit about trilobites, the segmented marine arthropods that dominated the seas in the Cambrian and Ordovician (figure 11.3). Typical Cambrian trilobites were quite unspecialized, with lots of segmentation and relatively simple head shields (cephalon) and tail shields (pygidium). However, in the Ordovician the trilobites evolved a variety of highly specialized forms, apparently to cope with new superpredators such as the giant nautiloids that could catch and eat almost any creature on the seafloor.

Some Ordovician trilobites turned to burrowing to survive, especially a group called the asaphids (figure 11.4A), which had smooth "snowplow" front and back ends to burrow beneath the seafloor. A few of these, for example *Isotelus rex*, were huge by trilobite standards, about 28 inches long, the largest trilobites ever found, and they vanished at the end of the Ordovician. Another distinctive group that arose in the Ordovician were the tiny trilobites known as trinucleids (figure 11.4B). They had a relatively large cephalon surrounded by an ornate border, giving them the nickname "lace-collar trilobites" for that reason. They had long spines protruding from the corners of their cephalon, but only a few segments in their thorax and a tiny pygidium—and they were typically only 1 to 2 centimeters long. Two other groups that did not survive into the Silurian were the olenids (figure 11.4C), a mostly Cambrian group that rediversified in the Ordovician, and the strange blind, floating arthropods known as agnostids (figure 11.4D). Agnostids were tiny (only a centimeter long) and had identical rounded front and back ends with no eyes and only two or three segments between their ends. They were thought to be highly aberrant trilobites, but scientists have questioned this idea recently. Agnostids had been dominant

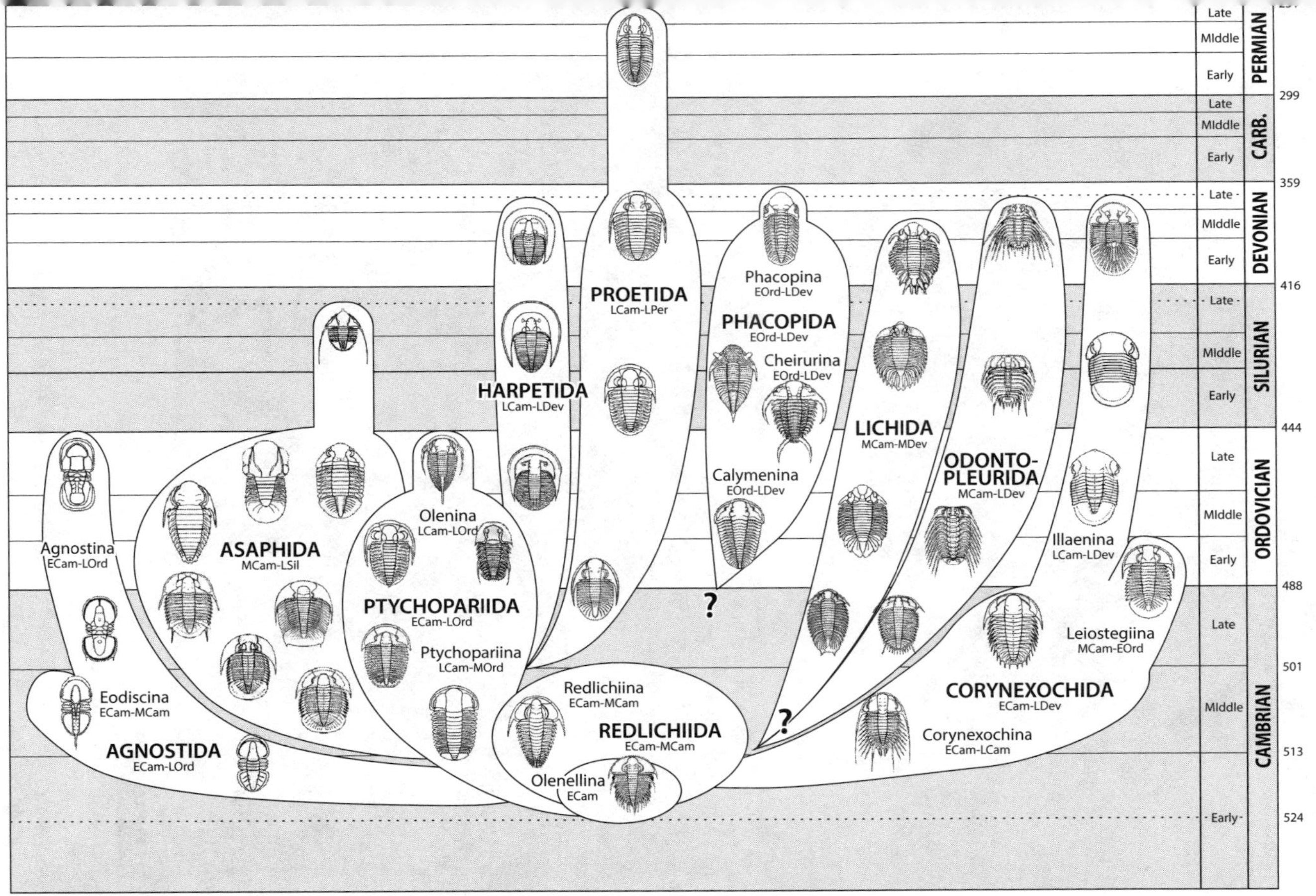

Figure 11.3 ▲

Evolution of the major groups of trilobites. (After D. R. Prothero, Prothero, *Bringing Fossils to Life*, 3rd ed. [Columbia University Press, New York, 2019])

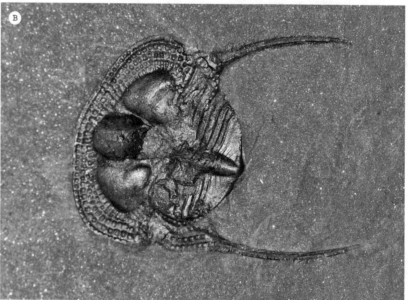

Figure 11.4 ▲ ▶

Typical groups of trilobites that were affected by the Ordovician extinction event: (*A*) "snow-plow" asaphid trilobites, with smooth front and back ends for shallow burrowing; (*B*) trincu-leids, with a broad "lace-collar" brim around the head shield, long spines, and tiny thorax and pygidium; (*C*) *Triarthrus*, a typical olenid trilobite; and (*D*) *Itagnostus*, a typical blind agnostid from the Middle Cambrian Wheeler Shale of Utah. (Courtesy of Wikimedia Commons)

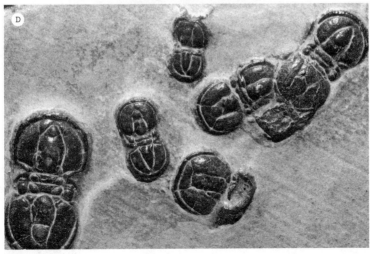

Figure 11.4 ▲
(continued)

in the Middle Cambrian but did not survive the end of the Ordovician. They made up most of the diversity of trilobites known in the Ordovician, and their extinction greatly reduced the total trilobite diversity in the Silurian.

For generations the area around Cincinnati, Ohio, and the adjacent states of Kentucky and Indiana, have been world famous for their incredibly rich fossil beds, which have been avidly collected by both professional paleontologists and amateurs alike. Almost any outcrop in the area produces huge numbers of the shelled creatures known as brachiopods or "lamp shells." They look superficially like clams. Like clams, brachiopods are enclosed in two hinged shells, and they open their shells to filter-feed the seawater that passes through. But in all their anatomical details and soft tissues, they are nothing like clams, which are mollusks and filter-feed using their gills. Brachiopods feed with a feather-like filtering fan-like structure inside their shell called a lophophore. In the Ordovician, brachiopods were so common and diverse that they literally paved the ground in some places. The two dominant groups of Ordovician brachiopods were the strophomenides, which were shaped like a capital "D" in profile, and the finely ribbed orthides (figure 11.5A). Both of these groups were incredibly abundant in Ordovician shallow marine shales and limestones, and both vanished in the late Ordovician. Side by side with all these brachiopods were another group most people have never noticed, the bryozoans, or "moss animals." Found in modern oceans, they are mostly inconspicuous encrustations on rocks and only skilled marine biologists notice them. In these encrusted colonies are hundreds of tiny animals that live in a hole the size of a pinhead and filter-feed seawater with a lophophore. Bryozoans first evolved in the Ordovician, and by the Late Ordovician the rocks (like those in the Cincinnati region) were loaded with them; they were mostly branching forms but some formed disk-like or mound-shaped colonies (figure 11.5B). A number of groups of corals that were dominant in the Ordovician vanished, especially certain horn corals. Nearly all of the groups that dominated the Ordovician seafloor had vanished by the Silurian.

Additional heavy losses occurred among a group of microscopic tooth-like fossils known as conodonts, which are now thought to be the denticles on the internal gill apparatus of an extinct group of jawless eel-like fishes. Floating in the waters above the Ordovician seafloor were tiny organisms known as graptolites, which lived in long branched colonies of tiny cup-like chambers that held their filter-feeding occupants (figure 11.6). Nearly all

Figure 11.5 ▲

Typical Ordovician fossils that were affected by the late Ordovician extinction: (A) a typical slab of Upper Ordovician limestone, covered by the D-shaped shells of strophomenid brachiopods (*Rafinesquina*) and orthid brachiopods (*Hebertella*); and (B) a diorama showing the reconstructions of the branching and massive bryozoans in life position on the late Ordovician seafloor. (Courtesy of Wikimedia Commons)

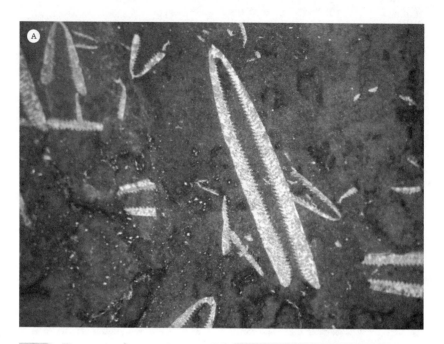

Figure 11.6 ▲

Graptolites were hemichordates (a close relative of vertebrates) and formed large colonies that floated on the sea surface, allowing the tiny animals inside each theca (cup-like structure) to filter feed. (*A*) Carbonized fossils of the graptolite *Monograptus* on a black shale; (*B*) a reconstruction of graptolites in life position. ([*A*] Courtesy of Wikimedia Commons; [*B*] drawing by Mary Persis Williams)

the typically Ordovician groups of graptolites, such as *Diplograptus, Didymograptus,* and *Climacograptus,* vanished, along with *Phyllograptus,* which was shaped like the four vanes on a dart.

The study of the Ordovician extinction is a hot topic in certain paleontological circles because it is so different from the patterns seen in other mass extinctions. There is no evidence of an extraterrestrial impact, so the glamorous impact models don't apply. Most important of all, one of the possible causes was a major glaciation event, which is puzzling and still controversial because the planet had been in a greenhouse world since the Early Cambrian and resumed greenhouse world conditions in the Silurian and right through the mid-Carboniferous (see chapters 10 and 13).

Something else that complicates the problem is that the Ordovician extinction is clearly a two-phase event (figure 11.7), with two separate pulses of extinction (something that also occurs in the Late Devonian extinction and in the Permian extinction). It is primarily focused on the boundary between the penultimate stage of the Ordovician, or Katian Stage (445.2 Ma) and the final stage of the Ordovician, the Hirnantian Stage, and a second event at the Ordovician-Silurian boundary (Hirnantian-Rhuddanian boundary) that occurred about 443.8 Ma.

The first pulse at the beginning of the Hirnantian corresponds with abundant geological evidence of huge glacial ice sheets on Gondwana, part of the famous Late Ordovician glacial event. The evidence of this glaciation is dramatic, with thick glacial deposits over what was then the South Pole in Gondwana (figure 11.8). (This is now the Sahara Desert region, ranging from North Africa to Arabia.) This implies that there was a dramatic cooling event on a global scale, which may have severely affected marine faunas adapted to the mild climates of the greenhouse planet of the Cambrian and most of the Ordovician.

Debate continues regarding the causes of the Late Ordovician glaciation; it is especially odd because it occurred during the warm greenhouse conditions of most of the early-middle Paleozoic. Geochemical evidence shows that there was a huge drop in carbon dioxide, from Katian levels of 4,000 to 7,000 ppm (extreme greenhouse) to much lower levels by the Hirnantian. Two main causes for this abrupt drop in greenhouse gases are suggested. One is that during the late Katian volcanic eruptions produced thousands of square kilometers of flood basalts (known as the "Katian large igneous province"). These might have darkened the skies and helped

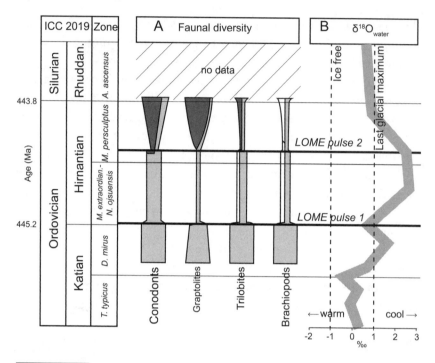

Figure 11.7 ▲

Global features of the Late Ordovician mass extinction (LOME) event. (*A*) Faunal diversity including pre-extinction biota (light gray), appearances post–LOME pulse 1 (white), and appearances post–LOME pulse 2 (dark gray). Width gives qualitative sense of within-group diversity changes. (*B*) δ¹⁸O curve. First LOME pulse occurs at the end phase of warming. Rhuddan = Rhuddanian. Zone refers to graptolite zones (genus abbreviations): *T.* = *Tangyagraptus*; *D.* = *Dicellograptus*; *M.* = *Metabolograptus*; *N.* = *Normalograptus*; *A.* = *Akidograptus*. (Redrawn by E. Prothero; modified from David P. G. Bond and Stephen E. Grasby, "Late Ordovician Mass Extinction Caused by Volcanism, Warming, and Anoxia, Not Cooling and Glaciation," *Geology* 48, no. 8 [2020]: 777–781)

trigger "nuclear winter" conditions, which have been postulated for the end-Cretaceous events and other mass extinctions. In addition, huge areas of black lava that were exposed on the earth's surface weathered rapidly, and weathering absorbs carbon dioxide as the rocks break down. As discussed elsewhere, weathering and absorption of carbon dioxide are major mechanisms for decreasing greenhouse gases.

The second possible trigger is the fact that vascular land plants began to diversify for the first time during Hirnantian. These would draw down

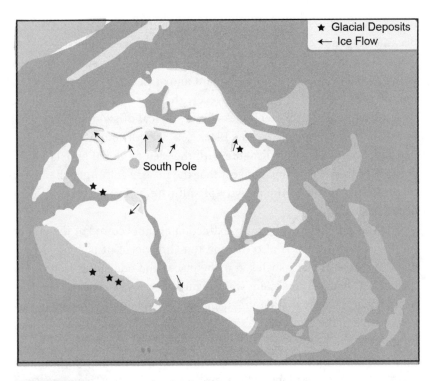

Figure 11.8 ▲
Paleogeographic map of the Late Ordovician Gondwana ice cap. (Redrawn by E. Prothero from several sources)

lots of carbon dioxide into their tissues and store it in the crust if they were not broken down by land plant decomposers (which had not evolved yet). In fact, probably the only land animals in the Late Ordovician were millipedes, who leave their characteristic burrows in ancient soils. Some scientists argue that the development of low-growing land plants is what triggered the rapid consumption of carbon dioxide and started the cooling that produced the Hirnantian glaciation.

With the rapid cooling as glaciers expanded across what is now North Africa, another important effect occurred. Glaciers lock lots of water on the land into ice caps and cause sea levels to drop as much as 80 meters (260 feet). This exposes huge areas of what had been shallow seafloor to drying conditions, which would have wiped out the marine communities that flourished in shallow epicontinental seas. Another consequence of the changed oceanographic conditions is that many marine basins became stagnant and anoxic. Study of the sulfur geochemistry of the world's oceans also points to something changing the sulfur isotopes in the global oceans, probably stagnant anoxic ocean basins that pulled lots of sulfur out of the sea. However, there are some doubts about this mechanism. Sulfur-rich black shales, typical of stagnant seas with anoxic bottom water, such as the Black Sea, are not common in the Hirnantian. Some scientists disagree with the idea of global anoxia in the oceans for this reason and suggest that the bizarre sulfur geochemistry might be the result of the activity of sulfur-reducing bacteria in local ocean basins.

The second phase of the mass extinction event occurred at the end of the Hirnantian, which is also the end the Ordovician. It is marked by a major increase in black shales in many places and odd sulfur isotopes in the oceans, which is consistent with stagnant deep ocean basins. The Hirnantian glaciers were in full retreat, and lots of evidence suggests that carbon dioxide in the atmosphere shot up abruptly as greenhouse conditions returned by the beginning of the Silurian. Indeed, the extinction of many cold-adapted fossil assemblages, including the *Hirnantia* brachiopod fauna and the *Mucronaspis* trilobite fauna, also suggests that the second phase of the Late Ordovician extinction was an abrupt global warming event after the world had adapted to cold glacial conditions.

However, in 2020 David Bond and Stephen Grasby argued that the large igneous province eruptions and ensuing anoxia of the Katian caused the early Hirnantian collapse of the deep-water ecosystem as oxygen became scarce and the waters became rich in toxic sulfides. Their data show that these geochemical signals coincide better with the first wave of extinction at the beginning of the Hirnantian, and that the cooling and glaciation that followed in the early Hirnantian took place after the first wave of mass extinction had occurred. Their data also showed the same volcanic signal with warming and anoxia at the end of the Ordovician. Thus they argue that both of the paired extinctions in the later Ordovician were triggered by volcanism, oceanic anoxia, and global warming, not by the Hirnantian glaciation.

Whatever the causes, after the second-largest mass extinction in Earth's history, most of the highly specialized, endemic Ordovician groups of fossils were gone, and the early Silurian is populated mostly by cosmopolitan organisms that could cope with a wide variety of oceanographic conditions. The warming at the end of the Ordovician brought the greenhouse conditions back by the Early Silurian, and they would continue without interruption until the middle of the Carboniferous Period, about 325 Ma.

FOR FURTHER READING

Algeo, Thomas J. "Terrestrial-Marine Teleconnections in the Devonian: Links Between the Evolution of Land Plants, Weathering Processes, and Marine Anoxic Events." *Philosophical Transactions of the Royal Society B: Biological Sciences* 353, no. 1365 (1998): 113–130.

Algeo, Thomas J., R. A. Berner, J. B. Maynard, S. E. Scheckler, and GSAT Archives. "Late Devonian Oceanic Anoxic Events and Biotic Crises: 'Rooted' in the Evolution of Vascular Land Plants?" *GSA Today* 5, no. 3 (1995).

Algeo, Thomas. J., Stephen E. Scheckler, and J. Barry Maynard. "Effects of the Middle to Late Devonian Spread of Vascular Land Plants on Weathering Regimes, Marine Biota, and Global Climate." In *Plants Invade the Land: Evolutionary and Environmental Approaches*, ed. Patricia G. Gensel and Dianne Edwards, 213–236. New York: Columbia University Press, 2001.

Bjerrum, Christian J. "Sea Level, Climate, and Ocean Poisoning by Sulfide All Implicated in the First Animal Mass Extinction." *Geology* 46 (2018): 575–576.

Bond, David P. G., and Stephen E. Grasby. "Late Ordovician Mass Extinction Caused by Volcanism, Warming, and Anoxia, Not Cooling and Glaciation." *Geology* 48, no. 8 (2020): 777–781.

Bond, David P. G., and Paul B. Wignall. "Large Igneous Provinces and Mass Extinctions: An Update." *GSA Special Papers* 505 (2014): 29–55.

Brannen, Peter. *The Ends of the World: Volcanic Apocalypses, Lethal Oceans, and Our Quest to Understand Earth's Past Mass Extinctions*. New York: Ecco, 2017.

Brenchley, P. J., Jim D. Marshall, and Charlie J. Underwood. "Do All Mass Extinctions Represent an Ecological Crisis? Evidence from the Late Ordovician." *Geological Journal* 36, no. 3–4 (2001): 329–340.

Caplan, Mark L., and R. Mark Bustin. "Devonian–Carboniferous Hangenberg Mass Extinction Event, Widespread Organic-Rich Mudrock and Anoxia: Causes and Consequences." *Palaeogeography, Palaeoclimatology, Palaeoecology* 148, no. 4 (1999): 187–207.

Hammarlund, Emma U., Tais W. Dahl, David A. T. Harper, David P. G. Bond, Arne T. Nielsen, Christian J. Bjerrum, Niels H. Schovsbo, et al. "A Sulfidic Driver for the End-Ordovician Mass Extinction." *Earth and Planetary Science Letters* 331–332 (2015): 128–139.

Hallam, Anthony, and Paul B. Wignall. *Mass Extinctions and Their Aftermath.* Oxford: Oxford University Press, 1997.

Harper, David A. T., Emma U. Hammarlund, and Christian M. Ø. Rasmussen. "End Ordovician Extinctions: A Coincidence of Causes." *Gondwana Research* 25, no. 4 (2014): 1294–1307.

Holland, Steven M., and Richard R. Davis. *A Sea Without Fish: Life in the Ordovician Seas of the Cincinnati Region.* Bloomington: Indiana University Press, 2009.

Jones, David S., Anna M. Martini, David A. Fike, and Kunio Kaiho. "A Volcanic Trigger for the Late Ordovician Mass Extinction? Mercury Data from South China and Laurentia." *Geology* 45, no. 7 (2017): 631–634.

Kravchinsky, Vadim A. "Paleozoic Large Igneous Provinces of Northern Eurasia: Correlation with Mass Extinction Events." *Global and Planetary Change* 86–87 (2012): 31–36.

Levi-Setti, Ricardo. *The Trilobite Book: A Visual Journey.* Chicago: University of Chicago Press, 2014.

MacLeod, Norman. *The Great Extinctions: What Causes Them and How They Shape Life.* London: Firefly, 2015.

McGhee, George. *The Late Devonian Mass Extinction: The Frasnian/Famennian Crisis.* New York: Columbia University Press, 1996.

——. *When the Invasion of the Land Failed: The Legacy of the Devonian Extinctions.* New York: Columbia University Press, 2013.

Melchin, Michael J., Charles E. Mitchell, Chris Holmden, and Peter Štorch. "Environmental Changes in the Late Ordovician-Early Silurian: Review and New Insights from Black Shales and Nitrogen Isotopes." *Geological Society of America Bulletin* 125, no. 11–12 (2013): 1635–1670.

Newell, Norman D. "Periodicity in Invertebrate Evolution." *Journal of Paleontology* 26 (1952): 371–385.

——. "Revolutions in the History of Life." In *Uniformity and Simplicity: A Symposium on the Principle of the Uniformity of Nature,* ed. Claude G. Albritton Jr., 63–91. Boulder, CO: Geology Society of America, 1967.

Prothero, Donald R. *The Evolving Earth.* New York: Oxford University Press, 2020.

Racki, Grzegorz. "Toward Understanding Late Devonian Global Events: Few Answers, Many Questions." In *Understanding Late Devonian and Permian-Triassic Biotic and Climatic Events,* ed. Jeff Over, Jared Morrow, and P. Wignall. London: Elsevier, 2005.

Raup, David M., and J. John Sepkoski Jr. "Mass Extinctions in the Marine Fossil Record." *Science* 215, no. 4539 (1982): 1501–1503.

Retallack, G. J. "Late Ordovician Glaciation Initiated by Early Land Plant Evolution, and Punctuated by Greenhouse Mass-Extinctions." *Journal of Geology* 123 (2015): 509–538.

Sallan, L. C., and M. I. Coates. "End-Devonian Extinction and a Bottleneck in the Early Evolution of Modern Jawed Vertebrates." *Proceedings of the National Academy of Sciences* 107, no. 22 (2010): 10131–10135.

Saltzman, Matthew R., and Seth A. Young. "Long-Lived Glaciation in the Late Ordovician? Isotopic and Sequence-Stratigraphic Evidence from Western Laurentia." *Geology* 33, no. 2 (2005): 109–112.

Sepkoski, J. John, Jr. "A Factor Analytic Description of the Phanerozoic Marine Fossil Record." *Paleobiology* 7, no. 1 (1981): 36–53.

Sheehan, Peter M. "The Late Ordovician Mass Extinction." *Annual Review of Earth and Planetary Sciences* 29, no. 1 (2001): 331–364.

Zou, Caineng, Zhen Qiu, Simon W. Poulton, Dazhong Dong, Hongyan Wang, Daizhao Chen, Bin Lu, et al. "Oceanic Euxinia and Climate Change 'Double Whammy' Drove the Late Ordovician Mass Extinction." *Geology* 46 (2018): 535–538.

12

THE DEVONIAN CRISES

> The inhabitants of each successive period in world's history have beaten
> their predecessors in the race for life, and are, insofar, higher on the
> scale of nature. If the Eocene inhabitants were put into competition
> with the existing inhabitants, the Eocene fauna or flora would certainly
> be beaten and exterminated, as would a secondary [Mesozoic] fauna by an
> Eocene, and a Palaeozoic fauna by a secondary fauna.
>
> —CHARLES DARWIN, *ON THE ORIGIN OF SPECIES* (1859)

THE LATE DEVONIAN EXTINCTIONS

Just as the Ordovician ended with a Big Five mass extinction, there was a
major mass extinction near the end of the Devonian that also occurred in
two pulses. The first, known as the Kellwasser Event, occurred 372 Ma on
the boundary between the last two stages (Frasnian and Famennian) of the
Devonian. The second pulse, the Hangenberg Event, occurred at the end of
the Devonian (end of the Famennian Stage), about 359 Ma. Other scientists
have suggested that it was not two distinct events but a series of as many
as 8 to 10 different pulses. This was the fourth or fifth largest mass extinction in Earth's history, wiping out 75 percent of the species, 55 percent of the
marine genera, and about 22 percent of the marine families in the marine
invertebrate community.

Who were the major victims of this extinction? Some of the groups were
typically Paleozoic fauna taxa such as the bryozoans, which had already

taken a hit during the Ordovician extinctions, and others were survivors of the Cambrian fauna. For example, the trilobites had recovered from the Ordovician extinctions and diversified with a number of new groups. These included the phacopids, with their complex protruding eyes, broad bumpy frog-like snouts, and their ability to roll up into a ball (figure 12.1A); the spiky odontopleurids (figure 12.1B); the lichad trilobites with their broad bodies and tail shield and their pointed snouts (figure 12.1C), including the gigantic 60-centimeter-long spiky lichid known as *Terataspis* (figure 12.1D); and the harpetids with a broad horseshoe-shaped brim around the front of the head (figure 12.1E). Other relatively unspecialized corynexochids, with their boxy pestle-shaped glabella and their large broad tail plate made of 12 or more segments, were also represented. This huge diversity of trilobites vanished at the end of the Devonian, leaving only a relatively simple and small group, the proetids, to straggle on in small numbers throughout the rest of the Paleozoic.

Another dominant Paleozoic fauna group was the brachiopods. In the Devonian, the seafloor was paved with huge numbers of brachiopods known as "spirifers" because their lophophore was shaped like a spiral. Genera such as *Mucrospirifer* are shaped somewhat like a pair of wings that a pilot might wear as a lapel pin, with a wide flaring shape and a very long straight hinge line (figure 12.2). In some Devonian localities, the ground is literally paved with hundreds of spirifer fossils including not only *Spirifer*, *Mucrospirifer*, *Delthyris*, and *Cyrtina* but also the odd-shaped but ubiquitous *Atrypa*. These genera largely vanished at the end of the Devonian except for *Spirifer* itself, which straggled on until the end of the Triassic.

The most impressive feature of shallow tropical seafloor marine assemblages before the end of the Devonian was the development of enormous coral reefs built by an assemblage of colonial animals. Some reefs were larger than the Great Barrier Reef off Australia, and stuck up from the seafloor in huge ramparts and mounds. Among the reef builders were a group of layered sponges known as stromatoporoids. There was also a remarkable range of corals, including tabulate corals persisting from the Ordovician, such as *Favosites* (figure 12.3A). The horn corals, or rugosids, were even more diverse. Some were solitary corals such as the distinctively rugose, wrinkled Devonian index fossil *Heliophyllum* (figure 12.3B). Some horn corals gave up the solitary life and clustered together in a colony that resembles a honeycomb or organ pipes. The best known of these is the distinctive

Figure 12.1 ▲ ▶

Typical Devonian trilobites that were victims of the Devonian extinction events. (A) *Phacops*, the common trilobite with huge eyes composed of large doublet lenses; (B) *Dicranurus*, a typical spiny odontopleurid trilobite; (C) *Arctinurus*, a typical lichad trilobite, with a big pygidium and a small cephalon with a pointed snout; (D) *Terataspis*, a giant spiny lichad; and (E) *Harpes*, a typical harpid trilobite with the broad horseshoe-shaped cephalon and tiny thorax and pygidium. (Courtesy of Wikimedia Commons)

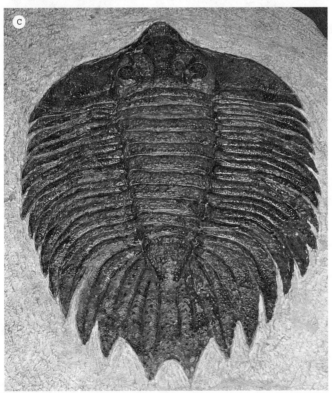

Figure 12.1 ▲

(*continued*)

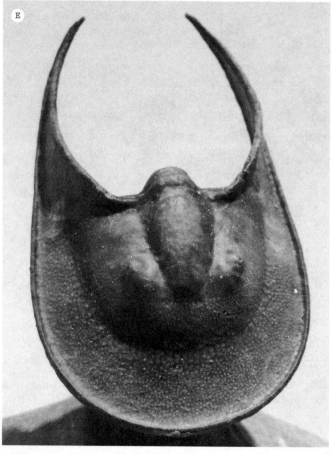

Figure 12.1 ▲
(continued)

Figure 12.2 ▲

The long-hinged Devonian *Mucrospirifer*, a common spirifer brachiopod of the Devonian. (Courtesy of Wikimedia Commons)

Devonian colonial rugosid *Hexagonaria* (figure 12.3C). Its name refers to the close packing of its cylindrical corallites that often have a hexagonal shape like the walls of a honeycomb. *Hexagonaria* fossils are common in many Devonian reef localities, such as the Falls of the Ohio near Louisville, Kentucky, and in many Devonian quarries in Iowa (especially near Coralville, Iowa). *Hexagonaria* fossils that eroded out of the Devonian rocks of Michigan and then abraded into smooth pieces on the shores of the Great Lakes are known as "Petoskey stones," which is the state rock of Michigan. Crinoids were also diverse and abundant on the flanks and lagoons of these Devonian reef complexes, and numerous Paleozoic localities in the Midwest have an incredible density of well-preserved crinoids. There were also relatives of the nautilus, with their shell coiled in a flat spiral, that soon became one of the dominant Devonian predators: the ammonoids.

Even more spectacular in the Devonian was the huge burst in the evolution of fish groups (figure 12.4). For the Cambrian and Ordovician, only a few fossils of tiny jawless fish have been found, and their skeletons are mostly made of cartilage without bone except for dermal scales. But during the Silurian and especially the Devonian, these early jawless ancestors diversified into a wide range of fish, including both extinct groups and most of the fish groups alive today. Indeed, fish fossils were so diverse and

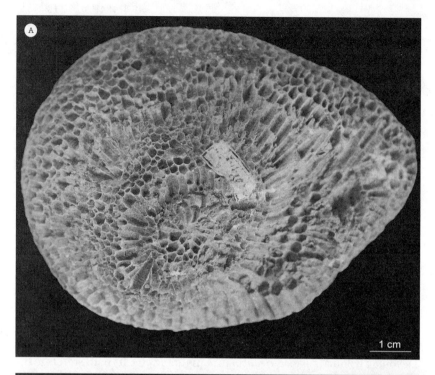

Figure 12.3 ▲ ▶

Typical common Devonian corals. (*A*) *Favosites*, with the tightly packed cylindrical corallites with tabulae along each tube; (*B*) *Heliophyllum*, a lumpy irregular horn coral, or rugosid; and (*C*) *Hexagonaria*, a colonial rugosid whose tightly packed cylindrical corallites produce a hexagonal "honeycomb" pattern. (Courtesy of Wikimedia Commons)

Figure 12.3 ▲
(*continued*)

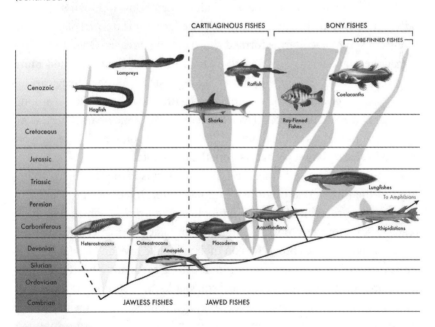

Figure 12.4 ▲
Family tree of early fish groups. (Redrawn from several sources)

abundant in some rocks that the Devonian has often been called the "Age of Fishes."

The family tree of fishes and their descendants shows that nearly every group was diverse and abundant in the middle Paleozoic. Jawless fishes continued to evolve rapidly and developed a variety of forms covered in bony armor (figure 12.5A). They were particularly common in Devonian river deposits such as the Old Red Sandstone of Scotland and England. Some had flat bodies and upward-facing eyes, suggesting that they sucked up food from the bottom. Others had rounded bodies, simple slit-like mouths, chain-mail bony armor, and tails that tended to propel their head upward. They probably wriggled along like tadpoles and filtered out food with their gills. But the major event that allowed fish to diversify so much was the invention of jaws. Without jaws, fish could only use their mouth to suck in food and water and filter out what they needed. Jaws gave them the power to bite and crush prey, and they had many other side benefits as well.

The largest and most impressive of the fish with jaws were the earliest group to obtain them, known as the placoderms (figure 12.5B). Like sharks, placoderms had no bones in their skeleton but were supported by a cartilage skeleton instead. Bone was found only in the dermal shields that covered their head and thorax; the back half of their body and their shark-like tails generally had no armor covering. The edges of the dermal plates in the mouth of some placoderms formed sharp cutting devices. One of the largest of the placoderms was *Dunkleosteus*, which may have reached more than 6 meters (20 feet) in length, weighed over a ton, and had heavy armor plates on its face and jaw that were hinged with a thick armor girdle around the thorax (recent estimates have reduced its size somewhat). These were the largest predators ever seen on Earth up to that point, and were not surpassed in size by larger predators until the Mesozoic.

There were dozens of other kinds of placoderms in the Devonian, filling most of the niches that other fish exploited later. In addition to *Dunkleosteus*, there were even bigger predators like *Titanichthys*, which may have reached 8 meters (26 feet) in length. Some placoderms (the rhenanids) were flat-bodied with broad fins on their side, like rays and skates. Others (the ptyctodonts) had heavy jaws with crushing teeth and resembled modern chimaeras ("ratfish") or Port Jackson sharks. Still others (the antiarchs) were small fossils (less than 30 centimeters [12 inches] long) with a heavy box of armor covering their entire body, and even hinged armor on their

Figure 12.5 ▲

The Devonian is often called the "Age of Fishes" because of the huge diversification of different groups of fish at that time, including the ancestors of all modern fish groups, plus extinct groups with no descendants, like most of the jawless fish, or the placoderms. (A) Images of the typical jawless fish of the Devonian. (B) In this painting, the gigantic 10-meter-long arthrodire placoderm *Dunkleosteus* is shown attacking a 2-meter-long shark *Cladoselache*. (Courtesy of Wikimedia Commons)

fins that gave them appendages that looked like crab legs. There were dozens of species of placoderms in eight separate orders during the Devonian; they were not only the largest but also the most abundant and diverse group in the Age of Fishes.

Not to be overlooked, however, is another important group: the sharks. Although not as abundant as placoderms, by the Late Devonian some early sharks had appeared. *Cladoselache* was about 2 meters (6 feet) long and had very primitive cartilaginous skulls with simple teeth composed of numerous tall conical cusps. Although their bodies were long and streamlined like most sharks, they had two dorsal fins with thick bony spines sticking out of the front of the fin. Their front (pectoral) fin was a broad, rigid triangle that served as a good stabilizer, but it was not as maneuverable as the pectoral fins in later sharks.

In addition to jawless fish, placoderms, and sharks, whose skeletons were made of cartilage with bone limited to their teeth, spines, and skin, the second big diversification of fishes are the bony fish. By the Devonian, the bony fish had split into the two groups that are both alive today: the ray-finned fishes (which make up 99 percent of living fishes) and the lobe-finned fishes (lungfish, coelacanths, and their amphibian descendants). Ray-finned fishes are the most familiar kinds of fish; their fins are supported by a number of thin bony rods or rays. Their Devonian ancestors already had evolved a bony skeleton, fin rays, and a skull with a bony braincase and powerful jaws. Although relatively rare and overshadowed by placoderms and other fish in the Devonian, ray-finned fish underwent a huge evolutionary radiation and eventually took over the entire realm of fishes (except for the sharks that are still alive). There are about 30,000 species of ray-finned fish today, which is more species than all the other groups of vertebrates (mammals, birds, reptiles, and amphibians) combined.

The other branch of bony fish contains the lobe-finned fishes, which have thick bony supports in their fins rather than thin rays or rods of bone. Those bony elements in their fins match exactly with the bones in your arm and in your leg, so they had the kinds of fins that could evolve into arms and legs. They include lungfish, which can breathe in air or water and were well represented in the Devonian as well. A second group of lobe-fins is the coelacanths, common in the Devonian but thought to be extinct until a single living species was found off the deep waters of southern Africa in 1938. The final branch of the lobe-fins is the group that leads to the amphibians, which finally crawled out onto land in the Frasnian, just before the Kell-wasser Event.

WHAT WERE THE CAUSES?

The waves of extinction struck different groups at different times. In the first wave in the late Frasnian (Kellwasser Event), the ammonoids, trilobites (most of the typical Devonian groups vanished), brachiopods (especially the diverse spirifers), armored jawless fish, most of the lobe-finned fish, and jawed placoderm fish were hit particularly hard, whereas the snails, clams, and bryozoans escaped with only minor extinctions. The second wave of extinction (the Hangenberg Event) at the end of the Devonian wiped out the algal community that produced acritarch fossils and finished off the placoderm fish completely (figure 12.6). From the huge radiation of different groups of fishes in the Age of Fishes, only sharks and bony fish survived into the Carboniferous. In fact, 97 percent of the vertebrate species were wiped out. The only survivors in the early Carboniferous were sharks less than a meter long and fish and amphibians less than 10 centimeters long.

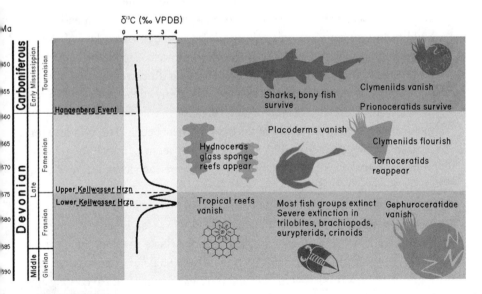

Figure 12.6 ▲
Diagram of the major events of the Devonian extinctions. (Diagram by K. Marriott)

The biggest victims of the Devonian extinction were the giant coral-stromatoporoid reefs that had dominated since the early Paleozoic. During the Hangenberg Event, stromatoporoid sponges vanished completely, and most of the tabulate and rugosan (horn coral) genera were wiped out as well. Only a few of these corals survived into the late Paleozoic. The complete obliteration of the warm water coral reef assemblage undoubtedly explains why many of the groups that lived in and around the reefs, such as the brachiopods and ammonoids, became extinct.

There is also a strong temperature signal in the extinction. The major group of brachiopods to vanish was the atrypids, which were tropical in their distribution. The tropical reefs of the world were decimated, but corals found in deep, colder water were unaffected, and reef communities in the polar latitudes (such as the Parana Basin of South America) also escaped severe extinction. Most striking of all are the organisms that replaced the tropical reefs. Giant reefs of the glass sponge *Hydnoceras* were particularly common in the Late Devonian of upstate New York (figure 12.7A). These were previously known from the colder, deeper waters and then spread to the shallows, apparently in response to the dramatic cooling of tropical waters (figure 12.7B).

The signal of tropical cooling can be seen in many other indicators, such as the geochemistry of oxygen in the oceans, which indicates a massive ice buildup. Indeed, Late Devonian glacial tills appeared in many places, including in the polar regions of Gondwana, and even as close to the ancient tropics as Pennsylvania, Maryland, and West Virginia (Rockwell Formation).

What caused the Devonian extinctions? Many ideas have been proposed, from asteroid or comet impacts, to massive volcanic events, to plate tectonic changes, to depletion of the oxygen in the ocean. The evidence for impacts has never been confirmed and doesn't seem to match the pattern of extinctions, and plate tectonic movements are too slow to trigger such abrupt events, so that is not supported as a cause.

The oceanic oxygen depletion is undoubtedly real. The typical Upper Devonian deposits (especially during the Kellwasser Event) are anoxic deep-water black shales, such as the Cleveland Shale in Ohio, the Chattanooga Shale in Tennessee, and the New Albany Shale in Illinois. However, many geologists think that anoxia and the clear signal of global cooling are related: as the ocean basins became dramatically cooled and stratified, their bottoms were starved of oxygen and became deadly for organisms.

Figure 12.7 ▲

(*A*) Fossils of the glass sponge *Hydnoceras*. (*B*) Diorama of the typical glass sponge reefs of *Hydnoceras*, common in the colder waters of the Late Devonian. (Courtesy of Wikimedia Commons)

Thus the only well-supported causes are dramatic cooling in the tropics and oceanic stagnation.

What could have caused this to happen? Volcanism has been a popular explanation since 2002. There were gigantic eruptions of basaltic flood lavas at both the Kellwasser and Hangenberg events, mostly in what is now Siberia and Russia. The Vilyuy lavas in the Siberian craton have been redated recently, and they match the dates for the Kellwasser Event. The Pripyat-Dneiper-Donets large igneous province, just north of the Caspian Sea, has been dated to just before the end of the Devonian, which could explain the Hangenberg Event. Such huge eruptions could have produced enough sulfur dioxide and dust in the atmosphere to contribute to global cooling, causing the Devonian glaciers to grow and sea level to drop. This would help produce stagnant anoxic ocean basins, and the high sulfur content of the gases would contribute to sulfur contamination of the oceans.

In recent years, a number of geologists have pointed to the fact that the planet developed large-scale forests for the first time in the Late Devonian. Until this time, only algae and a few low-growing land plants had affected the carbon dioxide balance in the atmosphere. Not only do large trees absorb and trap a lot more carbon dioxide, they also speed up deep weathering of the soil with their roots, which makes the soils absorb carbon dioxide as well. Earth was not completely done with the greenhouse world yet (it returns for its last hurrah in the Early Carboniferous), but this was the first significant pulse of global cooling, and it affected a planet that had been stable and had a huge tropical biomass for millions of years and was unprepared for such a severe cooling. In addition, the extinction was more severe than any previous extinction in important ways. Unlike the Ordovician extinction, which wiped out more genera and species but did not fundamentally change the ecological communities, the Late Devonian extinctions killed off much of the tropical fauna and changed the world by wiping out the reefs, as well as decimating the huge radiation of fish groups from the Age of Fishes. Life in the world's oceans would never look the same again.

FOR FURTHER READING

Algeo, Thomas J. "Terrestrial-Marine Teleconnections in the Devonian: Links Between the Evolution of Land Plants, Weathering Processes, and Marine

Anoxic Events." *Philosophical Transactions of the Royal Society B: Biological Sciences* 353, no. 1365 (1998): 113–130.

Algeo, Thomas J., R. A. Berner, J. B. Maynard, S. E. Scheckler, and GSAT Archives. "Late Devonian Oceanic Anoxic Events and Biotic Crises: 'Rooted' in the Evolution of Vascular Land Plants?" *GSA Today* 5, no. 3 (1995).

Algeo, Thomas J., Stephen E. Scheckler, and J. Barry Maynard. "Effects of the Middle to Late Devonian Spread of Vascular Land Plants on Weathering Regimes, Marine Biota, and Global Climate." In *Plants Invade the Land: Evolutionary and Environmental Approaches*, ed. Patricia G. Gensel and Dianne Edwards, 213–236. New York: Columbia University Press, 2001.

Bond, David P. G., and Paul B. Wignall. "Large Igneous Provinces and Mass Extinctions: An Update." *GSA Special Papers* 505 (2014): 29–55.

——. "The Role of Sea-Level Change and Marine Anoxia in the Frasnian-Famennian (Late Devonian) Mass Extinction." *Palaeogeography, Palaeoclimatology, Palaeoecology* 263, no. 3–4 (2008): 107–118.

Brannen, Peter. *The Ends of the World: Volcanic Apocalypses, Lethal Oceans, and Our Quest to Understand Earth's Past Mass Extinctions*. New York: Ecco, 2017.

Brenchley, P. J., Jim. D. Marshall, and Charlie J. Underwood. "Do All Mass Extinctions Represent an Ecological Crisis? Evidence from the Late Ordovician." *Geological Journal* 36, no. 3–4 (2001): 329–340.

Caplan, Mark L., and R. Mark Bustin. "Devonian–Carboniferous Hangenberg Mass Extinction Event, Widespread Organic-Rich Mudrock and Anoxia: Causes and Consequences." *Palaeogeography, Palaeoclimatology, Palaeoecology* 148, no. 4 (1999): 187–207.

Hallam, Anthony, and Paul B. Wignall. *Mass Extinctions and Their Aftermath*. Oxford: Oxford University Press, 1997.

Hammarlund, Emma U., Tais W. Dahl, David A. T. Harper, David P. G. Bond, Arne T. Nielsen, Christian J. Bjerrum, Niels H. Schovsbo, et al. "A Sulfidic Driver for the End-Ordovician Mass Extinction." *Earth and Planetary Science Letters* 331–332 (2015): 128–139.

Kravchinsky, Vandim A. "Paleozoic Large Igneous Provinces of Northern Eurasia: Correlation with Mass Extinction Events." *Global and Planetary Change* 86–87 (2012): 31–36.

Levi-Setti, Ricardo. *The Trilobite Book: A Visual Journey*. Chicago: University of Chicago Press, 2014.

MacLeod, Norman. *The Great Extinctions: What Causes Them and How They Shape Life*. London: Firefly, 2015.

Marriott, Katherine, Donald R. Prothero, and Alexander Bartholomew. *The Evolution of Ammonoids*. Boca Raton, FL: Taylor & Francis, 2023.

McGhee, George. *The Late Devonian Mass Extinction: The Frasnian/Famennian Crisis*. New York: Columbia University Press, 1996.

——. *When the Invasion of the Land Failed: The Legacy of the Devonian Extinctions*. New York: Columbia University Press, 2013.

Newell, Norman D. "Periodicity in Invertebrate Evolution." *Journal of Paleontology* 26 (1952): 371–385.

——. "Revolutions in the History of Life." In *Uniformity and Simplicity: A Symposium on the Principle of the Uniformity of Nature*, ed. Claude G. Albritton Jr., 63–91. Boulder, CO: Geology Society of America, 1967.

Prothero, Donald R. *The Evolving Earth*. New York: Oxford University Press, 2020.

Racki, Grzegorz. "Toward Understanding Late Devonian Global Events: Few Answers, Many Questions." In *Understanding Late Devonian and Permian-Triassic Biotic and Climatic Events*, ed. Jeff Over, Jared Morrow, and P. Wignall. London: Elsevier, 2005.

Raup, David M., and J. John Sepkoski Jr. "Mass Extinctions in the Marine Fossil Record." *Science* 215, no. 4539 (1982): 1501–1503.

Retallack, G. J. "Woodland Hypothesis for Devonian Evolution of Tetrapods." *Journal of Geology* 119 (2011): 235–258.

Retallack, G. J., and C.-M. Huang. "Ecology and Evolution of Devonian Trees in New York, USA." *Palaeogeography Palaeoclimatology Palaeoecology* 299 (2011): 110–128.

Sallan, L. C., and M. I. Coates. "End-Devonian Extinction and a Bottleneck in the Early Evolution of Modern Jawed Vertebrates." *Proceedings of the National Academy of Sciences* 107, no. 22 (2010): 10131–10135.

13

THE FIRST FORESTS

It's not as though we can keep burning coal in our power plants. Coal is a finite resource, too. We must find alternatives, and it's a better idea to find alternatives sooner than wait until we run out of coal, and in the meantime, put God knows how many trillions of tons of CO_2 that used to be buried underground into the atmosphere.

—ELON MUSK

THE COAL MEASURES

Throughout the Middle Ages and until about 1700, coal was a minor resource. It was difficult to mine, and there was abundant wood for charcoal and other types of fuel. All of this changed with the Industrial Revolution, which began to accelerate in the late 1700s. Although waterwheels and other sources of power were widely used, there were not enough rivers to power large factories—early industrialists would have run out of suitable river sites in England by 1830. With the invention of a workable steam engine in the late 1700s, steam became the most practical source for the power needed to run a factory or to propel a ship or a locomotive. Small steam engines could be heated with wood-burning furnaces, but a cheaper and more concentrated source of energy was needed to power larger machines. Coal became the first great fuel of the Industrial Revolution, and it made the entire industrial age possible (figure 13.1).

Figure 13.1 ▲

A nineteenth-century lithograph portraying the hard work of coal miners. (Courtesy of Wikimedia Commons)

The Industrial Revolution was launched in Great Britain, and it was among the first areas of the world to undertake large-scale coal mining. By 1700, 83 percent of the world's coal was mined in Britain, especially in the huge coalfields in southern Wales and in central and northern England from Manchester to Newcastle and into southern Scotland. At its peak in 1947, the British coal industry had about 750,000 miners in dozens of mines around the country.

Mining for coal was a dirty, dangerous, and often deadly business. In the early days of coal mining, the owners had all the power, and the workers had to accept whatever conditions were in the mine or starve—and those conditions were horrific. Coal mining releases fumes that can be toxic or explosive or both, and coal mine explosions were a common problem for years. Coal miners would take a caged bird (typically a canary) into the coal mine with them because the bird was more sensitive to the gasses and would react before the miners could detect the fumes (hence the phrase "Canary in a coal mine" indicating something that warns us of upcoming dangers). Coal mining also produced huge amounts of black coal dust, which filled the miners' lungs, and many died of black lung disease. Coal mines were also prone to caving in, often burying hundreds of miners alive.

Even more appalling was the fact that children as young as eight years old were expected to work in the coal mines in the nineteenth century. With their smaller size, they could work in more confined spaces and were particularly important for opening and closing trap doors to let the mine carts through while preventing the gases from building up. In the eighteenth and nineteenth century, children worked 12-hour shifts in the mine, just like the men, six days a week, with only Sunday off. They sat in the pitch darkness most of that time, only lighting their candle when necessary, and listened for the rumble of the mine carts so they could open the doors and let them through. During the cold short days of the winter, they would get up in the dark, work in the dark for 12 hours, and come home when it was still dark, so they only saw daylight on Sundays.

The casualties of mining were appalling. In the United States alone, more than 90,000 miners died between 1900 and 1950, with 3,200 dying in 1907 alone. Even with modern safety regulations, about 28 miners per year died between 2005 and 2014, making it one of the most hazardous jobs of all. If a miner did not die suddenly from an explosion, cave-in, or fire, he still died young from black lung disease. Thanks to the hard work of labor unions, in the twentieth century miners gradually began to win concessions from the coal barons, and laws were passed that mandated safety, reduced the hours of each shift, and outlawed child labor.

As the Industrial Revolution spread to other parts of the world, huge coal deposits were discovered to support their rapid rush to industrialization. In the United States, major coal deposits were found in the Appalachian region of western Pennsylvania, Virginia, West Virginia, and adjacent areas of Kentucky, Ohio, and Tennessee. By 1870 these fields were producing 40 million tons of coal, and production doubled every 10 years. By 1900 it had jumped to 270 million tons, and it peaked at 680 million tons in 1918 when huge demands for coal to power ships and factories during World War I pushed production.

The search for coal to power the Industrial Revolution was important, and it also provided the foundation for some of the earliest geological studies in Britain or anywhere in the world. When people began to study the major coal fields, they recognized that a certain sequence of rocks held most of Britain's coal. Called the "Coal Measures" in Britain, by the early 1700s this sequence became the basis for the geologic time term "Carboniferous" (meaning "coal bearing"), which was formally named by William Conybeare and William Phillips almost a century later in 1822.

THE MYSTERY OF THE CYCLOTHEMS

The best studied and most controversial examples of cyclic patterns in the rock record were called cyclothems. Geologists in Illinois or Pennsylvania looking for rich coal seams to be mined had noticed that the coal layer was part of a larger pattern. These rock sequences were found especially in the Upper Carboniferous (Pennsylvanian) rocks from the Appalachians to the Illinois Basin to central Kansas. In Illinois, the cyclic pattern started with a river-channel sandstone that was overlain by a floodplain shale, and on top of that was a gray layer full of root casts called the underclay, which was just beneath the coal (figure 13.2). The coal was thought to have formed

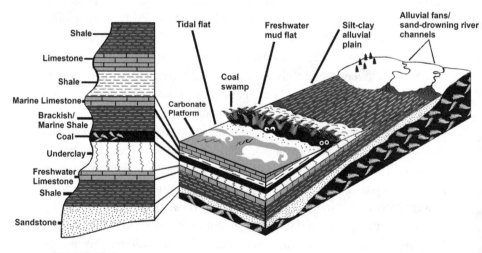

Figure 13.2 ▲

The classic diagram of the typical Illinois cyclothem, from the non-marine river channel sandstones and floodplain mudstones to the rooted gray muddy layer below the coal seam (the "underclay") and the alternating marine shales and limestones above the coal. (Redrawn by K. Marriott from several sources)

in gigantic swamps that covered much of the world in the Carboniferous. Above the coal were alternating layers of marine shale and limestone. This pattern was repeated over and over, sometimes dozens of times in coal-rich basins, and for decades geologists had long debates about the cause of this cyclic pattern.

Some suggested that this sequence represented the building out of ancient deltas and their coal swamps, followed by flooding of the delta sequence as it was abandoned and switched to a different location (something that has happened to the Mississippi Delta many times in the last 10,000 years). But this didn't explain why the same cycles could be traced all the way to Kansas, which was in the middle of a seaway that once flooded the continents and was not affected by deltas on distant shorelines. In addition, it didn't explain why there were dozens of these cyclothems in a row, all repeating themselves consistently over a wide region much larger than a delta such as the modern Mississippi Delta.

In another school of thought, some argued that global sea level had repeatedly risen and fallen during the Carboniferous, producing one identical cyclothem after another. At first this was ridiculed because no one could think of a mechanism for how sea level could go up and down dozens of times like a yo-yo, but in the 1970s two important pieces of evidence came together. First, if huge glaciers on the Gondwana continents in the Carboniferous and Permian had melted and frozen again in a regular fashion, they could cause sea level to fluctuate globally as well (see chapter 14). Second, the existence of the cycles of orbital variations (Croll-Milankovitch cycles) controlling the Pleistocene ice ages became well established (see chapter 22). Geologists who worked with the Carboniferous cyclothems eventually realized that the spacing of these cycles seemed to be roughly 100,000 years apart, the same Milankovitch periodicity that controlled the major Pleistocene cycles of climate. Finally, the idea that these cyclothems were controlled by global sea level fluctuations made sense, and the mystery of the cyclothems was solved.

COAL SWAMPS AND ICEHOUSE WORLDS

Why were there such extensive coal deposits in the Carboniferous around the world, and not so many in any other time period? A number of geological events interacted in a unique way. First, there were no tree-sized land

Figure 13.3 ▲
Diorama of a Carboniferous coal swamp, showing the enormous lycopsid trees ("club mosses") and sphenophtyes ("horsetails") that made up the bulk of the organic matter in coal. An enormous dragonfly is typical in such dioramas. (Courtesy of Wikimedia Commons)

plants before the Late Devonian, but during the Carboniferous they began to evolve into huge club moss trees, gigantic horsetails, and dense forests of ferns (figure 13.3). These plants grew in areas of dense swampy vegetation that formed on the floodplains, river deltas, and coastal lagoons along the fringes of the newly formed Appalachian Mountains in North America and similar mountains in Eurasia that were created by the collision of the various continents to form Pangea.

These giant swamps that grew in the tropical regions of Eurasia and North America were unlike any swamps that have lived since. In modern swamps, termites and other decomposers quickly break down the trees as they die and sink in the stagnant water, but no such wood-digesting insects had yet evolved in the Carboniferous. Enormous volumes of vegetation sank into the stagnant acidic muds of the coal swamps and became permanently buried in Earth's crust instead of rotting away as they do today.

A number of interesting consequences resulted from the huge amount of plant growth in the coal swamps of the Carboniferous. These plants both withdrew carbon dioxide from the atmosphere and also released a huge amount of oxygen. Numerous geochemical indicators suggest that the level

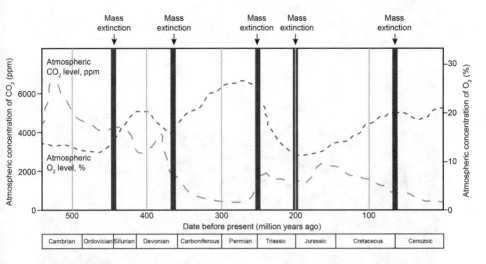

The history of free oxygen and carbon dioxide over the past 550 million years. The high carbon dioxide in the greenhouse world of the early Paleozoic is shown, along with the extremely low levels in the Carboniferous, when huge volumes of plant material absorbed the carbon and turned it into coal. Oxygen increased rapidly in the Carboniferous to almost 35 percent when huge volumes of plants from the coal swamps pumped out oxygen at higher rates than ever before. (Redrawn by E. Prothero based on several sources)

of oxygen in the atmosphere was extraordinarily high at that time (figure 13.4). Today our atmosphere contains 21 percent oxygen (the rest is mostly nitrogen gas), up from just a few percent in the Early Cambrian. But if the geochemical calculations are right, there was a spike of oxygen up to almost 35 percent, which would have made breathing Carboniferous air almost like breathing from an oxygen mask. It could not have been much higher than that because above 35 percent oxygen content the atmosphere becomes flammable and can combust easily, and we see no evidence of widespread wildfires at that time or since then. The gas mixture we consider normal today was reversed: carbon dioxide dropped to its lowest levels since the Proterozoic snowball earth conditions, and oxygen got so rich that the atmosphere was almost flammable.

Roaming these coal swamps were a wide variety of animals. Among the most spectacular were the arthropods (insects, spiders, and scorpions), which grew to enormous size. Dragonflies had wingspans of 75 centimeters (2.5 feet), similar to the size of an eagle's wings (figure 13.5A).

Figure 13.5 ▲

Thanks to the high oxygen levels, the insects and other arthropods of the Carboniferous could grow to gigantic sizes. (*A*) A fossil of a gigantic eagle-sized dragonfly *Meganeura*, from the Carboniferous. (*B*) A reconstruction of the 18-foot-long millipede *Arthropleura*, typical of the Carboniferous. (Courtesy of Wikimedia Commons)

There were cockroaches over 30 centimeters (1 foot) long. But the all-time largest terrestrial arthropod that has ever lived was the enormous millipede relative *Arthropleura*, which were almost 2.4 meters (8 feet) long (figure 13.5B)!

Why the arthropods got so large was a long-standing mystery until we discovered that these plants were pumping out high levels of atmospheric oxygen. One of the main factors that prevents arthropods from getting bigger is their inefficient respiratory system, which does not transfer oxygen to all parts of the body very well. In particular, there are narrow spots and bottlenecks in the trachea, especially in the joints for the limbs, that restrict how much oxygen reaches critical parts of the body. When atmospheric oxygen levels were around 35 percent, it was easier for an arthropod to get oxygen to all parts of its body, which would allow it to get big.

The increasing number of coal swamp beds pulled many tons of carbon dioxide out of the atmosphere through photosynthesis and locked up a significant volume of carbon in Earth's crust. The "greenhouse" climate of the Early Carboniferous (with no polar ice caps, high carbon dioxide, and high sea levels that drowned most of the continents) was slowly transformed into an "icehouse" planet by the Late Carboniferous (with polar ice caps on the South Pole, lower carbon dioxide, and a much lower sea level as all that polar ice pulled water out of the ocean basins). Earth was dominated by this icehouse planet setting for almost 200 million more years.

Another factor causing the change from greenhouse to icehouse is the widespread mountain building of the late Paleozoic. When mountains rise up and their surface rocks weather and erode, the chemical processes of weathering of minerals pulls carbon dioxide out of the atmosphere. Major mountain-building events, such as uplift of the Appalachians, Ozarks, Ouachitas, and Ancestral Rocky Mountains that dominated the Late Carboniferous, pulled a lot of greenhouse gases out of the atmosphere. Together with the huge volume of carbon locked into coals in Earth's crust, the planet's atmosphere lost a lot of carbon.

By the Late Carboniferous and Permian, Earth had become an icehouse world that would last until the Middle Jurassic. Huge deposits of ice built up on the poles in the Permian, and that is the subject of the next chapter.

FOR FURTHER READING

Freese, Barbara. *Coal: A Human History.* New York: Penguin, 2004.

Goodell, Jeff. *Big Coal: The Dirty Secret Behind America's Energy Future.* New York: Mariner, 2007.

Heckel, Philip H. "Sea-Level Curve for Pennsylvanian Eustatic Marine Transgressive-Regressive Depositional Cycles Along Midcontinent Outcrop Belt, North America." *Geology* 14, no. 4 (1986): 330–334.

Jacobson, Russell J. "Depositional History of the Pennsylvanian Rocks in Illinois." In *GeoNote 2.* Champaign: Illinois State Geological Survey, 2000.

Martin, Richard. *Coal Wars: The Future of Energy and the Fate of the Planet.* New York: St. Martin's, 2015.

Thomas, Larry J. *Coal Geology*, 3rd ed. New York: Wiley-Blackwell, 2020.

Wanless, Harold R., and J. Marvin Weller. "Correlation and Extent of Pennsylvanian Cyclothems." *Geological Society of America Bulletin* 43, no. 4 (1932): 1003–1016.

14

THE PANGEAN ICEHOUSE

A new theory is guilty until proven innocent, and the pre-existing theory innocent until proven guilty. . . . . Continental drift was guilty until proven innocent.

—DAVID M. RAUP

A SCIENTIFIC REVOLUTION IN GEOLOGY

In 1962, the famous philosopher of science Thomas Kuhn wrote an influential book entitled *The Structure of Scientific Revolutions*. He said that for a long time scholars and the general public thought of science as a constant search for the "right answer." Kuhn pointed out that if you look at the history of science it's not a slow and steady march to final truth. Instead, most science operates within a given set of assumptions that Kuhn called a "paradigm." Most research is undertaken to work out the small details of the paradigm, but researchers do not challenge its fundamental assumptions. Eventually, however, lots of small problems that cannot be easily explained within the existing paradigm accumulate, and someone suggests that there is a better way to solve these problems with a totally new set of assumptions. Science then undergoes conflict between the old and new paradigms during a scientific revolution. If the new paradigm explains the observations better than the old one, eventually it replaces the old paradigm and becomes the fundamental belief system that dictates research for the future.

Kuhn derived this insight from his earlier research on the Copernican revolution in astronomy. Copernicus was bothered about the old Earth-centered cosmology because it could not account for things like the retrograde motion of planets like Mars and Jupiter without the clumsy explanations of epicycles that were proposed in ancient times by Ptolemy. Copernicus changed one simple assumption: that the sun, not Earth, was the center of our planetary system. From this heliocentric model, all of modern astronomy was born. That revolution did not happen all at once when Copernicus published *De Revolutionibus* in 1543, or even when Galileo first observed the skies with a telescope in 1610 and published his famous *Dialogue Concerning the Two Chief World Systems* in 1632. Copernican astronomy wasn't fully accepted until Isaac Newton worked out the mechanics of gravity in the late 1600s.

Newton was the architect of a different scientific revolution in physics with his *Principia Mathematica* in 1687. Newtonian mechanics remains the prevailing paradigm for most of physics. But if you are dealing with things operating at near the speed of light, the scientific revolution of Einsteinian mechanics explains physics better. Biology underwent its own revolution with Darwin's 1859 book *On the Origin of Species*, which transformed biology from natural theology to the new paradigm of natural selection and evolution. Whether chemistry has had a similar scientific revolution is still debated, although the discovery of the periodic table by Dmitri Mendeleev was certainly revolutionary and groundbreaking.

In contrast to these events of more than a century ago, geology experienced a scientific revolution in my own lifetime. I am among the first generation of students that were trained in the new paradigm of plate tectonics in the early 1970s. The old paradigm reigned until a series of discoveries at the bottom of the ocean in the 1950s and 1960s produced a mountain of new data that led to plate tectonics. However, the revolution had started years earlier.

In 1910, the 30-year-old German meteorologist Alfred Wegener happened to glance through a world atlas a friend had received as a Christmas gift. Flipping through the pages, he saw the amazing match between the coastlines of South America and Africa. It was like a light bulb going off in his head. Why did these two continents, separated by the huge South Atlantic Ocean, seem to fit together so well? In fact, Wegener was not the first to notice this. People had commented on this match as early as the 1500s, almost as soon as the first decent maps of the Atlantic were available. In Wegener's own words:

> The first concept of continental drift first came to me as far back as 1910, when considering the map of the world, under the direct impression produced by the congruence of the coast lines on either side of the Atlantic. At first I did not pay attention to the ideas because I regarded it as improbable. In the fall of 1911, I came quite accidentally upon a synoptic report in which I learned for the first time of palaeontological evidence for a former land bridge between Brazil and Africa. As a result I undertook a cursory examination of relevant research in the fields of geology and palaeontology, and this provided immediately such weighty corroboration that a conviction of the fundamental soundness of the idea took root in my mind.

But Wegener did not casually notice this pattern and then move on to other things, as everyone else had done for 400 years. He was building a career as a budding scientist and had many years' worth of experience in meteorology and climate. That same year he wrote the standard German textbook in the science, *Thermodynamics of the Atmosphere*, so he clearly knew his stuff. Nor was he green and inexperienced. At the young age of 26, he had organized and led the first of four expeditions to Greenland to study polar climate and weather (figure 14.1).

Despite his heavy commitment of time to meteorological and polar research and teaching classes at the University of Marburg, Wegener continued to look for evidence that the continents once fit together. In 1908, he began to work with the legendary climatologist Wladimir Köppen, who created the standard classification of climate regions and zones that is still used today. Wegener and Köppen searched the scientific literature and found that the climate zones of the Permian Period (250 to 300 Ma) don't make any sense with the way the continents are distributed today. They began to compile evidence that these climatically sensitive rocks suggested that the continents had moved. By 1912, Wegener gave a few lectures on his evidence for continental drift, and he published three short papers on the evidence in German geographical journals. In 1913, Wegener married Köppen's daughter—and then led a second expedition to Greenland. He spent the winter on the ice and almost died there before he and his companion were rescued.

Figure 14.1 ▲

Alfred Wegener on his last polar expedition in 1930. (Courtesy of Wikimedia Commons)

On June 28, 1914, Archduke Franz Ferdinand was assassinated, and World War I soon broke out. Along with every other able-bodied man in Germany at the time, Wegener was called to serve in the Kaiser's army. He was wounded twice (once in the neck) before the German High Command decided that he was more useful to them in their weather stations than as trench fodder because he was a trained meteorologist. As he worked in one German weather station after another, he continued to document his ideas. *On the Origin of Continents and Oceans* was published late in 1915, but few people read it or even saw it due to the wartime restrictions. In it, he published the first-ever maps of the continents drifting apart since the Permian (250 Ma), showing their configuration in the supercontinent of Pangea (figure 14.2), with the southern half known as Gondwana and the northern part called Laurasia (Laurentia plus Eurasia). Wegener found that being an army meteorologist was a good posting, and he had published another 20 papers in meteorology and climatology by the war's end.

Once the war was over, Wegener held several jobs in Hamburg before accepting a permanent post at the University of Graz. He used this time to

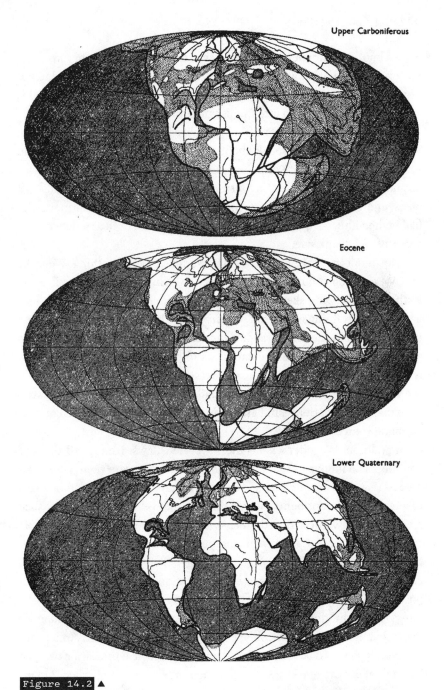

Upper Carboniferous

Eocene

Lower Quaternary

Figure 14.2 ▲

Wegener's (1915) map of all the continents from Pangea to present. (Courtesy of Wikimedia Commons)

write a book with Köppen about climates in the geological past based on the evidence he had originally collected to support his ideas about continental drift. Wegener remained mostly unknown outside of Germany because his works were not translated into English until 1925.

After more than 14 years of being ignored by mainstream geologists, Wegener was invited to speak about his ideas at the 1926 meeting of the American Association of Petroleum Geologists in New York City. This symposium had been organized by his opponents as a chance to ridicule his ideas, and Wegener was walking into a lion's den. Only the chairman, who had invited him, gave Wegener a fair hearing; the rest of the audience were scornful and dismissive.

Why didn't they take him or his ideas seriously? For one thing, Wegener was not a geologist but a meteorologist and a climatologist. There is some justification in being skeptical of outsiders dabbling in your field without having a full training in it. I run into crackpot ideas about the earth on the internet all the time. They range from flat earthers to hollow earthers to geocentrists to young-earth creationists to people who believe the earth is expanding. Anyone with the basic first few courses in geology can easily see why they are wrong, and it is especially obvious to those geologists who have real field experience and are not just dreaming up ideas based on secondhand sources.

In addition, Wegener's ideas did have their flaws. He argued that the continents had moved around the globe. If Wegener were right, geologists argued that there should be huge areas of oceanic crust crumpled up like a carpet when the continents plowed through them—and there are few such places. (We now know that the oceanic crust is nothing like what people thought at that time. Rather than crumpling up, plates usually slide beneath other continents in subduction zones.) Wegener could not explain how the continents moved, or what powered their motion, and the ideas he did propose (such as centrifugal force) were geophysically impossible. Also, the rates of plate motion were too high (250 centimeters per year); today we know that most plates move only about 1/100 as fast as that (2.5 to 6 centimeters per year). To be fair, in 1915 dating the geological time scale was just beginning, and no one knew how long ago the Permian was when Pangea was assembled.

Finally, the best evidence came from the southern hemisphere. Nearly all geologists were living and working in North America and Europe then, with only a handful coming from the less developed countries around the world. In the early twentieth century, traveling on an ocean liner down to Brazil or South Africa was very slow and expensive, so very few geologists had actually been to these regions and seen the rocks for themselves. Most geologists had only seen written descriptions in journals and books with a handful of muddy black and white photographs that don't do justice to the brilliant colors and striking similarity of rocks in the southern hemisphere. The South African geologist Alexander du Toit had seen the rocks firsthand and was among the strongest supporters of continental drift, but these geologists were also outsiders whose ideas were seldom presented at the meetings of North American or European geologists. One of the few European supporters of the idea, Arthur Holmes, had worked in Africa, so he also knew the rocks from personal experience. As a result, the entire idea of continental drift remained a crackpot notion for another 30 to 40 years.

Meanwhile, Wegener did not sit in a corner and sulk over the poor reception of his grand ideas. He continued his meteorological research as a polar explorer, and in 1929 he led his third expedition to Greenland. The next year he led his fourth and largest expedition—and also his last. It was equipped with a wide range of weather equipment, plus a propeller-driven snow sled, and other devices. There was one remote station in the middle of the Greenland ice sheet known as *Eismitte* ("mid-ice" in German) that is one of the coldest locations in the Northern Hemisphere. It has an average mean temperature of -30°C (-22°F), and routinely reaches temperatures of -80°F in winter. Thanks to being near the Arctic Circle, the sun does not rise from November 23 to January 20. Station *Eismitte* was so remote that it was a dangerous trip of many days to reach for resupply, and then return.

In November 1930, Wegener and his partner Rasmus Villumsen were on their way back from a supply run when they got caught in horrific blizzards and extremely cold temperatures. There Wegener died, possibly from a heart attack (he was a heavy smoker), or possibly because hypothermia overcame him and he froze to death. Villumsen buried him in the snow with skis to mark his grave, and then was never seen again. Later a team found Wegener's grave, and reburied it with a cross to mark it. There his body still lies under 100 m (330 feet) of ice, moving along with the flowing ice of the Greenland ice sheet. Wegener died just past his fiftieth birthday, unnoticed and unmourned by the geological community. If he had lived another 30 years, he might have witnessed the vindication of his ideas—but luck and fate were not with him. Instead, he is yet another example of a genius with a great idea who died scorned and unappreciated, never living to see his work make it from crackpot to scientific paradigm.

PANGEA AND CLIMATE

If you travel to South Africa or Brazil or Antarctica or India or Australia, you will see a strikingly similar sequence of rocks. They include a distinctive Carboniferous sandstone sequence with coal beds, followed by the Lower-Middle Permian glacial till, overlain by a thick sequence of Permian-Triassic red beds full of fossils of reptiles and protomammals, and finally huge Jurassic lava flows capping the entire sequence. I've heard several geologists remark that if you didn't read the formation names or hear the local geologists speaking either Afrikaans or Portuguese you could not tell whether you were looking at rocks from South Africa or Brazil.

Even more revealing is where these distinctive deposits were found (figure 14.3). The coal deposits of the Permian, which should be found only in the equatorial rain forest belt, are far outside that region today. Likewise, the Permian desert dune deposits are not located in the subtropical high-pressure belts between 10° and 40° north and south of the equator as they are in the modern world's deserts. But if you put the continents back into the Pangea configuration, all of the Permian coal beds are in the tropical rain forest belt, and all of the Permian sand dunes are in the subtropical high-pressure desert belt where they belong.

Most impressive of all are thick glacial beds found on many of the Gondwana continents, such as the Dwyka tillite of South Africa (figures 14.4A and 14.4B), which is more than 1,000 meters (3,300 feet) thick! There are equivalent glacial deposits on other continents, including the Itararé Group in the Parana Basin of Brazil (more than 1,400 meters or a mile thick) and the Lyons Formation (3,500 meters thick) in the Carnarvon Basin of eastern Australia. Similar sediments are found in India and Antarctica as well. These only make sense in the Permian Gondwana location of these land masses (figure 14.5A). If you tried to plot the distribution of those ice sheets on the

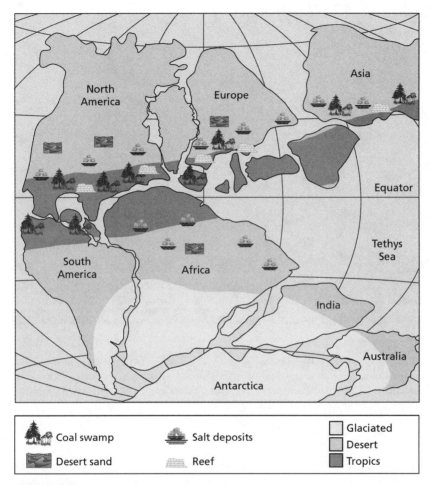

Coal swamp
Salt deposits
Glaciated
Desert
Desert sand
Reef
Tropics

Figure 14.3 ▲

Distribution of climatically sensitive rock types, from coals in the tropical swamps, to desert deposits in the arid subtropical high-pressure belts, to the polar glaciers. (Redrawn from several sources)

modern continents, a Permian ice sheet would straddle the South Atlantic, most of the Indian Ocean, and even reach across the equator to some parts of India (figure 14.5B). Clearly, this makes no sense paleoclimatically.

Even more remarkable were the scratches and gouges created as the glaciers dragged huge rocks in their base across the landscape. In some cases, the gouges originate in South Africa and line up with similar scratches in

Figure 14.4 ▲ ▶
Glacial deposits of the Permian Dwyka tillite in South Africa. (*A*) A typical random mixture of pebbles and fine sand and clay, a classic glacial till. (*B*) A giant dropstone, or large rock embedded in a floating iceberg which was dropped into fine-grained deep sea muds as the iceberg melted. (Courtesy of Wikimedia Commons)

South America. For this to make sense on the modern globe, the Permian glaciers would have had to jump into the Atlantic and cross it in a straight line, then jump right back onto land in that same straight course. Clearly, this is absurd as well. Only if the continents were in their Gondwana configuration, with no Atlantic Ocean to cross, would these aligned scratches make any sense.

In addition to evidence from the late Paleozoic and early Mesozoic Gondwanan rocks, the fossils found in these deposits are even more revealing (figure 14.6A). Nearly every Permian deposit in Gondwana contains leaves of the primitive extinct seed fern *Glossopteris*—including even Antarctica, Australia, and Madagascar (figure 14.6B). Then there is the small, 50 cm long reptile *Mesosaurus* (figure 14.6C). This reptile is only known from Permian lake beds in South Africa and Brazil; it was too small to swim

Figure 14.4 ▲
(*continued*)

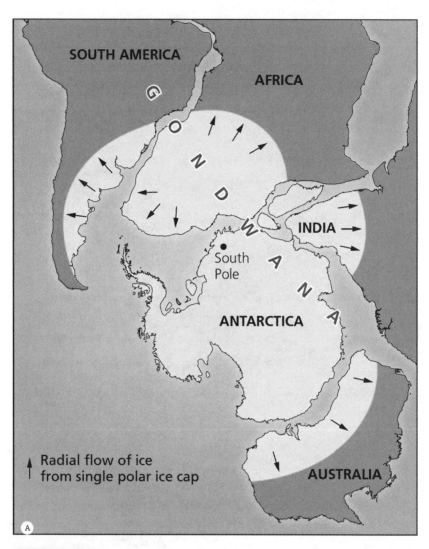

Figure 14.5 ▲ ▶

(A) The glacial scratches found in Permian rocks match up only if the southern continents are aligned in a Gondwana configuration. (B) Those same Permian glacial deposits plotted on their modern position of the continents would require such absurdities as ice caps crossing oceans, and part of the ice cap straddling the equator to reach India. (Redrawn from several sources)

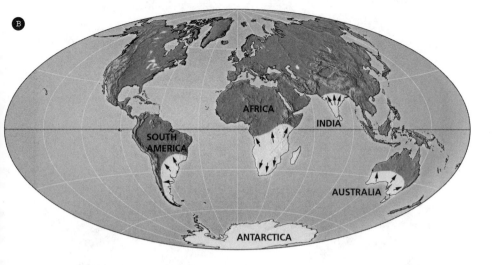

Figure 14.5 ▲
(*continued*)

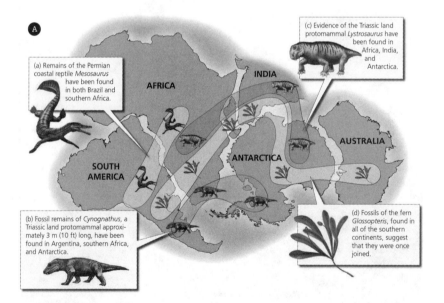

Figure 14.6 ▲ ▶

The occurrence of certain Permian fossils tie the Gondwana continents together. (*A*) Map showing the distribution of key fossil types. (*B*) The tongue-shaped leaf of the Permian seed fern *Glossopteris*. (*C*) The small aquatic reptile *Mesosaurus*, found in lake deposits of Brazil and South Africa. (*D*) The dog-sized protomammal *Lystrosaurus*, found on most of the Gondwana continents. (*E*) The bear-sized protomammal *Cynognathus*. ([*A–D*] Redrawn from several sources; [*E*] courtesy of Wikimedia Commons)

Figure 14.6 ▲

(*continued*)

Figure 14.6 ▲
(*continued*)

across the modern South Atlantic. Almost every Gondwana continent yields fossils of the beaked protomammal (formerly but incorrectly called a "mammal-like reptile") known as *Lystrosaurus* (figure 14.6D). It was already known from Africa, South America, and India, and its discovery in Antarctica in 1969 was considered by some to be the clinching evidence for continental drift. Finally, there is the bear-sized predatory protomammal *Cynognathus*, which has been found not only in South Africa and South America but also in Russia in the Late Permian (figure 14.6E).

All of this evidence convinced geologists such as Du Toit and Holmes of the possibility of continental drift, but skeptical northern hemisphere geologists tried to dismiss it or explain it away. The peculiar distribution of Permian Gondwana plants and animals was attributed to land bridges or to animals rafting across oceans. The matches in the rocks themselves were either ignored (perhaps because they were not that convincing in tiny black and white journal photos) or dismissed as circumstantial. In short, almost none of this evidence, which seems overwhelming to geologists today, was given a fair hearing by geologists in 1915, and almost 50 years of appalling blindness by the majority of the world's geologists ensued.

Continental drift theory was considered crackpot by the most powerful American and European geologists of the late 1940s and early 1950s. The famous biologist and documentary filmmaker David Attenborough, who went to university in the late 1940s, recalled this exchange: "I once asked one of my lecturers why he was not talking to us about continental drift and I was told, sneeringly, that if I could prove there was a force that could move continents, then he might think about it. The idea was moonshine, I was informed."[1] In 1949, the American Museum of Natural History held a symposium attacking the evidence for drift, talking about land bridges and dismissing the similarity between rocks on different continents (even though most geologists who did so had never seen these rocks firsthand). The symposium was not published until three years later, but in retrospect it is an amazing monument to being shortsighted and wrong just as the world was about to change beneath our feet.

In the late 1940s, oceanographic institutes around the world began to send out numerous research vessels to map and sample the ocean floor. After World War II, the U.S. Navy realized that they needed to better understand the oceans if they were to deal with submarines in the future. Many federal agencies poured millions of dollars into oceanographic institutes such as

Woods Hole in Massachusetts, Lamont-Doherty Geological Observatory of Columbia University in New York, and Scripps Institute of Oceanography in San Diego, California. This effort continued for more than a decade. In the late 1950s, Marie Tharp and Bruce Heezen produced the first detailed map of the ocean floor, which makes up more than 70 percent of the earth's surface. Other oceanographers and marine geologists had sampled the ocean bedrock and taken cores of the sediments that sat on top of the bedrock. The key breakthrough, however, came from shipboard magnetometers, which had been used to detect submarines during World War II and were surplus on Navy ships. The magnetometers detected a striking pattern of magnetic signals on the ocean floor that by 1963 were shown to be caused by seafloor spreading and the birth of a new oceanic crust. At that point, the revolution swung into high gear, and when I took my first college geology classes in 1972 many of the professors considered it worth talking about.

THE PANGEA ICEHOUSE

By the Permian, the Pangea supercontinent was fully assembled, with huge ice caps over its southern end on the Gondwana portion (mainly Antarctica, South America, and southern Africa) and smaller ice caps in Siberia near the Permian North Pole. It had dense coal swamps across its tropical belt, but it also suffered the effects of being a supercontinent. When continents are smaller and lower in elevation, they have lots of coastlines, and the cooling, moderating effect of the ocean along most of their edges prevents extreme temperature fluctuations (this is called the maritime climate effect). But the interior regions of gigantic continents are not cooled by nearby oceans, and they experience extremes in what is known as the continental climate effect. Without large volumes of water from nearby oceans to buffer the changes in temperature, continental interiors can fluctuate from one extreme to another, often in a single day. For example, in the Great Plains of North America or in the center of Asia, it is not unusual to start out the day with temperatures over 104°F (40°C), and then have a cold front move through and drop the temperature to near freezing by the same afternoon. Likewise, isolation from nearby oceans and their life-giving moisture in clouds means that continental centers tend to have drier climates. The great deserts of the central Asian steppe—from the Gobi Desert in Mongolia through the deserts of southern Siberia and western China, as well as the huge Sahara Desert—are all consequences of this isolation. Finally, the collision of supercontinents tends to uplift the entire land surface much higher than when the continents were smaller. The effect of this uplift can be seen in the Tibetan Plateau and the Himalayas, which are very high and extremely cold and dry as well.

Putting this all together, we expect Permian landscapes (except in the tropical coal swamps and polar ice caps) to reflect this drier, more extreme continental climate. Indeed, that is what we find for North America in particular, which was still tropical to subtropical during the Permian. The entire eastern half of the continent was still uplifted by the Appalachian-Ouachita orogeny, and there was nothing but erosion and almost no deposits of Permian age in that region. In the basins of the Ancestral Rockies and adjacent areas, however, there are extensive Permian beds. The most famous of these occur in the Colorado Plateau of Arizona, New Mexico, Utah, and Colorado. Most of these sediments are thick, cross-bedded sandstones formed in ancient desert dunes (such as the Coconino Sandstone in the Grand Canyon or the De Chelley Sandstone in Monument Valley). These are interbedded with red shales representing arid floodplains, such as the upper Supai and Hermit formations in the Grand Canyon and the Organ Rock Shale in Monument Valley. In some places, the aridity was so extreme that huge deposits of salt and gypsum were formed. The most famous of these was the Paradox Basin of south-central Utah, where more than 4,600 meters (15,000 feet) of salt, gypsum, and other deposits accumulated. This area is an important commercial source of salt even today. When you think "Permian" in North America, think of red beds, dune sands, and thick deposits of evaporites.

NOTE

1. Robin McKie, "David Attenborough: Force of Nature," *The Guardian*, October 27, 2012, https://www.theguardian.com/tv-and-radio/2012/oct/26/richard-attenborough-climate-global-arctic-environment.

FOR FURTHER READING

Cox, Allan, and Robert B. Hart. *Plate Tectonics: How It Works*. New York: Wiley-Blackwell, 1986.

Greene, Mott T. *Alfred Wegener: Science, Exploration, and the Theory of Continental Drift*. Baltimore, MD: Johns Hopkins University Press, 2015.

Kearney, Philip, Keith A. Klepeis, and Frederick J. Vine. *Global Tectonics*. New York: Wiley-Blackwell, 2009.

McCoy, Roger M. *Ending in Ice: The Revolutionary Idea and Tragic Expedition of Alfred Wegener*. Oxford: Oxford University Press, 2006.

Molnar, Peter. *Plate Tectonics: A Very Short Introduction*. Oxford: Oxford University Press, 2015.

Oreskes, Naomi, ed. *Plate Tectonics: An Insider's History of the Modern Theory of the Earth*. Boulder, CO: Westview, 2003.

——. *The Rejection of Continental Drift: Theory and Method in American Science*. New York: Oxford University Press, 1999.

Prothero, Donald R. *The Story of the Earth in 25 Rocks: Tales of Important Geological Puzzles and the People Who Solved Them*. New York: Columbia University Press, 2018.

Roberts, Peter. *Tectonic Plates: How the World Changed*. New York: Russet, 2016.

Wegener, Alfred. *The Origin of Continents and Oceans*, trans. John Biram. New York: Dover, 1915.

15

THE "GREAT DYING"

Extinction is the rule. Survival is the exception.

—CARL SAGAN

THE MOTHER OF ALL MASS EXTINCTIONS

The general public is familiar with the end-Cretaceous extinctions because they involved glamorous, sexy topics such as dinosaurs and an asteroid impact. But at best, that was only the third largest mass extinction in Earth's history. The Late Ordovician extinctions were even bigger, but few people other than paleontologists have heard of it. The largest of all was the extinction that ended the Paleozoic altogether. In fact, the Permian extinction may have wiped out as much as 95 percent of the species in the ocean, 83 percent of the genera, and 57 percent of the marine families. It is estimated that 70 percent of land vertebrates also vanished by the end of the Permian. Paleontologists call it "the mother of all mass extinctions," or "the great dying," because it was far more severe than any other extinction event.

Who were the victims of the great dying? In the oceans, the extinction afflicted nearly every group: some vanished entirely, others were reduced to just a few lineages, and only a few had more survivors. The extinction reached to the very bottom of the food chain. Among the most abundant and diverse creatures of the Late Permian were the fusulinids, a group of single-celled amoeba-like creatures from a group of protistans known as foraminiferans (figure 15.1). Other groups of foraminiferans are

Figure 15.1 ▲

The single-celled amoebas known as foraminiferans secrete a calcite shell. These are Permian fusulinids, which secrete a shell about the size and shape of a grain of rice, but sometimes they are much larger. (Courtesy of Wikimedia Commons)

still abundant in modern oceans, both in the plankton and in the marine sediment, but in the late Paleozoic they were mostly seafloor dwellers. Shaped like a large grain of rice, fusulinids grew in a tight spiral around a central axis, so their shells ended up somewhat spindle-shaped. The shell of the fusulinid can be up to 1 centimeter (2 inches) or more long, but the most incredible fact is that it was produced by a single-celled organism! (Bottom-dwelling foraminiferans have gotten this large several times in the history of the oceans.) Most important, they were incredibly abundant in the late Paleozoic seas (especially during the Permian), and huge thick units of limestone with millions of fusulinids have been found in some places. Their other great value is that they evolved very rapidly, so they are the best time marker for the Late Carboniferous and Permian.

Another group typical of the Paleozoic was the brachiopods, or "lamp shells." In the Permian, the major group of brachiopods were known as productids, which first flourished in the Carboniferous (figure 15.2). Typical late Paleozoic productids had one shell shaped like a teacup, which was propped up on the muddy seafloor with long spines that acted like stilts or stabilizers. Their other shell was a flat lid hinged over the opening of the tea-cup. When they fed, they opened the lid and let the plankton-rich currents flow through their feather-like filter-feeding device called a lophophore.

Many of these productids were truly bizarre in shape. Some of them, known as richthofenids, had modified the simple round teacup of a typical productid into a cone-like shape, with the other shell forming a tiny lid inside the cone. But the weirdest of all the brachiopods was a group of productids that had one shell shaped like a shallow soap dish, and the other shell shaped like a simple comb that covered part of the soft tissues. How these strange creatures lived is still a mystery, but it is clear that brachiopods were becoming more and more specialized. This incredibly weird and specialized community vanished completely when the greatest mass extinction in Earth's history occurred.

Brachiopods were the most common shelled animals on the Permian seafloor, but primitive clams and snails were abundant there too. Archaic sponges and horn corals and tabulate corals were still around, but they no longer formed the giant coral reefs that had been so typical in the Silurian and Devonian. The "reefs" of the Permian were, technically speaking, not true reefs held up by massive cemented corals like the reefs of the Silurian and Devonian or those found today. Instead, they were a series of mounds

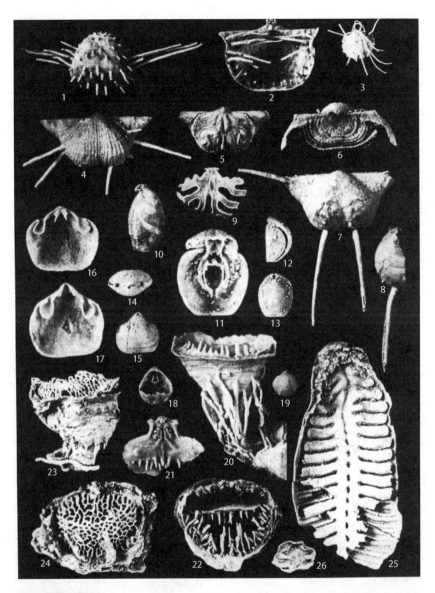

Figure 15.2 ▲

The weird, spiny, cup-shaped productid brachiopods dominated the Late Permian seas. The cone-shaped specimens 21–24 (lower left) are the richthofenids, and specimen 25 (bottom right) is *Leptodus*, a productid shaped like a soap dish with a comb making up its upper shell. (Courtesy of Wikimedia Commons)

of small bead-like sponges, algae, bryozoans, and other less massive organisms that provided a hard-wave-resistant core and a back-reef lagoon (figure 15.3). This "Permian reef complex" is beautifully exposed in the Guadalupe Mountains National Park in southeastern New Mexico and across the Texas border. Instead of sponges and small corals, most of the seafloors of the Permian were dominated by lots of lacy bryozoans and "sea lilies" or crinoids—a stalked group of echinoderms related to sea stars and sea urchins and their kin. In addition, another group of stalked echinoderms, the blastoids, were also numerous. They looked vaguely like crinoids with a stalk and a "head" with arms, but their heads were shaped like flower buds, and they had distinctive features within their armored "flower bud."

In addition to groups typical of the Paleozoic fauna, there were still relicts from the Cambrian in the Late Permian. The trilobites were mostly decimated by the Devonian extinctions, but one group of small, unspecialized trilobites, the proetids, still hung on (see chapter 11). They had persisted

Figure 15.3 ▲

Diorama showing the Permian "reef" as it might have looked, with lots of spiny productids, goniatite ammonoids, and bead-like calcareous sponges. (Photo by author)

in the background as a minor component of late Paleozoic marine communities. They are rare and hard to collect, but they were hanging on nonetheless.

The waters above these seafloor communities were dominated by a variety of swimming predators, both invertebrate and vertebrate. The nautilus-like ammonoids, known as goniatites, had been evolving rapidly since their near-extinction at the end of the Devonian, but they did survive and diversified throughout the Carboniferous and Permian. These archaic creatures were nowhere near as fast-swimming or as voracious as the evolutionary explosion of sharks and bony fish that took place after the Devonian extinctions. The most common types of fish were primitive armored bony fish known as palaeoniscoids, which are distantly related to the living sturgeons and paddlefish. One of the peculiar looking sharks of the Permian were the xenacanths, which were up to 2 meters (6.6 feet) long, had wicked double-pronged teeth, long ribbon-like fins down their back, and a symmetrical pointed tail more like that of an eel (figure 15.4A). Their pelvic and pectoral

Figure 15.4 ▲ ▶

(A) The weird, spiky-headed Permian xenacanth shark; (B) The "buzz-saw toothed" shark relative *Helicoprion*. (Courtesy of Wikimedia Commons)

Figure 15.4 ▲
(*continued*)

fins had a unique arrangement of supporting cartilages not seen in any other shark group. And most peculiar of all, they had a long straight spike on the top of their head. The function of this spike is still a mystery. Even stranger was the shark relative called *Helicoprion* (figure 15.4B). It is only known from their fossilized spiral whorls of teeth. Newly discovered specimens demonstrate that they were part of a buzz-saw lower jaw that sliced rapidly through prey.

TERRESTRIAL LIFE

The Ordovician extinction had almost no effect on life on the land. The only terrestrial organisms then were primitive mosses, liverworts, and simple vascular plants, plus millipedes. The Devonian extinctions did not seem to make a very big difference on land either because only primitive amphibians and a variety of insects, spiders, and scorpions lived on land in the Late Devonian. But by the Permian the land communities were much more complex and developed. Huge forests of primitive relatives of conifers covered the drier landscapes, and the swampy areas were covered by ferns and mosses as well as gigantic tree-sized horsetails, tree ferns, and club mosses. A wide range of amphibians that had evolved from the primitive forms that crawled out of the water in the Late Devonian were living in these swamps as well. The largest and most impressive of these were the flat-headed, flat-bodied amphibians known as temnospondyls. In the Permian of Brazil, there was a huge narrow-snouted crocodile-shaped temnospondyl called *Prionosuchus*, which reached 9 meters (30 feet) in length. The Lower Permian red beds of north-central Texas are famous for their huge temnospondyls, best known from the sprawling wide-skulled form *Eryops* (figure 15.5A), which at 3 meters (10 feet) in length, was the largest animal ever to walk on land at that time. The second great group of amphibians was the lepospondyls, which came in a range of sizes and shapes, from

Figure 15.5 ▲ ▶

Permian land animals: (A) *Eryops*, a giant flat-bodied amphibian; (B) *Diplocaulus*, a small amphibian with a set of horns on its skull that gave it a boomerang shape; and (C) the huge hippo-like reptiles known as *Bradysaurus*. (Courtesy of Wikimedia Commons)

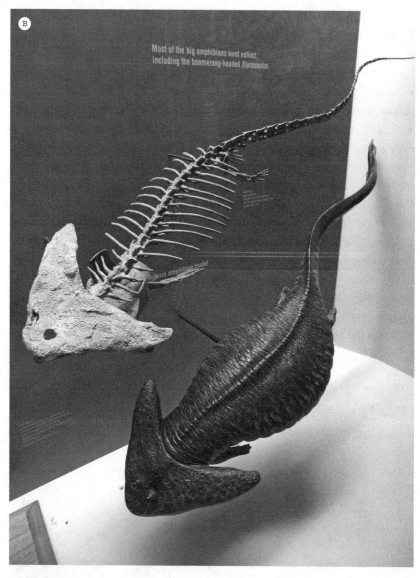

Most of the big amphibians went extinct, including the boomerang-headed *Diplocaulus*.

Figure 15.5 ▲ ▶

(*continued*)

the tiny lizard-shaped microsaurs, to the snake-like aistopods, to the truly weird *Diplocaulus*, which had broad horns on the side of its head that made its skull look like a boomerang (figure 15.5B). By the Permian, primitive relatives of frogs had also appeared.

Figure 15.5 ▲
(*continued*)

Reptiles also flourished during the late Paleozoic. In the Late Permian, reptiles had truly begun to diversify. They came in a wide variety of shapes and sizes, from the huge hippo-like pareiasaurs that weighed up to 600 kilograms (1,300 pounds) (figure 15.5C), to the swimming *Mesosaurus* found in Permian lake beds in Brazil and Africa (figure 14.6C), to the earliest turtle ancestors that had a shell on its belly but not on its back, and still had teeth in its beak. By the Late Permian, the ancestors of the lineage that led to dinosaurs, crocodiles, snakes, and lizards were present, although those groups would not evolve until the Triassic.

But reptiles did not rule the world yet. Far more important was the wide range of synapsids, or protomammals. These creatures used to be called "mammal-like reptiles," but that term is misleading and has been abandoned. Synapsids are a separate lineage from true reptiles, and they originated at the same time and evolved side by side with true reptiles.

In the Permian, protomammals took over the terrestrial realm. By the Late Permian, a great variety of protomammals had evolved. Found in red beds from South Africa, South America, and Russia, these protomammals included some of the first herbivores known. One group, the dinocephalians ("terrible heads" in Greek), had huge hippo-like bodies up to 2.7 meters (9 feet) long and thick bony skulls, often with weird flanges and protuberances projecting from around their eyes (figure 15.6A). The other main group of herbivorous protomammals was the dicynodonts ("two dog teeth" in Greek), which had a large plant-slicing beak with no teeth except two huge canine tusks (figure 15.6B). These ranged from the small, dog-sized *Lystrosaurus* (figure 14.6D) to a variety of forms that were pig-sized and even

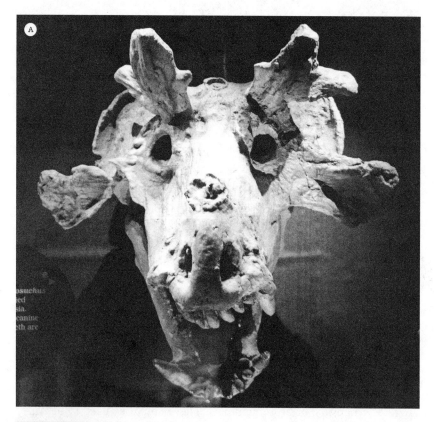

Figure 15.6 ▲ ▶

Protomammals of the Permian: (*A*) The weird dinocephalian *Estemmenosuchus*, with a face covered with bumps and ridges; (*B*) the herbivorous dicynodonts had two tusks but an otherwise toothless beak; and (*C*) the largest predators of the late Permian, the huge tusked gorgonopsians. (Photos by author)

Figure 15.6 ▲

(continued)

hippo-sized. In addition to these early herbivores, there were a wide variety of predatory protomammals in the Late Permian, most notably the tiger-sized predators with saber-like teeth known as the gorgonopsians (figure 15.6C).

The Permian landscape was forested with primitive swamp plants, such as horsetails, ferns, and tree-sized club mosses in the swamps, and dense forests of conifers in the uplands. A wide range of amphibians, many with the size and shape of large crocodiles, lived in the ponds and rivers. True reptiles of many different sizes and shapes were found on the land and in the lakes and rivers. Ruling over all of them were the synapsids or protomammals, which began with the finbacks of the Early Permian and climaxed with the huge predatory gorgonopsians and a variety of the first herbivorous land animals, the dinocephalians and dicynodonts.

EXTINCTION PATTERNS

What does the fossil record reveal about what happened in the Late Permian? In the marine realm, the extinction was catastrophic (figure 15.7), and probably 95 percent of the species in the ocean vanished. The end of the Permian marked the end of the line for five groups of animals that had survived every other extinction event of the Paleozoic. This included not only the trilobites, which were down to just two genera, and the tabulate and rugose corals, which were nearly gone after the Devonian extinction, but also the stalked echinoderms known as blastoids, which had been slowly declining since their heyday in the Early Carboniferous. None of these groups was very diverse when the catastrophe came, but the fifth group, the fusulinid foraminiferans (figure 15.1), were still covering the seafloor and evolving faster than ever before.

In addition to groups that vanished completely, many other groups were nearly wiped out, with only a few lineages straggling through. The goniatite ammonoids, which had survived the Devonian extinction, were so badly affected that only two or three lineages survived. These became the ancestors of another great radiation of ammonoids in the Mesozoic. The major groups of Paleozoic bryozoans vanished, resulting in 80 percent of the genera going extinct, as did 98 percent of the genera of Paleozoic crinoids, leaving only a few minor families that would repopulate the world in the Mesozoic. The extinction of clams was not nearly so severe (only 59 percent of the genera), but 98 percent of the marine snail genera vanished.

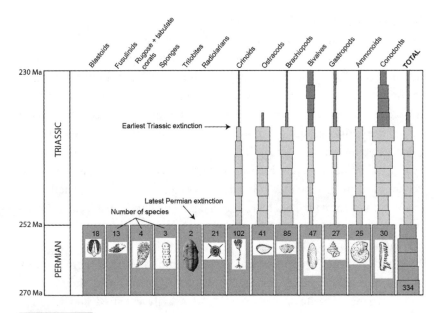

▲

Summary of the groups that vanished (blastoids, fusulinids, rugose and tabulate corals, trilobites) or nearly vanished (sponges, radiolarians) at the end of the Permian, and groups that were decimated but managed to hang on into the Triassic (on the right side of the diagram). (Redrawn by E. Prothero from several sources)

Most significant, the ubiquitous brachiopods that had dominated the shelly fauna of the seafloor were nearly wiped out. About 96 percent of their genera vanished, marking the end of the group known as the orthids (dominant in the Ordovician) and especially the productids, which were the most common group of the late Paleozoic. Only four lineages of brachiopods managed to survive into the Mesozoic: the spirifers (they hung on until extinction at the end of the Triassic), and the three groups that are alive today, the primitive brachiopods known as lingulids, the lamp-shell-shaped terebratulids, and the corrugated rhynchonellids.

This devastating event completely transformed the nature of marine faunas. We saw how the typical Paleozoic fauna arose after extinction of most of the Cambrian fauna in the Ordovician (figure 11.1). This Paleozoic fauna dominated from the Ordovician to the Permian, and you can collect typical fossils in any Paleozoic locality after the Cambrian, especially brachiopods, bryozoans, corals, crinoids, and cephalopods. The individual genera of corals and brachiopods changed over the Paleozoic, so each time interval has its own index fossils, but the seafloor fossils assemblage is very similar from about 500 million years ago until the Permian crisis 252 million years ago. If you look at the ocean floor today, or any modern collection of shells on a beach, you will see almost no brachiopods or crinoids or cephalopods or any of the extinct groups of corals and bryozoans. The Permian event completely decimated these groups and ended their domination of the marine realm. When life began to recover in the Mesozoic, the marine realm was recolonized by groups that still rule the oceans today: clams, snails, and sea urchins (among the shell producers), plus fish and crustaceans. This is often called the Modern fauna, although it has ruled the earth for the past 250 million years.

What did these Permian victims have in common? Most of them had very simple inefficient respiratory mechanisms, suggesting that they needed well-oxygenated waters. They were also heavily calcified and required oceans with normal amounts of carbon dioxide to create their shells. In contrast, the survivors all had active control of their respiration and elaborate gas exchange mechanisms, which helped them endure conditions of low oxygen. Their shells were also lightly calcified, and they did not need to expend a lot of energy to crystallize out the calcite from seawater to make their shells. All of the victims had low metabolic rates as well. These victims had many things in common: weak respiratory systems, low metabolic rates, and required normal calcium carbonate conditions to make their shells. The survivors all had well-developed respiratory systems, higher metabolic rates, and did not require a lot of carbon dioxide in the seawater to make their hard body parts. All of this points to low oxygen conditions (hypoxia) and odd carbon dioxide conditions in seawater at the same time, such as an oversaturation with carbon dioxide (hypercapnia), which is toxic to most marine invertebrates.

The extinctions on land were slightly less severe, but this was still the worst land extinction Earth had ever seen. It wiped out 70 to 80 percent of all creatures on land, including most of the lineages of protomammals that had long ruled the Permian (only the dicynodonts and some of the predatory groups survived), nearly all the reptilian groups, and most of the archaic amphibians (lepospondyls and anthracosaurs vanished, and only a few temnospondyls survived). The severity of this extinction can be seen in places like the Karoo beds of South Africa, where the huge diversity of protomammals in the Late Permian is depleted to just a few dicynodonts (such as *Lystrosaurus*), a few small predatory forms, and just a few small reptiles in the Early Triassic.

These survivors had a few things in common: most were small-bodied forms that could burrow and could survive with fewer food resources. They could deal with bad atmospheric conditions such as low oxygen or high carbon dioxide. The ones with inefficient respiratory systems (like skin respiration found in the amphibians) were particularly hard hit, and those with advanced respiration systems, such as the diaphragm breathing of protomammals like *Lystrosaurus* and *Thrinaxodon*, had an advantage. These common characteristics of survivors suggests that they could adapt to an atmosphere that was low in oxygen but high in carbon dioxide.

ALL HELL BREAKS LOOSE

At one time, scientists blamed the great Permian extinction on the assembly of the Pangaea supercontinent (which wiped out the shallow seas crushed between the colliding continental blocks), but that event occurred more than 50 million years earlier, in the Early Permian. Others have blamed the extinction on the growth of the great Gondwana ice sheet, but that too was already in place 100 million years earlier and was actually declining by the Late Permian. A few scientists claimed Earth was hit with a giant impact from a meteor or a comet (such as what happened at the end of the Mesozoic), but no claim of this impact has ever withstood the scrutiny of other scientists. None of the typical signs of impact (such as droplets of melted crustal rock splattered around the earth, or rare elements such as iridium that come from space rocks) have been proven to exist.

Instead, the cause of the mother of all mass extinctions seems to be even more frightening: the largest volcanic eruptions in all of Earth's history, which produced lava beds today known as the Siberian traps (figure 15.8A). These ancient lava flows are still visible in many places in northern Siberia (figure 15.8B). In this event, almost 4 million cubic kilometers (1 million cubic miles) of lava erupted, covering 2 million square kilometers (770,000 square miles). Such eruptions from deep mantle sources would have released enormous volumes of greenhouse gases, especially carbon

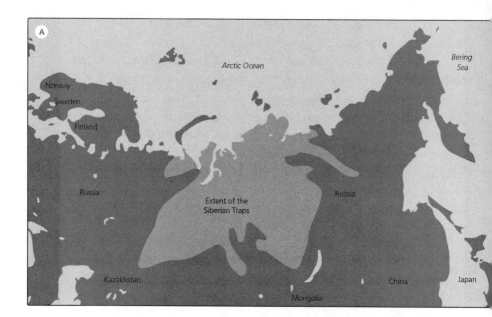

Figure 15.8 ▲

The Siberian traps: (*A*) map showing the areal extent of the eruptions; (*B*) outcrop of Siberian traps in central Russia in the plateaus (in the background) with broken blocks of basaltic lava eroding (in the foreground). ([*A*] Courtesy of Wikimedia Commons; [*B*] courtesy of the U.S. Geological Survey)

dioxide and sulfur dioxide. In 2013, scientists estimated that these eruptions released 8.5×10^{13} metric tonnes of carbon dioxide, 4.4×10^{11} metric tonnes of carbon monoxide, 7.0×10^{12} metric tonnes of hydrogen sulfide, and 6.8×10^{13} metric tonnes of sulfur dioxide—staggering amounts of volatile gases that completely transformed the atmosphere into a super greenhouse full of toxic sulfates and sulfides, which quickly changed the chemistry of the oceans. In addition, these lavas ignited the thick Carboniferous coal seams in the bedrock below them, releasing even more carbon dioxide into the atmosphere.

The oceans then became supersaturated with carbon dioxide, making them too hot and acid and killing nearly everything that lived there. Ocean temperatures are estimated to have reached over 40°C (104°F), far hotter than even most tropical life can stand. The warming of the sea-floor may have released immense quantities of frozen methane from the bottom sediments, producing a huge burst of methane that is an even more potent greenhouse gas than carbon dioxide. There are many places where the black shales and geochemical evidence suggests that the waters became depleted of oxygen, and perhaps even poisoned by hydrogen sulfide. The atmosphere was also low in oxygen and full of excess carbon dioxide. Nearly all land animals above a certain size vanished, and only a few smaller lineages of synapsids, reptiles, amphibians, and other land creatures made it through the hellish planet of the latest Permian and survived the aftermath into the world of the earliest Triassic.

Only a few lineages of clams, snails, sea urchins, ammonoids, and a few other groups managed to survive into the world of the Early Triassic (figure 15.9). Thick deposits of bacterial and algal mats known as stromatolites, typical of the Precambrian, which had nearly vanished when snails evolved in the Cambrian, made their reappearance in the Early Triassic as most of the grazing snails died out. The animals found in the Triassic oceans were mostly hardy opportunistic "weedy" species that could survive extreme conditions and then exploded in numbers onto a seafloor empty of all their former competitors. The tabulate and rugose corals (horn corals) of the Paleozoic were gone, and it would not be until the Middle Triassic that a new group of corals, the modern Scleractinia that make up the modern coral reefs, evolved from a soft-bodied polyp ancestor. It would take 5 to 10 million years for the marine ecosystems of the world's oceans to recover and rediversify. When they did so, the Paleozoic fauna rich in

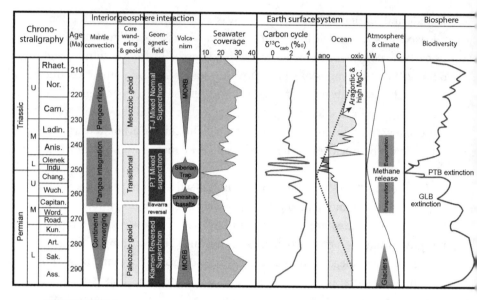

Diagrammatic summary of the major tectonic, volcanic, geophysical, and climatic events at the Permian-Triassic transition. (Modified by E. Prothero from several sources)

brachiopods, bryozoans, crinoids, and corals was gone, and a new world with the Modern fauna had taken over. These animals still dominate the seafloor today.

FOR FURTHER READING

Benton, Michael J. *When Life Nearly Died: The Greatest Mass Extinction of All Time*. London: Thames & Hudson, 2003.

Benton, Michael J., and Richard J. Twitchett. "How to Kill (Almost) All Life: The End-Permian Extinction Event." *Trends in Ecology & Evolution* 18, no. 7 (2003): 358–365.

Berner, Robert A. "Examination of Hypotheses for the Permo-Triassic Boundary Extinction by Carbon Cycle Modeling." *Proceedings of the National Academy of Sciences* 99, no. 7 (2002): 4172–4177.

Brannen, Peter. *The Ends of the World: Volcanic Apocalypses, Lethal Oceans, and Our Quest to Understand Earth's Past Mass Extinctions*. New York: Ecco, 2017.

Dickens, Gerald R., James R. O'Neil, David K. Rea, and Robert M. Owen. "Dissociation of Oceanic Methane Hydrate as a Cause of the Carbon Isotope Excursion at the End of the Paleocene." *Paleoceanography* 10, no. 6 (1995): 965–971.

Erwin, Douglas H. "The End-Permian Mass Extinction." *Annual Review of Ecology and Systematics* 21 (1990): 69–91.

——. *Extinction: How Life on Earth Nearly Ended 250 Million Years Ago.* Princeton, NJ: Princeton University Press, 2006.

Hallam, Anthony, and Paul B. Wignall. *Mass Extinctions and Their Aftermath.* Oxford: Oxford University Press, 1997.

Jin, Y. G., Y. Wang, W. Wang, Q. H. Shang, C. Q. Cao, and D. H. Erwin. "Pattern of Marine Mass Extinction Near the Permian–Triassic Boundary in South China." *Science* 289, no. 5478 (2000): 432–436.

Jurikova, Hana, Marcus Gutjahr, Klaus Wallmann, Sascha Flögel, Volker Liebetrau, Renato Posenato, et al. "Permian–Triassic Mass Extinction Pulses Driven by Major Marine Carbon Cycle Perturbations." *Nature Geoscience* 13, no. 11 (2020): 745–750.

Kaiho, Kunio, Md. Aftabuzzaman, David S. Jones, and Li Tian. "Pulsed Volcanic Combustion Events Coincident with the End-Permian Terrestrial Disturbance and the Following Global Crisis." *Geology* 49, no. 3 (2020): 289–293.

Kamo, Sandra L., Gerald K. Czamanske, Yuri Amelin, Valeri A. Fedorenko, D. W. Davis, and V. R. Trofimov. "Rapid Eruption of Siberian Flood-Volcanic Rocks and Evidence for Coincidence with the Permian–Triassic Boundary and Mass Extinction at 251 Ma." *Earth and Planetary Science Letters* 214, no. 1–2 (2003): 75–91.

Knoll, A. H., R. K. Bambach, D. E. Canfield, and J. P. Grotzinger. "Comparative Earth History and Late Permian Mass Extinctions." *Science* 273, no. 5274 (1996): 452–457.

Leighton, Lindsey R., and Chris L. Schneider. "Taxon Characteristics That Promote Survivorship Through the Permian–Triassic Interval: Transition from the Paleozoic to the Mesozoic Brachiopod Fauna." *Paleobiology* 34, no. 1 (2008): 65–79.

MacLeod, Norman. *The Great Extinctions: What Causes Them and How They Shape Life.* London: Firefly, 2015.

Marriott, Katherine, Donald R. Prothero, and Alexander Bartholomew. *The Evolution of Ammonoids.* Boca Raton, FL: Taylor & Francis, 2023.

Ogdena, Darcy E., and Norman H. Sleep. "Explosive Eruption of Coal and Basalt and the End-Permian Mass Extinction." *Proceedings of the National Academy of Sciences of the United States of America* 109, no. 1 (2011): 59–62.

Payne, Jonathan L., Daniel J. Lehrmann, Jiayong Wei, Michael J. Orchard, Daniel P. Schrag, and Andrew H. Knoll. "Large Perturbations of the Carbon Cycle During Recovery from the End-Permian Extinction." *Science* 305, no. 5683 (2004): 506–509.

Payne, Jonathan, L., Alexandra V. Turchyn, Adina Paytan, Donald DePaolo, Daniel J. Lehrmann, Meiyi Yu, and Jiayong Wei. "Calcium Isotope Constraints on the End-Permian Mass Extinction." *Proceedings of the National Academy of Sciences of the United States of America* 107, no. 19 (2010): 8543–8548.

Reichow, Marc K., M. S. Pringle, A. I. Al'Mukhamedov, M. B. Allen, V. L. Andreichev, M. M. Buslov, et al. "The Timing and Extent of the Eruption of the Siberian Traps Large Igneous Province: Implications for the End-Permian Environmental Crisis." *Earth and Planetary Science Letters* 277, no. 1–2 (2009): 9–20.

Retallack, Gregory J. "Multiple Permian-Triassic Life Crises on Land and at Sea." *Global and Planetary Change* 198 (2021): 103415.

Retallack, Gregory J., and Giselle D. Conde. "Deep Time Perspective on Rising Atmospheric CO_2." *Global and Planetary Change* 189 (2020): 103177.

Retallack, Gregory J., and A. Hope Jahren. "Methane Release from Igneous Intrusion of Coal During Late Permian Extinction Events." *Journal of Geology* 116, no. 1 (2008): 1–20.

Retallack, Gregory J., Roger M. H. Smith, and Peter D. Ward. "Vertebrate Extinction Across Permian–Triassic Boundary in Karoo Basin, South Africa." *Bulletin of the Geological Society of America* 115, no. 9 (2003): 1133–1152.

Rothman, David H., Gregory P. Fournier, Katherine L. French, Eric J. Alm, Edward A. Boyle, Changqun Cao, and Roger E. Summons. "Methanogenic Burst in the End-Permian Carbon Cycle." *Proceedings of the National Academy of Sciences* 111, no. 15 (2015): 5462–5467.

Sahney, Sandra, and Michael J. Benton. "Recovery from the Most Profound Mass Extinction of All Time." *Proceedings of the Royal Society B* 275, no. 1636 (2008): 759–765.

Saunders, Andy, and Marc Reichow. "The Siberian Traps and the End-Permian Mass Extinction: A Critical Review." *Chinese Science Bulletin* 54, no. 1 (2009): 20–37.

Tang, Qingyan, Mingjie Zhang, Chusi Li, Ming Yu, and Ii Liwu. "The Chemical Compositions and Abundances of Volatiles in the Siberian Large Igneous Province: Constraints on Magmatic CO_2 and SO_2 Emissions Into the Atmosphere." *Chemical Geology* 339 (2013): 84–91.

Twitchett, Richard J., Cindy V. Looy, Ric Morante, Henk Visscher, and Paul B. Wignall. "Rapid and Synchronous Collapse of Marine and Terrestrial Ecosystems During the End-Permian Biotic Crisis." *Geology* 29, no. 4 (2001): 351–354.

Ward, Peter D. *Gorgon: Paleontology, Obsession, and the Greatest Mass Extinction in Earth History*. New York: Viking, 2004.

——. *Rivers in Time: The Search for Clues to Earth's Mass Extinctions*. New York: Columbia University Press, 2000.

——. *Under a Green Sky: Global Warming, the Mass Extinctions of the Past, and What They Can Tell Us About Our Future*. New York: HarperCollins, 2007.

Ward, Peter, and Joseph Kirschvink. *A New History of Life: The Radical New Discoveries About the Origin and Evolution of Life on Earth*. New York: Bloomsbury, 2015.

White, Rosalind V. "Earth's Biggest 'Whodunnit': Unravelling the Clues in the Case of the End-Permian Mass Extinction." *Philosophical Transactions of the Royal Society of London* 360, no. 1801 (2002): 2963–2985.

Wignall, Paul B., Yadong Sun, David P. G. Bond, Gareth Izon, Robert J. Newton, Stéphanie Vedrine, Mike Widdowson, et al. "Volcanism, Mass Extinction, and Carbon Isotope Fluctuations in the Middle Permian of China." *Science* 324, no. 5931 (2009): 1179–1182.

Wignall, Paul B., and Richard J. Twitchett. "Extent, Duration, and Nature of the Permian-Triassic Superanoxic Event." *Geological Society of America Special Papers* 356 (2002): 395–413.

——. "Oceanic Anoxia and the End Permian Mass Extinction." *Science* 272, no. 5265 (1996): 1155–1158.

FROM ICEHOUSE TO GREENHOUSE

Of all the stream-cut gorges, Zion Canyon is one of the most conspicuous. Niagara Gorge seems impressive as we gaze into its depths, but fifteen Niagaras might be placed in Zion Canyon, one above the other, before they reached the rim. The splendidly colored rocks carved by natural forces in impressive outlines; the great walls rising in astonishing grandeur; the deep, narrow trench cut by the little stream working through untold ages—they cannot be described, they must be seen.

—"STORIES IN STONE," NATIONAL PARK SERVICE BROCHURE

CLIMATE CHANGE IN AN OUTCROP

Any visit to Zion National Park in southwestern Utah is an inspiration. You can travel up Zion Canyon and see towering walls of white sandstone more than 610 meters (2,000 feet) above your head. This white rock is the Lower Jurassic Navajo Sandstone, which forms not only the striking "White Cliffs" in Zion National Park but is also found in Capitol Reef National Park, Glen Canyon National Recreational Area, and many other places in Utah, Colorado, Arizona, and New Mexico (figure 16.1A). In the foreground in figure 16.1B, just above the trees, are the red shales of the Triassic Moenkopi Formation, capped by a ledge of the Shinarump Sandstone. Behind it is another set of slopes of red shale of the Triassic Chinle Formation, capped by Vermilion Cliffs of the Lower Jurassic Kayenta-Moenave formations.

In the middle distance are the White Cliffs of the Lower Jurassic Navajo Sandstone. In the far distance, the lower slopes are Lower Cretaceous Mancos Shale. The furthest outcrops are the Pink Cliffs of the Eocene Claron Formation, famously exposed at Bryce Canyon National Park.

The White Cliffs were formed by a blanket of sand dunes across the southwestern United States with about 40,000 cubic kilometers in volume. In many places they show incredibly thick (up to 25 meters or 80 feet) sets of giant cross-beds, which were formed on the back slopes of enormous sand dunes (figure 16.1C). The Early Jurassic dune field stretched from near Las Vegas (Red Rock Canyon State Park), where the white sandstone is called the Aztec Sandstone and is more than 760 meters (2,500 feet) thick, to the Four Corners region, where the Navajo Sandstone and the underlying red river sandstones of the Kayenta Formation are 700 meters (2,300 feet) thick, and even to central Wyoming, where it is represented by the Nugget Sandstone, which reaches about 655 meters (2,150 feet) in thickness. The enormous Navajo-Aztec-Nugget dune field stretched over a wide area parallel to the Jurassic coastline. Careful measurements of the cross-bedding show that the winds blew the sand from northeast to the southwest, perpendicular to the coastline that trended northwest to southeast. Although it was a virtual "sea" of giant sand dunes, it was not as big as the continent-sized Sahara Desert dune fields of North Africa. Instead, the Navajo dunes are probably more like the huge coastal dune fields on the shores of Namibia today. Dune environments tend not to preserve many fossils, but some reptile and pterosaur trackways and a few skeletons of small dinosaurs are known.

The Navajo Sandstone is also significant in that it is one of the last indicators of the dry, harsh, interior continental climates of the icehouse world that began in the Permian and dominated Triassic landscapes (see chapter 14). Driving east out of Zion National Park are the uppermost Navajo beds, with impressive cross-bedding in every outcrop. Just past the East Entrance Gate to Zion, the Navajo vanishes beneath your feet (you can see it below in the surrounding washes), and you are on top of a flat surface eroded into the Navajo Sandstone. Immediately above the Navajo east of Zion are small roadcuts of a soft yellowish or pale creamy colored limestone known as the Carmel Formation (figure 16.1D). It is full of the star-shaped stem pieces of a crinoid (sea lily) known as *Pentacrinus*, and it tells us that the Lower Jurassic desert dune surface had been flooded by the ocean. This suggests that sea levels had risen and the planet was back in a greenhouse world again.

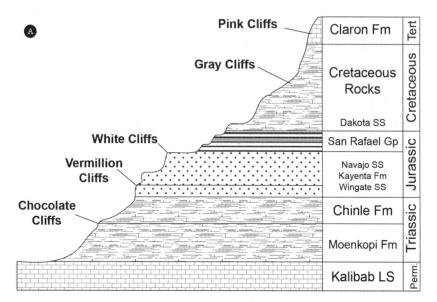

Figure 16.1 ▲ ▶

Mesozoic facies of the Colorado Plateau area. (*A*) Diagram of the sequence of Mesozoic rocks that make up the "Grand Staircase," including the White Cliffs of Navajo Sandstone. (*B*) The entire Grand Staircase from the North Rim of the Grand Canyon, looking north. (*C*) Typical cliff of Navajo Sandstone, Zion National Park, with its enormous cross-beds. (*D*) Above the Navajo Sandstone near Mt. Carmel is a shallow marine limestone full of crinoids; it is known as the Carmel Formation. ([*A*] Modified by E. Prothero from several sources; [*B–D*] photos by author)

Figure 16.1 ▲
(*continued*)

In fact, over the whole Rocky Mountains lies the first epicontinental seaway deposit since the greenhouse days of the Devonian and the Mississippian (figure 16.2). Often called the Sundance Seaway, it flooded the Great Plains from the Gulf of Mexico to the Arctic Ocean, and also drowned much

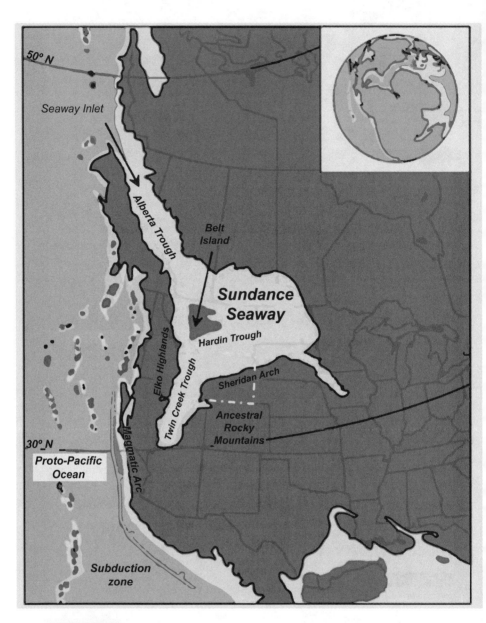

Figure 16.2 ▲

Paleogeographic map of the Late Jurassic Sundance Seaway. During the peak of the middle-late Jurassic transgression, the Sundance Seaway and the Sundance Formation covered much of the northern Rockies. (Redrawn by E. Prothero from several sources)

Labels on the map:

50° N

Seaway Inlet

Alberta Trough

Belt Island

Sundance Seaway

Hardin Trough

Elko Highlands

Twin Creek Trough

Sheridan Arch

Ancestral Rocky Mountains

30° N

Proto-Pacific Ocean

Magmatic Arc

Subduction zone

of the region of the future Rockies and Colorado Plateau. Represented by the rocks of the San Rafael Group in the Arizona-Utah borderland around Lake Powell, or the Sundance Formation over much of the northern Rockies, these thick deposits of shales with rare limestones represent shallow seas drowning most of what would become the Great Plains and Rocky Mountain region.

Major transgressions that flooded the middle of the continent corresponded to the transition of Earth from its former icehouse state to a greenhouse planet once again. The rise of sea level tells us that the arid icehouse conditions of the Permian were over and there were no more polar icecaps by the Jurassic. The shallow seaways of the Sundance and the San Rafael group were home to a wide variety of marine life, including bizarre oysters on the sea bottom and other marine snails and clams. Swimming above them were many kinds of ammonites, as well as marine reptiles such as ichthyosaurs and plesiosaurs.

EARLY MESOZOIC ICEHOUSE WORLD

The Navajo Sandstone is one of the last North American rock units formed in an icehouse world that began in the Late Carboniferous and spanned the Permian and the entire Triassic. These Triassic rocks resembled the rocks of the Permian: red shales and sandstones formed on hot, sunny floodplains, with dune sands and occasional evaporites (figure 16.3A). This is not surprising because the extreme continental interior climates that dominated the Permian in much of the world continued into the Triassic while Pangea was still together. If you travel around the Colorado Plateau in northern Arizona, southern Nevada, eastern and southern Utah, western New Mexico, or western Colorado, there is a consistent Triassic sequence above the gray Permian rocks such as the Kaibab Limestone that forms the rim of the Grand Canyon. The lowest unit above the Kaibab is the Moenkopi Formation, a floodplain shale that comes in many hues—from brick red to gray to brown to lavender to green—giving color to the region known as the Painted Desert. In many places, the Moenkopi Formation is capped by a resistant ledge of sandstone and conglomerate known as the Shinarump Formation, which represents a widespread river channel sequence that once covered the area. The uppermost Triassic unit is the famous Chinle Formation, a brick-red shale with minor sandstones that forms red cliffs in many places.

Figure 16.3 ▲ ▶

Triassic rocks of the western United States. (*A*) Gigantic petrified logs protrude from the Upper Triassic Chinle Formation, Petrified Forest National Park, Arizona. (*B*) Red shales with white gypsum from the Triassic Spearfish Shale, behind Evans Plunge, Hot Springs, South Dakota. (Photos by author)

Figure 16.3 ▲
(*continued*)

Its most famous exposures, however, are in Petrified Forest National Park in Arizona where huge fossil logs are steadily eroding out of the soft shales (figure 16.3A).

Elsewhere in the western United States the names of the formations are different, but the distinctive red shale is the signature of the Triassic. In Wyoming, the Triassic red shale is called the Chugwater Formation. In the Black Hills of South Dakota, the red shale bed full of gypsum is called the Spearfish Formation (figure 16.3B). It forms an easily eroded oval shale valley nicknamed "the Racetrack" that completely surrounds the Black Hills. It lies between resistant layers of Mississippian Pahasapa Limestone on the inside track and the resistant Cretaceous Dakota Sandstone on the outside rim.

THE REAL "JURASSIC PARK"

The fossil plants of the Triassic in Petrified Forest National Park, as well as fossil plants from the Jurassic, present a type of forest very different from the one seen in the movie *Jurassic Park*. The real Jurassic Park had no flowering plants at all (99 percent of the plants that we see around us in most parts of the world today) because they did not evolve until the Early Cretaceous. Instead, the plants that dominated the Triassic and Jurassic were mostly gymnosperms, or "naked seed" plants, which keep their seeds in cones rather than in flowers. They first evolved in the Carboniferous and Permian and had been abundant in the drier upland habitats of the Permian. Their cones and wind-blown pollen enabled them to reproduce without being immersed in water, so they didn't require living in swampy wetlands, or even a wet climate. By the later Triassic, a whole range of gymnosperms had evolved. These included the familiar pine trees along with the "sego palms," or cycads, that superficially look like palm trees but are actually gymnosperms with cone-bearing male and female plants (figure 16.4A). Another familiar gymnosperm is the ginkgo tree, with its distinctive duck-foot-shaped leaves (figure 16.4B). Most abundant of all were the trees of the genus *Araucaria*, which today include such plants as the Norfolk Island pine (figure 16.4C), the monkey puzzle tree, and the Kauri pine. *Araucaria* trees very like these were the source of the giant tree trunks found in the Upper Triassic rocks of the Petrified Forest National Park, which have been given the scientific name *Araucarioxylon*. Ferns were still abundant in the wetter areas, and ferns and tree-sized "club mosses" (lycopsids) survived in the few tropical swampy areas left on the dry, hot desert latitudes of an icehouse planet. This gymnosperm flora would continue to dominate the landscape until the Early Cretaceous, when flowering plants first diversified. When you walk in a pine forest or a tree fern forest—or any other places with few flowering plants—you are walking in forests similar to those of the Triassic or Jurassic.

Living in these forests during the later Triassic was a land fauna transitional between the typical Permian and Jurassic assemblages, a mixture

Figure 16.4 ▲ ▶

Gymnosperms that dominated Triassic and Jurassic forests. (*A*) Cycads or "sego palms,"
with the long phallic male cone and the shorter, rounder female cones. (*B*) Leaves of the
ginkgo or "maidenhair tree," another survivor of the Jurassic forests. (*C*) The distinctive
short needles and long flexible branches of the Norfolk Island pine, genus *Araucaria*, one of
the dominant conifers of the Triassic and Jurassic. (Photos by author)

Figure 16.4 ▲
(*continued*)

of holdovers from the Permian plus new forms. These holdovers included the huge, flat-bodied temnospondyl amphibians, which enjoyed their last successful period before their final extinction at the end of the Triassic, and some of the earliest true frog fossils that come from the Triassic. The modern groups had already appeared along with their ancient relatives. The main holdover, however, was the diversity of the protomammals (figure 16.5A), which ruled the land in the Permian and still dominated in the Early and Middle Triassic. These included the beaked herbivores, called dicynodonts, like the pig-sized *Lystrosaurus* (figure 14.6D) and the cow-sized *Kannemeyeria*, and numerous predatory protomammals from the weasel-like *Thrinaxodon* to the bear-sized killer *Cynognathus* (figure 14.6E). These had all survived the low-oxygen greenhouse planet of the Permo-Triassic extinction and had begun to get larger as oxygen levels rose again in the Middle Triassic.

Battling for supremacy with these relics of the Permian were several new groups. The most important were primitive members of the reptilian group Archosauria, which includes crocodiles, dinosaurs, pterosaurs, and birds (figure 16.5B). Archosaurs ("ruling reptiles" in Greek) had more advanced limbs and posture than most other sprawling reptiles, and many had additional adaptations for mobility. In addition, they may have had lots of air sacs throughout their bodies as aids for breathing (as their bird descendants have today), which would have given them an advantage in the low-oxygen world of the Triassic and Jurassic. The typical Late Triassic archosaurs (formerly lumped in the wastebasket group "thecodonts") included the crocodile-like phytosaurs (figure 16.6), the long-bodied deep-snouted predators known as erythrosuchids ("bloody crocodiles" in Greek), the armored aetosaurs, the beaked pig-sized herbivores known as rhynchosaurs, and the strange long-necked reptile *Tanystropheus*, whose 7 meter (23 foot) long body had a rigid neck more than 3 meters (10 feet) long. Some of these lineages were related to the crocodile branch of archosaurs, and others were closely related to dinosaurs and pterosaurs. But they all began to decline in the Late Triassic as the two living groups of archosaurs replaced them. One was the true crocodilians, which were slender-legged, dog-sized, hopping reptiles in the Late Triassic. Only later when the phytosaurs died out and vacated that niche did they become the large aquatic ambush predators we think of today.

The most important group of archosaurs, however, was the dinosaurs. They first appeared in the Late Triassic as small, bipedal, fast-running predators such as *Eoraptor* and *Herrerasaurus*. They soon began to evolve into a variety of body forms, and before the Triassic was over there were large ancestors of the long-necked sauropods. Known as sauropodomorphs, they

Figure 16.5 ▲

(A) Two groups of Triassic protomammals from South Africa. The bear-sized cynodont *Cynognathus* are attacking the cow-sized herbivorous dicynodont *Kannemeyeria*. (B) Display of typical land animals of the Triassic Petrified Forest in Arizona. On the left is the dicynodont *Placerias*, with its toothless beak and tusks. On the right is the armored aetosaur, with spikes in its shoulder armor. Between them is a predatory archosaur known as a rauisuchid. (Courtesy of Wikimedia Commons)

Figure 16.6 ▲

The huge crocodile-like phytosaur known as *Redondasaurus*. Phytosaurs dominated the crocodile niche while true crocodiles were still small, long-legged runners and jumpers. (Courtesy of Wikimedia Commons)

were typified by the largest herbivore of the Triassic, *Plateosaurus*, which was about 10 meters (33 feet) long. Although it was still fully bipedal, it already showed the lengthening neck and tail that characterized the Jurassic sauropods. It had simple, leaf-shaped teeth suitable for chopping vegetation, unlike the sharply pointed teeth of most archosaurs that were still predators.

RAIN, RAIN, GO AWAY: THE CARNIAN PLUVIAL EPISODE

One of the strangest climatic events of the Mesozoic (and indeed of all global climatic history) was the Carnian pluvial episode, when the world's climate got very wet and humid, with heavy rains, for about 2 million years, from 234 to 232 Ma. These events spanned the boundary between two sub-stages of the Carnian stage of the Triassic, the Julian and the Tuvalian, which ran from 237 to 228 Ma.

The first mention of something peculiar was described in a 1974 article by Wolfgang Schlager and Walter Schöllnberger, who noticed a dramatic pulse of muds and sands right in the middle of the shallow marine limestones of the Carnian sequence in the northern Alps. Other people soon began to notice sandy mudstones—such as the Schilfsandstein of Germany and the Mercia Mudstone of Wales, which were rich in tropical kaolinite clays and plant debris and rain forest pollen—punctuating a sequence of deposits with pollen indicating arid climates. This idea was further developed in a 1989 paper by Michael Simms and Alastair Ruffell, who were working on the extinction of Triassic crinoids (Simms) and the geology of the Mercia Mudstone (Ruffell). They described geological evidence of an episode of high humidity and abundant rainfall in the mid-Carnian in many places in the world. Soon, geochemical studies of carbon and oxygen isotopes also indicated an abrupt warming of climate and abundant carbon from unusual sources. In 2008 this obscure idea was again revived in a series of studies of the Triassic rocks of the Italian Alps, and by 2018 there were major international workshops on the event as the evidence became overwhelming.

The evidence came from many sources. There were soil horizons in many Carnian terrestrial rock sequences that suggested tropical weathering, even in areas far from the tropics, and the fossil pollen were also consistent with wet tropical conditions over much of the globe. In terrestrial sequences, thick deposits of floodplain mudstones and sandstones were found over wide areas, even in regions that normally produced only dry climate sediments throughout the Triassic. There were also widespread occurrences of amber, which tend to form mostly from tropical and subtropical rain forest trees. In the ocean, normal shallow marine limestones were almost always interrupted by a thick deposit of sands and muds, suggesting lots of erosion of sediments from the land. The geochemical evidence kept getting stronger as well: oxygen isotopes gave a clear signal of warming of 3°C to 4°C (37°F to 39°F), the carbon isotopes seemed to indicate that a lot more plant matter was being produced, and osmium isotopes suggested a volcanic source for most of the unusual signals. Today the Carnian pluvial episode is being intensively studied by scientists all over the world, and the consensus is that it was a global event of extremely warm and wet climate that lasted almost 2 million years, with only a few minor episodes of drying during this interval.

Millions of years of wet climates also had a major effect on life on Earth. The Carnian pluvial episode triggered the first wave of extinctions in the oceans, where the ammonoids, the "moss animals" or bryozoans, and green algae were all hit with severe extinctions. But the more remarkable consequence of this global warm and wet climate was how it triggered the evolution of many different groups that would flourish in the rest of the Mesozoic. The oldest known dinosaurs were found in Brazil and Argentina and are all dated from 233 to 231 Ma. The first primitive relatives of crocodilians, phytosaurs, lizards, turtles, and mammals all date to the end of the

Carnian pluvial. The plants typical of the Late Triassic that lived in the Petrified Forest diversified during these warm wet conditions, including major families of ferns, podocarps, araucarias, and several types of conifers.

What could have caused global climate to become suddenly warm and wet, despite the cold dry icehouse conditions of the late Paleozoic and early Mesozoic? Two major ideas have been suggested. The distinctive geochemical signal from osmium isotopes and carbon isotopes suggests a massive carbon dioxide injection in the oceans. Both of these geochemical signals strongly point to a volcanic source releasing huge amounts of greenhouse gases. The best candidate for these eruptions is the Wrangellia Large Igneous Province in southern Alaska, dated 231 Ma and older. These eruptions were caused when a block of exotic crust from the central Pacific, the Wrangellia terrane, slammed into the North American plate in the Triassic, triggering gigantic eruptions of basaltic lavas that released many millions of metric tonnes of carbon dioxide and sulfur dioxide into the atmosphere.

Other scientists pointed to a huge mountain-building event in what is now central Asia, called the Cimmerian orogeny, which was caused by closure of the northern branch of the Tethys Seaway. These mountains were close to where the Himalayas are today, and they would have created a monsoonal weather pattern with intense summer flooding around the Indian Ocean region, just as India and the Himalayas do today. This might work well for Eurasia and the Tethys margin, but it does not explain why the monsoonal conditions were global.

THE TRIASSIC-JURASSIC EXTINCTION

Of the Big Five major mass extinction events first recognized by Raup and Sepkoski in 1982, the fourth or fifth largest is the one at the end of the Triassic, about 201 Ma. It is only slightly smaller than the end-Cretaceous event that wiped out the dinosaurs. About 35 percent of marine genera and possibly 50 percent of marine species vanished, and an equally severe crash in the land animals occurred (at least 42 percent of the terrestrial vertebrates vanished, according to one estimate).

Details of the Late Triassic extinctions are complicated because the last stage of the Triassic (the Rhaetian stage) is poorly known in many parts of the world. It is not clear whether many animals died out at the very end of the Triassic, or near the end of the Triassic at the Norian-Rhaetian boundary. As Tanner, Lucas, and Chapman pointed out in 2004, some organisms thought to have vanished at the end of the Triassic now appear to have been significantly affected by the Carnian pluvial episode or the end-Norian extinctions. Most scientists see the Triassic extinctions as a series of severe events, similar to the two pulses in the Ordovician extinctions, the Devonian extinctions, and the Permian extinction. In fact, four of the Big Five extinctions show this pattern of at least two pulses, and only the end-Cretaceous event does not.

In the marine realm, the most obvious victims of this extinction were the ceratitic ammonoids, which had been the dominant shelled cephalopods of the entire Triassic. They had a first wave of extinction at the end of the Norian Stage but vanished completely at the end of the Rhaetian, leaving just a few survivors that radiated into an evolutionary explosion of new lineages in the Early Jurassic, such as the Lytoceratina, Phylloceratina, and Ammonitina. Other molluscs, such as the clams and oysters, suffered as well, but marine snails did not seem much affected. The brachiopods that had survived the Permian extinction were hit again, especially the spirifers. This group had been dominant in the Devonian and Carboniferous, barely survived the Permian extinction, but finally vanished in the Early Jurassic after the Triassic extinctions decimated them. Even the coral reefs, which had been destroyed in the Permian but had finally recovered in the Middle Triassic, were wiped off the face of the earth again. They did not return until the Sinemurian, the second stage of the Early Jurassic.

As for marine vertebrates, the fossil record of bony fish is not complete enough to know which events, other than the Carnian pluvial episode, caused extinctions among them. However, one group of marine vertebrates, the eel-like conodonts that had survived all the extinction events of the Paleozoic, were severely affected by the Carnian pluvial episode, and they met their final demise at the end of the Triassic. Of the marine vertebrates in the Triassic, such as the dolphin-like ichthyosaurs, a lot of primitive groups became extinct but recovered with more advanced groups in the Jurassic. The only marine reptiles to vanish completely were the weird, turtle-shaped mollusk eaters known as the placodonts.

In the terrestrial realm, there were some dramatic changes in the dominant vertebrate groups (figure 16.7). Due to the poor record of the Rhaetian Stage, the last interval of the Triassic, it is difficult to know which animals

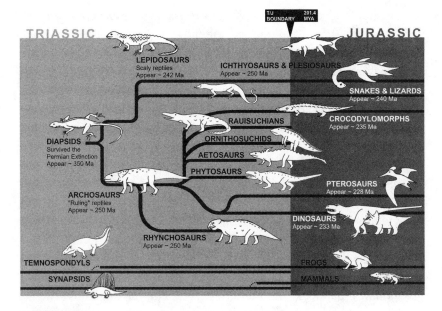

Figure 16.7 ▲

Diagram showing the relationships and survival of different branches of vertebrates during the Late Triassic extinctions. (Drawing by K. Marriott)

died out at the end of the Triassic, and how many were affected by the Carnian pluvial episode or by the extinctions at the end of the Norian Stage. Nevertheless, the cast of characters that had once dominated the Triassic, such as the gigantic flat-bodied temnospondyls, died out in stages at the end of the Norian, with only a few lineages surviving until the end of the Rhaetian.

A wide range of archosaurian reptiles vanished. The pig-sized beaked herbivores known as rhynchosaurs were gone by the end of the Norian or the earliest Rhaetian, but most of the others were still alive in the Rhaetian. These included the crocodile-like phytosaurs, the armored aetosaurs, the long-necked tanystropheids, the big quadrupedal terrestrial predators such as the rauisuchids and the erythrosuchids, the "flying lizard" kuehneosaurs, and many other groups of typically Triassic archosaurs reptiles. All vanished at the end of the Norian or at the end of the Rhaetian, and only the delicate, small, running crocodilians survived to radiate into new lineages in the Early Jurassic.

The temnospondyls and the archaic archosaurs were two dominant Triassic groups that were wiped out near the end of the Triassic. The third group was the last of the synapsids, or protomammals, which were often lumped into the wastebasket taxon group Cynodontia. The bear-like *Cynognathus* (figure 14.6E), the weasel-like *Thrinaxodon*, and the huge beaked herbivores known as dicynodonts were all present near the end of the Triassic but vanished at the end of the Norian or the end of the Rhaetian. In their place, their direct descendants, tiny shrew-sized mammals, appeared in the Late Carnian, and they became much more diverse by the Early Jurassic.

This was the general pattern of all the extinctions on land around the end of the Triassic. Archaic groups vanished, and their modern relatives survived to form new lineages. The archaic temnospondyls vanished, but their frog and salamander relatives were evolving fast in the Jurassic. The archaic protomammals were phased out, but their mammalian descendants survived. In place of the archaic archosaurs—phytosaurs, aetosaurs, and the like—a new group took over the terrestrial realm: the dinosaurs. Finally, soaring in the air were the first true flying vertebrates: the pterosaurs. These groups—dinosaurs, pterosaurs, mammals, and frogs and salamanders—would dominate the terrestrial ecosystems until the end of the Mesozoic.

What caused the Late Triassic extinctions? Like many other mass extinction events, a wide variety of culprits have been blamed. After the 1980 publication of the Cretaceous asteroid impact model, many scientists assumed there would be signs of an impact at the end of the Triassic as well. One of the prime candidates was the huge Manicouagan impact crater in Quebec, which is 100 kilometers (62 miles) across, the largest impact event known during the Mesozoic and second only to the end-Cretaceous Chicxulub crater in size (figure 16.8). Naturally, in the late 1980s, a lot of scientists were pointing at this event as the Triassic killer. However, in 1992, the crater was precisely dated, and its age was 214 Ma —far too young to have anything to do with extinctions at the end of the Norian (227 Ma) and far too old to have anything to do with the end of the Triassic (201 Ma). Today, the lack of effects from this huge impact event stand as an object lesson on not getting ahead of the data or jumping to the conclusion that impacts always causing extinctions. Other smaller impacts have been dated closer to the end of the Triassic, but they are too small to cause all of the extinctions of the entire Late Triassic interval. Gradual sea level drops were also blamed for the extinctions. But the fact that the extinction was far more severe in the

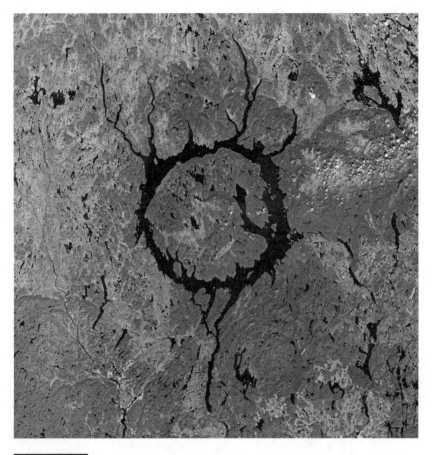

Figure 16.8 ▲
Satellite view of the Manicougan crater in northern Quebec. (Courtesy of Wikimedia Commons)

land realm, which was unaffected by any changes in sea level, seems to rule that idea out.

The likeliest culprit is volcanism and associated climate change, the same killer suggested for the Permian extinction (Siberian lavas), the Devonian extinction (Vilyuy lavas), the late Ordovician (Pripyat-Dneiper-Donets lavas), the Carnian pluvial event (Wrangell lavas), and the end-Cretaceous extinctions (Deccan lavas). During the Late Triassic, huge eruptions called the Central Atlantic Magmatic Province (CAMP) formed as North America and South America ripped away from Africa when

Pangea split apart and the North Atlantic began to open (figure 16.9A). The first pulses of eruption ran from Nova Scotia to Morocco at the end of the Norian and climaxed during the Rhaetian, with the largest pulse at the end of the Triassic. Eventually it formed a blanket of 11 million square kilometers of lava, with the volume estimated at 3 million cubic kilometers of basaltic rock. Based on the latest radiometric dating, the eruptive pulses

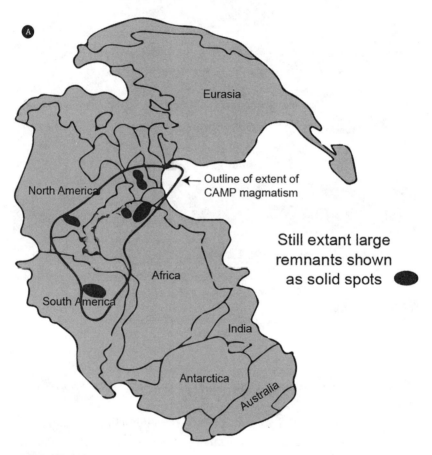

Figure 16.9 ▲ ▶

The Central Atlantic Magmatic Province (CAMP) is a leading candidate for causing the climate changes and extinctions at the end of the Triassic. (A) Map of the CAMP eruptions. (B) The Palisades Sill, on the west bank of the Hudson in northwestern New Jersey and southern New York, was one of these Late Triassic CAMP basalts that intruded parallel to bedding in the Triassic. ([A] Redrawn by K. Marriott from several sources; [B] photo by author)

Figure 16.9 ▲
(*continued*)

spanned at least 600,000 years. The flows eventually included the area from Brazil to Scandinavia, making them the most widespread eruptions in Earth's history. The Moroccan lavas alone formed a layer more than 300 meters (1,000 feet) thick.

Similar to other mass extinctions, the main killer from such huge basaltic lava flows over a wide landscape (known as "flood basalts") was the gigantic volume of greenhouse gases, especially carbon dioxide and sulfur dioxide. Most of the volcanic rocks in the Palisades Sill, which lies along the western shore of the Hudson River northeast of New York City (figure 16.9B), and the Watchung lava flows, which are in northern New Jersey in the Newark Basin, are part of the CAMP. Only eruptions of this scale explain the carbon isotope signal found in marine rocks. Such a huge volume of carbon would have made the oceans more acidic, as happened in the Permian, so shells that dissolved easily would have been most affected. As in the case of some of the Paleozoic extinctions, the most vulnerable organisms were those with easily dissolved shells made of aragonite ("mother of pearl") such as the goniatitic ammonoids, or those with little direct control of the mineralization of their calcite skeletons (such as corals and calcified sponges).

To summarize, the Jurassic world was dominated by the survivors of a severe mass extinction, including the dinosaurs, crocodilians, pterosaurs, and small mammals and frogs on land, and a new radiation of ammonites in the oceans. By the Late Jurassic, these groups would have reached their peak of diversity. By the Late Jurassic, gigantic sauropods dominated the landscape, and a wide variety had evolved.

Meanwhile, the Navajo Sandstone marks the last gasp of the icehouse world of the Early Jurassic. As shown by the Middle and Upper Jurassic marine limestones and shales that cover it, the icehouse was replaced by a greenhouse planet in the Middle Jurassic. This greenhouse continued throughout the entire Cretaceous and did not vanish until about 45 Ma. That subject is addressed in the next chapters.

FOR FURTHER READING

Bardet, Nathalie. "Extinction Events Among Mesozoic Marine Reptiles." *Historical Biology* 7, no. 4 (1994): 313–324.

Bernardi, Massimo, Piero Gianolla, Fabio Massimo Petti, Paolo Mietto, and Michael J. Benton. "Dinosaur Diversification Linked with the Carnian Pluvial Episode." *Nature Communications* 9, no. 1 (2018): 1499.

Blackburn, Terrence J., Paul E. Olsen, Samuel A. Bowring, Noah M. McLean, Dennis V. Kent, John Puffer, Greg McHone, et al. "Zircon U-Pb Geochronology Links the End-Triassic Extinction with the Central Atlantic Magmatic Province." *Science* 340, no. 6135 (2013): 941–945.

Brannen, Peter. *The Ends of the World: Volcanic Apocalypses, Lethal Oceans, and Our Quest to Understand Earth's Past Mass Extinctions.* New York: Ecco, 2017.

Dal Corso, Jacopo, Massimo Bernardi, Yadong Sun, Haijun Song, Leyla J. Seyfullah, Nereo Preto, Piero Gianolla, et al. "Extinction and Dawn of the Modern World in the Carnian (Late Triassic)." *Science Advances* 6, no. 38 (2020): eaba0099.

Dal Corso, Jacopo, Paolo Mietto, Robert. J. Newton, Richard D. Pancost, Nereo Preto, Guido Roghi, and Paul B. Wignall. "Discovery of a Major Negative $\delta^{13}C$ Spike in the Carnian (Late Triassic) Linked to the Eruption of Wrangellia Flood Basalts." *Geology* 40, no. 1 (2012): 79–82.

Deenen, M. H. L., Micha Ruhl, N. R. Bonis, Wout Krijgsman, Wolfram Kuerscher, Mariel Reitsma, and M. J. van Bergen. "A New Chronology for the End-Triassic Mass Extinction." *Earth and Planetary Science Letters* 291, no. 1 (2010): 113–125.

Fox, Calum P., J. H. Whiteside, P. E. Olsen, X. Cui, R. E. Summons, E. Idiz, and Kilti Grice. "Two-Pronged Kill Mechanism at the End-Triassic Mass Extinction." *Geology* 50, no. 4 (2022): 448–452. https://doi.org/10.1130/G49560.1.

Greene, Sarah E., Rowan C. Martindale, Kathleen A. Ritterbush, David J. Bottjer, Frank A. Corsetti, and William M. Berelson. "Recognising Ocean Acidification in Deep Time: An Evaluation of the Evidence for Acidification Across the Triassic-Jurassic Boundary." *Earth-Science Reviews* 113, no. 1 (2012): 72–93.

Hallam, Anthony, and Paul B. Wignall. *Mass Extinctions and Their Aftermath.* Oxford: Oxford University Press, 1997.

Hautmann, Michael, Michael J. Benton, and Adam Tomašových. "Catastrophic Ocean Acidification at the Triassic-Jurassic Boundary." *Neues Jahrbuch für Geologie und Paläontologie—Abhandlungen* 249, no. 1 (2008): 119–127.

Hodych, J. P., and G. R. Dunning. "Did the Manicougan Impact Trigger End-of-Triassic Mass Extinction?" *Geology* 20, no. 1 (1992): 51–54.

MacLeod, Norman. *The Great Extinctions: What Causes Them and How They Shape Life.* London: Firefly, 2015.

Mcelwain, Jennifer C., D. J. Beerling, and Ian Woodward. "Fossil Plants and Global Warming at the Triassic-Jurassic Boundary." *Science* 285, no. 5432 (1999): 1386–1390.

McHone, Greg. "Volatile Emissions of Central Atlantic Magmatic Province Basalts: Mass Assumptions and Environmental Consequences." In *The Central Atlantic Magmatic Province: Insights from Fragments of Pangea*, ed. W. E. Hames, Greg McHone, P. R. Renne, and Carolyn Ruppel, 241–254. Washington, D.C.: American Geophysical Union, 2003.

Ogg, James G. "The Mysterious Mid-Carnian 'Wet Intermezzo' Global Event." *Journal of Earth Science* 26, no. 2 (2015): 181–191.

Payne, Jonathan L., Daniel J. Lehrmann, Jiayong Wei, Michael J. Orchard, Daniel P. Schrag, and Andrew H. Knoll. "Large Perturbations of the Carbon Cycle During Recovery from the End-Permian Extinction." *Science* 305, no. 5683 (2014): 506–509.

Preto, Nereo, Evelyn Kustatscher, and Paul B. Wignall. "Triassic Climates: State of the Art and Perspectives." *Palaeogeography, Palaeoclimatology, Palaeoecology* 290, no. 1 (2010): 1–10.

Racki, Grzegorz. "The Alvarez Impact Theory of Mass Extinction: Limits to Its Applicability and the 'Great Expectations Syndrome.' " *Acta Palaeontologica Polonica* 57, no. 4 (2010): 681–702.

Retallack, Gregory J. "Greenhouse Crises of the Past 300 Million Years." *Geological Society of America Bulletin* 121, no. 9–10 (2009): 1441–1455.

Ruffell, Alastair, Michael J. Simms, and Paul B. Wignall. "The Carnian Humid Episode of the Late Triassic: A Review." *Geological Magazine* 153, no. 2 (2016): 271–284.

Schaller, Morgan F., James D. Wright, and Dennis V. Kent. "Atmospheric Pco$_2$ Perturbations Associated with the Central Atlantic Magmatic Province." *Science* 331, no. 6023 (2011): 1404–1409.

Schlager, Wolfgang, and Walter Schöllnberger. "Das Prinzip stratigraphischer Wenden in der Schichtfolge der Nördlichen Kalkalpen." *Österreichische Geologische Gesellschaft* 66–67 (1974): 165–193.

Simms, Michael J., and Alastair H. Ruffell. "The Carnian Pluvial Episode: From Discovery, Through Obscurity, to Acceptance." *Journal of the Geological Society* 175, no. 6 (2018): 989–992.

——. "Climatic and Biotic Change in the Late Triassic." *Journal of the Geological Society* 147, no. 2 (1990): 321–327.

——. "Synchroneity of Climatic Change and Extinctions in the Late Triassic." *Geology* 17, no. 3 (1989): 265–268.

Tanner, Lawrence H., John F. Hubert, Brian P. Coffey, and Dennis P. McInerney. "Stability of Atmospheric CO$_2$ Levels Across the Triassic/Jurassic Boundary." *Nature* 411, no. 6838 (2001): 675–677.

Tanner, Lawrence H., Spencer G. Lucas, and M. G. Chapman. "Assessing the Record and Causes of Late Triassic Extinctions." *Earth-Science Reviews* 65, no. 1 (2004): 103–139.

Thorne, Philippa M., Marcello Ruta, and Michael J. Benton. "Resetting the Evolution of Marine Reptiles at the Triassic-Jurassic Boundary." *Proceedings of the National Academy of Sciences* 108, no. 20 (2011): 8339–8344.

Whiteside, Jessica H., Paul E. Olsen, Timothy Eglinton, Michael E. Brookfield, and Raymond N. Sambrotto. "Compound-Specific Carbon Isotopes from Earth's Largest Flood Basalt Eruptions Directly Linked to the End-Triassic Mass Extinction." *Proceedings of the National Academy of Sciences of the United States of America* 107, no. 15 (2010): 6721–6725.

Xu, Guangping, Judith L. Hannah, Holly J. Stein, Atle Mørk, Jorunn Os Vigran, Bernard Bingen, Derek L. Schutt, and Bjørn A. Lundschien. "Cause of Upper Triassic Climate Crisis Revealed by Re–Os Geochemistry of Boreal Black Shales." *Palaeogeography, Palaeoclimatology, Palaeoecology* 395 (2014): 222–232.

GREENHOUSE OF THE DINOSAURS

But the slice of chalk presents a totally different appearance when placed under the microscope. The general mass of it is made up of very minute granules; but, imbedded in this matrix, are innumerable bodies, some smaller and some larger, but, on a rough average, not more than a hundredth of an inch in diameter, having a well-defined shape and structure. A cubic inch of some specimens of chalk may contain hundreds of thousands of these bodies, compacted together with incalculable millions of the granules.

—THOMAS HENRY HUXLEY, *ON A PIECE OF CHALK* (1868)

OCEANS OF KANSAS

In the early 1870s, the western parts of Kansas were still unexplored, but construction of the Union Pacific Railway was underway throughout the region. There were few settlers because hostile Comanche, Lakota, Cheyenne, and other tribes were still roaming the area. The only permanent settlements were cavalry forts, such as Fort Wallace, which protected travelers on the trail along the Smoky Hill River on their way to Colorado. As early as 1858, pioneering explorers and naturalists reported chalk deposits along the Smoky Hill River (figure 17.1) and the Republican River in northwest Kansas. They noted that the chalk was made of the "exuviae of microscopic

Figure 17.1 ▲
Finely bedded chalk beds of the Smoky Hill Member of the Niobrara Formation, Monument Rocks, Gove County, Kansas. (Photo by author)

animals." In that same year, the pioneering geologist Ferdinand Vandiveer Hayden led an expedition across the region and published the first geologic map of Kansas, but he did not map the Cretaceous rocks of western Kansas in much detail. It was not until 1868 that the geologist Joseph LeConte formally described the chalk beds and reported on a number of fossil shells of oysters that had been recovered.

In his 1919 memoir, *Life of a Fossil Hunter*, the famous paleontologist Charles H. Sternberg described the rocks this way:

Both sides of my ravine are bordered with cream-colored, or yellow, chalk, with blue below. Sometimes for hundreds of feet the rock is entirely denuded and cut into lateral ravines, ridges, and mounds, or beautifully sculptured into tower and obelisk. Sometimes it takes on the semblance of a ruined city, with walls of tottering masonry, and only a near approach can convince the eye that this is only another example of that mimicry in which nature so frequently indulges. The chalk beds are entirely bare of vegetation, with the exception of a desert shrub that "finds a foothold in the rifted rock" and sends its roots down every crevice. . . . Sometimes I come upon gorges only two feet wide and fifty feet deep; sometimes for five miles or more the sides of the ravine will be only a few feet high.[1]

Paleontologists were not far behind the original settlers in exploiting these fossil beds. In 1867, an army surgeon, Theophilus Hunt Turner, was exploring the rocks around Fort Wallace and recovered some neck vertebrae of an "extinct monster," which he gave to LeConte, who forwarded them to the legendary naturalist Edward Drinker Cope of Philadelphia. Cope immediately recognized them as the neck vertebrae of a plesiosaur, which had been known in Europe since the 1823 discovery of the first complete plesiosaurs by Mary Anning—but the Kansas specimens were much larger. Cope asked Turner to collect more specimens and send them to him. When the shipment arrived on the train, Cope immediately described them at the May 1868 meeting of the Philadelphia Academy of Sciences and gave them the name *Elasmosaurus*. By 1869, Cope had enough bones to publish a formal description and for reconstruction of the whole skeleton.

In 1871, Othniel Charles Marsh of Yale University (Cope's archrival) led one of several expeditions to collect fossils out west, and his crew recovered spectacular marine reptiles and the first known specimens of a Cretaceous bird with teeth called *Hesperornis*. In 1872, his crew went back to the Smoky Hill River region and found more important fossils, including a much more complete *Hesperornis* with a good skull. Marsh continued to work on the specimens collected from Kansas, and he even bribed Cope's favorite collector, George Hazelius Sternberg, to sell some of his latest finds to Yale. This angered Cope and became part of their feud, which eventually exploded into all-out "bone wars." Their warfare escalated even further in 1870, when both Marsh and Cope's mentor Joseph Leidy reexamined Cope's bones and realized that Cope had put the skull of *Elasmosaurus* at the end of its short tail rather than at the end of its long neck. Their feud escalated through the 1870s and 1880s, and by the early 1890s they had destroyed each other's careers. But other collectors followed in their footsteps, and the chalk beds of Kansas were soon yielding spectacular marine reptiles and giant fish fossils to a number of collectors from different museums.

In addition to the beds of western Kansas, Cretaceous chalks are common all over North America, from the Austin Chalk in central Texas to the Selma Chalk of Alabama and Mississippi. But its greatest extent is in the Great Plains of North America, which were drowned by shallow seas throughout most of the Cretaceous. This formed a permanent marine barrier, the Western Interior Seaway (figure 17.2), that lasted

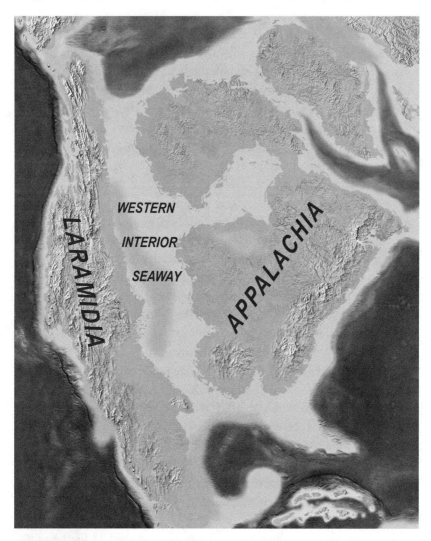

Figure 17.2 ▲

During the Cretaceous, the high sea levels drowned parts of most continents and formed the Western Interior Seaway, cutting North America in half from the Gulf of Mexico to the Arctic Ocean. (Courtesy of Wikimedia Commons)

more than 110 million years, from the Middle Jurassic (about 180 Ma) until just before the end of the Cretaceous (about 70 Ma). It isolated the land and the dinosaur faunas, which resulted in the dinosaurs of Montana having more in common with those from Mongolia than those from New Jersey.

The Cretaceous chalks and all the shales and other marine rocks of the Cretaceous are a product of a much bigger phenomenon: the global greenhouse world of the Late Cretaceous. During the latter half of the Age of Dinosaurs, the climate was so warm that dinosaurs and crocodilians roamed above the Arctic and Antarctic circles. There was virtually no ice or snow anywhere at that time because the atmospheric carbon dioxide levels were over 1,000 ppm (today it is over 415 ppm and climbing). The melting of all those ice caps resulted in extremely high sea level, and shallow seas covered many of the continents. The shallow chalk-rich seas of the Cretaceous drowned most of Europe as the distribution of chalk not only in Britain but in Belgium and France attests.

These shallow inland seas in what are now the western Great Plains of Kansas, Oklahoma, Texas, Nebraska, and the Dakotas were filled with an amazing variety of marine life. Coccolithophorid algae were at the base of the planktonic food chain, and the rapid evolution of zooplankton fed on these tiny algae, especially the first foraminiferans to live in the plankton. Feeding on this rich planktonic food supply was an incredible diversity of marine invertebrates on the seafloor. These included a wide variety of marine snails, many still represented in today's oceans. Even more impressive was the extreme evolution of the bivalves. In addition to the familiar types of clams and scallops, there was a wide range of odd-shaped oysters. Taking things to an extreme was the spiral oyster called *Exogyra*, whose shell was built like the coiled cup with a flat lid but spiraled up an axis like a snail's shell (figure 17.3A). Even more extreme were the huge flat clams known as inoceramids; shaped like a dinner plate, they were up to 1.7 meters (5 feet) across and rested flat on the sea bottom (figure 17.3B). Many of them were found with numerous fish fossils inside them, suggesting that they harbored a lot of symbiotic fish in their mantle. But the strangest of all were the weird oysters known as rudistids; these bivalves had one shell shaped like an ice-cream cone, and the other shell was a cup-shaped lid (figure 17.3C). They were found in densely packed colonies in the tropical seas over the

▲ ▶

Typical invertebrate fossils of the Western Interior Seaway. (*A*) The spirally coiled oyster *Exogyra*. (*B*) The gigantic flat "dinner plate" clam *Inoceramus*, which reached up to 5 feet in diameter. (*C*) The bizarre cone-shaped oysters known as rudistids. (*D*) A selection of ammonites from the Cretaceous. The two long tusk-shaped shells on top are *Baculites*. The weirdly coiled specimen in the lower right is *Didymoceras*. (Photos by author)

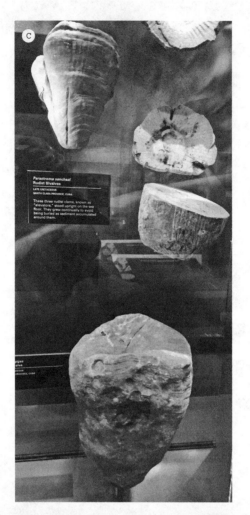

Parastroma sanchezi
Rudist Bivalves

LATE CRETACEOUS
SANTA CLARA PROVINCE, CUBA

These three rudist clams, known as "elevators," stood upright on the sea floor. They grew continually to avoid being buried as sediment accumulated around them.

Figure 17.3 ▲
(*continued*)

entire Cretaceous and formed gigantic reefs that apparently took the role of coral reefs (which were unimportant during most of the Mesozoic). Some were even weirder in shape, forming twisted shells like ram's horns, and many other peculiar forms.

If the clams were strange looking, the ammonites were even more so. The Jurassic radiation of ammonitic ammonoids continued into the Cretaceous as they swam the extensive seas of the greenhouse world. But the oddest of all were the ammonites that abandoned the simple spiral pattern in a single plane that had been standard for coiled cephalopods since the Devonian. These were the heteromorph ammonites ("heteromorph" means "different shape" in Greek). They varied from slightly uncoiled forms, such as the scaphitids, to a variety of ammonites with shells coiled like a paper clip (*Diplomoceras* and *Hamites*), a question mark (*Didymoceras*), or a corkscrew spiral like a snail *Turrilites*, to the truly weird *Nipponites*, whose coils formed a tight knot (figure 17.3D). With these odd arrangements of coils, the squid-like creature inside could not have used jet propulsion with its water nozzle, as a nautilus does; instead it used jet propulsion to spin erratically to avoid predators. Most of the time they must have floated slowly in the open water or just above the sea bottom, grabbing prey that came within reach of their tentacles or filter-feeding the plankton. The most common heteromorph of all was the straight-shelled group known as baculitids, which had a long conical shell that was completely straight except for a tiny coil at the end. Baculitid fossils are abundantly preserved in the deposits of the Western Interior Cretaceous Seaway, and they evolved so rapidly that they are one of the principal index fossils of the Cretaceous. Their mode of life is curious. They had no dense calcite deposits inside the shell to act as a counterweight and hold the shell horizontal, like Ordovician nautiloids did. Instead, the shell must have floated with the point upward and the squid-like creature inside dangled below, using its tentacles to move along as it floated above the seafloor to filter-feed on plankton.

Sailing above these bottom dwellers and ammonites were the marine vertebrates of the Cretaceous seaways. These include some enormous bony fish, such as the gigantic fossil *Xiphactinus*, which is known from many complete skeletons in the chalk beds of western Kansas and reached up to 6 meters (20 feet) long (figures 17.4A and 17.4B). There were huge sharks up to 10 meters (33 feet) long as well. But the largest sea creatures were among

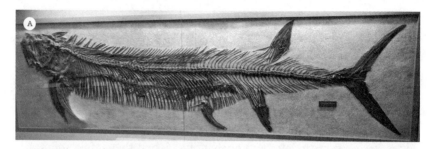

Figure 17.4 ▲

(A) A complete specimen of the enormous fish *Xiphactinus*, found with its last meal—another complete fish—in its stomach. (B) George Sternberg and crew with a *Xiphactinus* fossil during excavation, giving a sense of its size. ([A] Photo by author; [B] courtesy of Wikimedia Commons)

the great diversification of marine reptiles (figure 17.5). The immense sea turtle *Archelon* was more than 4 meters (13 feet) long—the largest sea turtle that ever lived. The dolphin-like ichthyosaurs were declining in the Cretaceous, and they vanished about 95 Ma in the Early Cretaceous. Two other groups took over the role as dominant marine predators. The paddling plesiosaurs were in their heyday. These included the forms with extremely long necks and tiny heads (Cope's elasmosaurs) that reached up to 10.3 meters (35 feet) in length. But the biggest of all were the short-necked, long-snouted plesiosaurs known as pliosaurs. The largest of these was the immense *Kronosaurus* from the Cretaceous of Australia, which was almost 11 meters (37 feet) long.

Figure 17.5 ▲

The Western Interior Seaway waters were full of marine reptiles, such as huge mosasaurs, long-necked plesiosaurs, and gigantic sea turtles. (Courtesy of Wikimedia Commons)

Finally, a new group of marine reptiles took to the Cretaceous seaways and evolved rapidly throughout the entire Cretaceous. These were the mosasaurs, a group of monitor lizards related to Komodo dragons that became so specialized for swimming that their feet were modified into paddles, their tails had a tall narrow fin for propulsion, and their snouts became elongated for catching fast-swimming prey such as fish and squid. They reached about 17 meters (56 feet) in length, and most were much smaller, but they were fast, agile, fearsome marine predators that were able to out-swim and out-maneuver just about any other marine animal in the oceans of Kansas. Most had long snouts with sharp conical teeth for catching fast prey, such as fish, squid, and ammonites. Indeed, some ammonite specimens have a V-shaped row of conical bite marks that match the mosasaur tooth pattern. One mosasaur was found with a diving bird, several fish, a shark, and a smaller mosasaur in its stomach, so they were opportunistic feeders. A few of them had globular crushing teeth for eating mollusks, so mosasaur feeding habits were adaptable. They diversified into dozens of genera and species, and during the last 20 million years of the Cretaceous, the plesiosaurs vanished, leaving the role of dominant marine predator to the mosasaurs.

THE WHITE CLIFFS OF DOVER

Chalk is typical of deposits found all over Europe as well, especially in southeastern England, and across the English Channel in France and Belgium. Famous as the bedrock that makes up the White Cliffs of Dover, the chalk cliffs are often a symbol of Britain as an island nation (figure 17.6A). When most people hear the word "chalk," they think of the sticks of white powdery material used to write on blackboards for generations. Even though real chalk was once used for that purpose, most modern blackboard "chalk" is not made of real chalk at all; it is powdered gypsum compressed into sticks.

Real chalk is soft, white, porous limestone composed of the mineral calcite (calcium carbonate or $CaCO_3$). It forms under deep marine conditions from the gradual accumulation of minute calcite shells (coccoliths) shed from micro-algae known as coccolithophores (figure 17.6B). Flint (a type of chert typical of chalk) is very common as bands parallel to the bedding or as nodules embedded in chalk (figure 17.6C). The silica in the

Figure 17.6 ▲ ▶

The Chalk beds of southeast England were the product of shallow warm seas during the greenhouse world of the Cretaceous. (A) The famous White Cliffs of Dover. (B) Chalk is made of the shells of trillions of tiny planktonic algae known as coccolithophores. These specimens are only a few tens of microns across. (C) The Chalk is full of Cretaceous marine invertebrate fossils and also nodules of chert that grew after the soft sediment turned to stone. ([A–B] Courtesy of Wikimedia Commons; [C] photo by author)

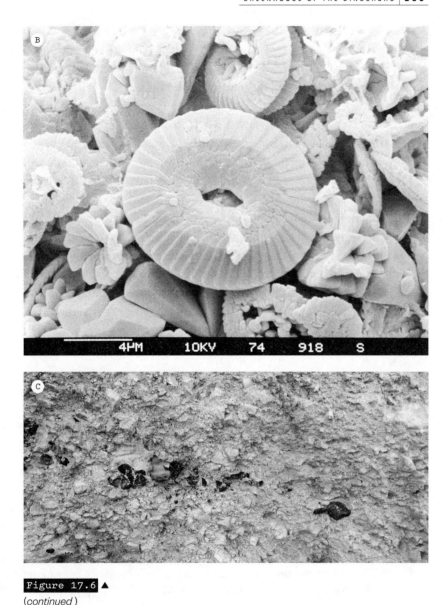

Figure 17.6 ▲
(*continued*)

flint is probably derived from sponge spicules or other siliceous organisms as water is expelled upward during compaction. Flint is often deposited around larger fossils, and they may be silicified (i.e., the calcite is replaced molecule by molecule by silica).

The Chalk Cliffs are rich in fossils that can be collected on the beaches of England in many places. It is famous for its heart urchins, sea biscuits, and other sea urchins, which have been studied by many paleontologists. It also yields a variety of clams and oysters, especially the weirdly coiled oysters *Gryphaea* and *Exogyra* (figure 17.3A), which have been the subject of many famous studies by paleontologists as well. It also yields a few poorly preserved ammonites, as well as shark teeth and various fish fossils.

The chalk outcrops are not restricted to southeastern England. The belt of Chalk extends to the Alabaster Coast of Normandy and to Cape Blanc Nez in France across the Dover Strait. There are chalk beds beneath the Champagne region of France, which produces the soils planted in Champagne grapes, and the erosion of that soft soluble chalk also creates natural caverns for wine storage. The Chalk extends further east through Belgium and countries to the northeast as well, including Jasmund National Park in Germany and Mons Klint in Denmark. The Chalk was so distinctive and widespread across so much of Europe that in 1822 Jean D'Omalius D'Halloy named the Cretaceous Period after it (from the Latin word *creta*, which means "chalk").

CRETACEOUS GREENHOUSE

What could have caused such an extensive flooding of the continents in the Cretaceous? Several things were happening at this time, all of which contributed to the rising sea levels and greenhouse atmosphere, but it is difficult to know which effects were most important.

BREAKUP OF PANGEA

During the 80 million years of the Cretaceous, nearly all the remaining pieces of Pangea broke up and separated into their modern continents (figure 17.7). The North Atlantic began to open in the Late Jurassic, but most of its widening occurred in the Cretaceous. However, most the opening of the South Atlantic occurred in the Cretaceous, as did the separation of Africa from Antarctica and the beginning of the split between Australia from Antarctica. Most impressive is how India was ripped away of its former position in Gondwana and raced across the Indian Ocean; by the end of the Cretaceous, India was close to colliding with Asia.

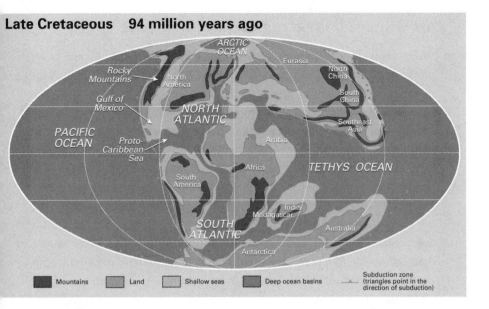

Late Cretaceous 94 million years ago

Mountains | Land | Shallow seas | Deep ocean basins | Subduction zone (triangles point in the direction of subduction)

Figure 17.7 ▲

In the Cretaceous, most of the Pangaea continents broke up, so the North Atlantic grew wider, the South Atlantic opened for the first time, Africa split from Antarctica, and India split from Gondwana to race across the Indian Ocean. At the end of the Cretaceous, Antarctica and Australia began to pull apart as well. (Courtesy of Wikimedia Commons)

This breakup of all the Pangaea continents into their modern configurations had several important effects. The most important among them is that, after rifting, the edges of the continents are wider and more stretched out, so their average continental elevations are lower, allowing the seas to transgress across them. The seas leave thick sedimentary wedge deposits on the edges of the rifting continents. For example, the bulk of the enormous passive margin wedge more than 6,100 meters (20,000 feet) thick on the Atlantic Coast and the Gulf Coast of North America was formed during the 80 million years of the Cretaceous. These sinking continental edges were continuously drowned by shallow seas, producing enormous piles of marine shales and limestones that are today found all over the Atlantic and Gulf Coast Plains. Much of the drowning of the continents was a product of the sinking of passive margin wedges.

RAPID SEAFLOOR SPREADING AND RIDGE VOLUME

When seafloor spreading is very rapid, it produces much thicker, taller profiles of rock on the midocean ridge complex (see chapter 10). This probably happened when the Pannotia supercontinent broke up in the Cambrian, and the record rates of seafloor spreading at that time would have produced large midocean ridges that would have displaced much of the ocean's water onto the land, causing the global transgression (figure 10.4). The same would have been true in the Cretaceous, when all the Pangea continents were pulling apart at a record rate of seafloor spreading. The increased volume of all those rapidly spreading ridges would have been enormous, and all that extra rock volume made the ocean shallower and pushed the water up onto the land. Numerous studies have shown extraordinarily high rates of spreading beginning in the Early Cretaceous (Aptian Stage, about 126 Ma) and continuing at lesser rates until nearly the end of the Cretaceous.

GREENHOUSE GASES

High rates of seafloor spreading are important in terms of the depth of the ocean, and a lot more greenhouse gases are released from the mantle during all these intense submarine eruptions. This would be a major reason atmospheric carbon dioxide levels may have reached values as high as 1,000 ppm (only 415 ppm today, even with all the greenhouse gases we are introducing due to global warming). Another contributor was the indirect effect of oceans covering so much of the land, with limited areas of mountain ranges uplifting at the time. Rapid rates of mountain uplift cause rapid weathering, which is a major absorber of carbon dioxide. When no large areas of mountains are rising around the world, the limited land area also restricts the amount of weathering in soils, regulating how fast the soil can absorb carbon dioxide. If any significant polar ice caps had been left in the Jurassic or the Early Cretaceous, this extreme warming would have melted them all away, further contributing to the amount of water in the oceans. Those dinosaur movies that show glacial ice in the mountains in the background are wrong—there was probably little or no ice anywhere in most of the world during the Cretaceous.

MANTLE SUPERPLUME ERUPTIONS

Recent research has shown that the Cretaceous was a time of extraordinarily large eruptions of huge hot spots, or mantle superplumes, coming up from the lower mantle and erupting beneath the oceans. These rocks were formed by gigantic mantle eruptions and were first discovered in the 1950s and 1960s, when oceanographers mapped huge submarine plateaus, especially in the Pacific Ocean. When scientists drilled and analyzed and dated the rocks, they proved to be the result of huge mantle lava eruptions that spilled enormous volumes of magma into the Cretaceous oceans—and pumped huge volumes of greenhouse gases from the mantle at the same time. The biggest of these is the Ontong-Java Plateau in the southwestern Pacific, which was formed about 126 Ma by the eruption of 1.5 million cubic kilometers (360,000 cubic miles) of lava across the ocean floor that lasted less than a million years. This timing coincides with the huge spike in seafloor volcanism in the Aptian Stage of the late Early Cretaceous and the rapid rise in sea level around the globe at that time. The Hess Plateaus and the Shatsky Rise also erupted during the Cretaceous, and their huge volumes of lava undoubtedly contributed to the Cretaceous greenhouse gases as well.

Many factors may have contributed to the extraordinarily intense greenhouse world of the Cretaceous. Which was the most important is difficult to determine, but they were all working at the same time and all contributed to greenhouse warming.

DINOSAURS OF DARKNESS

The warming of the "Greenhouse of the Dinosaurs" was so extreme that ice-free polar regions were relatively balmy and mild, even if they did experience almost six months of darkness during the winter. In both the North Slope of Alaska (above the Cretaceous Arctic Circle) and the Antarctic and southern Australia (both near the South Pole then), we find evidence of temperate and even subtropical plants such as breadfruit along with tree ferns and conifers and other plants adapted to mild climates. This tells us that temperatures were about 10°C (18°F) warmer at the poles than they are now.

The first evidence for warm poles came from Alaska in the 1960s. The Arctic was a region with cool but mild climates and abundant forests and dinosaurs during the Cretaceous. As early as 1961, geologists were finding bones of duckbill dinosaurs in places like the Colville River of Alaska. When major excavations were undertaken by the University of Alaska, thousands of bones were found representing 12 species of dinosaurs. Most abundant were the common duckbills from Alberta, similar to *Edmontosaurus*, which have now been renamed *Ugrunaaluk*. There are also Albertan duckbills such as *Kritosaurus* and *Lambeosaurus*. There were horned dinosaurs including *Pachyrhinosaurus*, with the flat bony boss on its snout (star of the computer-generated movie *Walking with Dinosaurs*), and the three-horned *Anchiceratops*. Much less common were pachycephalosaurs, the dinosaurs with a thick boss of bone over their tiny brains. Other herbivores included *Thescelosaurus*, an ornithopod dinosaur. Their predators include the Albertan tyrannosaur *Albertosaurus*, plus typical Alberta and Montana theropods such as *Troodon*, *Dromaeosaurus*, and *Saurornitholestes*.

These dinosaurs lived in a world with cool temperatures suitable for bald cypress, cycads, ginkgoes, ferns, horsetails, and conifers. Crocodilians and pond turtles were also common. The temperatures in Cretaceous Alaska (based on plants and on geochemistry) averaged about 5°C (41°F) with summer means in the 10°C (50°F) and winter temperatures around freezing. The dinosaurs lived within the Arctic Circle, so they experienced many months of near darkness. It is possible that many of the dinosaurs (especially the duckbills) migrated down to warm Alberta to escape the cold dark winters.

In the Antarctic, the Early Jurassic dinosaurs including *Cryolophosaurus* (a theropod with a peculiar crest that looks like the hair comb that señoritas used to wear) and the theropod *Glacialisaurus* have been discovered, but the best evidence of the Cretaceous warming comes not from Antarctica but from Australia, which was also polar at that time. These Cretaceous creatures of the darkness included a small ornithopod named *Leaellynasaura amicagraphica* and *Atlascopcosaurus loadsi*, as well as a small predatory coelurosaur *Timimus hermani*, and other small theropods. The meter-long (39 inch) *Leaellynasaura* is the most complete fossil of the bunch. It had unusually large eye sockets and large optic lobes in the brain, suggesting that it had huge eyes for seeing in the many months of darkness or semi-darkness of the Antarctic winter. Its bones show no growth lines, so it did not hibernate in the winter. Even though polar Australia was dark or dimly lit during the winter, this little herbivore must have found enough food to survive. The little coelurosaur *Timimus* did show strong growth lines in its bone, suggesting it did hibernate during the polar winters when conditions were dark and prey was scarce. In addition to dinosaurs, fish, turtles, pterosaurs, birds, and amphibians were also recovered, indicating that an entire ecosystem was adapted to months of darkness and cool temperatures.

This was the greenhouse world of the Age of Dinosaurs. The planet was warm to the poles with no ice, high sea levels, and lots of different animals living in conifer and fern cooled temperate forests within a few miles of the North Pole.

NOTE

1. Charles H. Sternberg, *Life of a Fossil Hunter* (Bloomington: Indiana University Press, 1990).

FOR FURTHER READING

Everhart, Michael J. *Oceans of Kansas: A Natural History of the Western Interior Sea.* Bloomington: Indiana University Press, 2005.

Huxley, Thomas H. *On a Piece of Chalk.* New York: Scribner, 1967.

Marriott, Katherine, Donald R. Prothero, and Alexander Bartholomew. *The Evolution of Ammonoids.* Boca Raton, FL: Taylor & Francis, 2023.

Retallack, Gregory J. "Greenhouse Crises of the Past 300 Million Years." *Geological Society of America Bulletin* 121 (2009): 1441–1455.

Retallack, Gregory J., and Giselle D. Conde. "Deep Time Perspective on Rising Atmospheric CO_2." *Global and Planetary Change* 189 (2020): 103177.

Rich, Thomas H., and Patricia Vickers-Rich. *Dinosaurs of Darkness: In Search of the Lost Polar World.* Bloomington: Indiana University Press, 2000.

Skelton, Peter W., Robert A. Spicer, Simon P. Kelley, and Iain Gilmour. *The Cretaceous World*, ed. Peter W. Skelton. Cambridge: Cambridge University Press, 2003.

Smith, Andrew B., and David J. Batten, eds. *The Palaeontological Association Field Guide to Fossils, Fossils of the Chalk*, 2nd ed. London: Wiley-Blackwell, 2002.

THE DEATH OF THE DINOSAURS

The Age of Reptiles ended because it had gone on long enough and it was all a mistake in the first place. A better day was dawning at the close of the Mesozoic Era. There were some little warm-blooded animals around which had been stealing and eating the eggs of the dinosaurs, and they were gradually learning to steal other things, too. Civilization was just around the corner.

—WILL CUPPY, *HOW TO BECOME EXTINCT* (1941)

DINOMANIA

Dinosaurs are incredibly popular in our culture these days, especially since the wave of Dinomania triggered by the *Jurassic Park* franchise has turned almost every kid into a dinosaur lover. (I was crazy about dinosaurs when I was a kid, but when I grew up in the 1950s and early 1960s, that was very rare—I was the only kid I knew through all of my school years who loved dinosaurs.) Today, most kids seem to go through a "dino phase" from about five years old until puberty hits; then they shift their attention to sports and pop stars and video games and the opposite sex. Dino merchandise is a huge business with million dollar profits—*none* of which supports the research and discovery of dinosaur fossils that fueled it in the first place. Dinosaurs are everywhere in the media. If you put a dinosaur on the cover or in the title of something, it gets more hits or sells more copies than most other similar media without the dinosaur "hook." If you publish an article or post a story about the extinction of the dinosaurs, it's guaranteed to get attention—especially if it's simple and easy to summarize in one sentence.

Naturally, the bulk of the attention goes to simplistic explanations of the extinction of the dinosaurs that most of the public has heard, namely, "a rock from space came down and killed the dinosaurs." This is sexy and flashy enough that it made the cover of *Time* magazine in the early 1980s, and many magazines and internet articles continue to promote this simplistic story. After 44 years of promoting the asteroid impact model for the death of the dinosaurs, it's pretty well ingrained in the public consciousness. But is that the entire story?

DINO-CENTRISM

Let me clarify one thing right away: the end-Cretaceous extinction was a global event. It affected the base of the food chain from the phytoplankton, through most of the marine predatory groups including ammonites and marine reptiles, and land floras and some of the land animals, including the nonbird dinosaurs. Thus any explanation that is too "dinocentric" is irrelevant because the dinosaurs are only one part of a much larger picture.

Before 1980, a pointless and inconclusive debate about what killed the dinosaurs at the end of the Cretaceous had been going on for decades. Some said the climate got too hot; others said it was too cold. Some blamed it on the evolution of flowering plants—except that occurred in the Early Cretaceous, 80 million years earlier, and actually might have helped spur the evolution of herbivorous dinosaurs such as the duckbills and horned dinosaurs. Some suggested that mammals ate their eggs—but both mammals and dinosaurs originated together in the Late Triassic, about 200 Ma, and they coexisted for 135 million years without mammals suddenly developing a taste for dinosaur eggs. There were even wilder and less scientifically testable ideas—epidemics and diseases, widespread depression and psychological problems, and even the notion touted in the tabloids that aliens kidnapped them or killed them off! As the paleontologist Glenn Jepsen wrote in 1964:

Why Become Extinct? Authors with varying competence have suggested that dinosaurs disappeared because the climate deteriorated (became suddenly or slowly too hot or cold or dry or wet), or that the diet did them in (with too much food or not enough of such substances as fern oil; from poisons in water or plants or ingested minerals; by bankruptcy of calcium or other necessary elements). Other writers have put the blame on disease, parasites, wars, ana-tomical or metabolic disorders (slipped vertebral discs, malfunction or imbal-ance of hormone and endocrine systems, dwindling brain and consequent stupidity, heat sterilization, effects of being warm-blooded in the Mesozoic world), racial old age, evolutionary drift into senescent overspecialization, changes in the pressure or composition of the atmosphere, poison gases, volcanic dust, excessive oxygen from plants, meteorites, comets, gene pool drainage by little mammalian egg-eaters, overkill capacity by predators, fluc-tuation of gravitational constants, development of psychotic suicidal factors, entropy, cosmic radiation, shift of Earth's rotational poles, floods, continen-tal drift, extraction of the moon from the Pacific Basin, draining of swamp and lake environments, sunspots, God's will, mountain building, raids by lit-tle green hunters in flying saucers, lack of standing room in Noah's Ark, and *palaeoweltschmerz*.[1]

Without any independent evidence to test these ideas, they are just speculation and guesswork, not science. Moreover, they all focused only on dinosaurs and ignored the much more important picture: the end-Cre-taceous extinction was a global event that affected both the marine realm as well as plants on land. Any mass extinction that was this widespread and pervasive needed a broader explanation than one specific to the dinosaurs. In fact, if the extinction was this far-reaching, and killed off so many other organisms at every level in the food chain, the dinosaurs are just a side effect, an epiphenomenon, not the most important piece of the puzzle.

But the public only cares about dinosaurs and has no interest in any of the other organisms that existed and died out at the same time, and they don't realize that these other forms of life are much more important to our understanding of the big picture. Raup and Sepkoski argued that this extinction killed off about 75 percent of the marine species, so it was the third or fourth largest mass extinction in Earth's history. More powerful extinction events include the end-Permian extinction (see chapter 15) and the late Ordovician extinction (see chapter 11).

ACCIDENTAL DISCOVERY IN THE APENNINES

Explaining the extinction of the dinosaurs was largely in the realm of unscientific, untestable speculation in the 1850s and 1860s when dinosaurs were first described and their extinction was accepted. All this changed at the end of the 1970s when the geologist Walter Alvarez was working in the Apennine Mountains of Italy. (I was a graduate student at the Lamont-Doherty Geological Observatory at Columbia University when I first met Walter; he was just a humble post-doc then.) Walter Alvarez, William Lowrie, and their colleagues were studying a long sequence of Upper Cretaceous and lower Paleocene limestones, which recorded the last few million years of the Mesozoic and the beginning of the Cenozoic in a roadcut and quarry near Gubbio, Italy. In the middle of this long sequence of marine limestones formed by planktonic microfossils, they noticed a distinct dark clay layer right at the boundary between the uppermost Cretaceous and lowest Paleocene rocks (figure 18.1).

In 1978, Alvarez brought home samples of this layer to the University of California, Berkeley, where he was on the geology faculty. His father, the Nobel Prize-winning physicist Luis Alvarez (who had worked on the Manhattan Project to develop the first atomic bombs and watched them dropped over Nagasaki) was on the physics faculty. Walter was looking for a way to determine how long it had taken for this boundary clay layer to be deposited. His father suggested that they measure the amount of cosmic dust that had rained down on the ocean floor. If there was relatively little dust, then the deposition might have been quick; if there was a lot of dust, then it had been a long slow accumulation of clays and cosmic particles. The best proxy they knew to estimate the amount of extraterrestrial dust was the platinum-group metal iridium, which is extremely rare in earth rocks but slightly more abundant in space rocks and in the Earth's interior. They gave their samples to the geochemists Frank Asaro and Helen Michel at the Lawrence Berkeley National Laboratory, which has its own nuclear reactor, so they could look for the trace elements with neutron activation analysis (NAA).

To everyone's surprise, when Asaro and Michel got their results, the amount of iridium was hundreds of times more than what they might expect, even if there had been thousands of years of slow clay accumulation on the ancient seafloor. Putting their minds to the problem, they tried

Figure 18.1 ▲

The geologist Walter Alvarez (right) and his father, the physicist Luis Alvarez (left), at the Cretaceous-Paleogene boundary layer in Gubbio, Italy. Walter's fingers are in the boundary clay layer, and the Cretaceous limestones are in the lower right and Paleocene limestones above them on the left. (Courtesy of Wikimedia Commons)

to find an explanation for the high iridium levels. They finally hit upon the idea that the Cretaceous extinctions occurred because a 10 kilometer (6 mile) diameter asteroid had hit Earth, creating global wildfires and gigantic amounts of dust in the stratosphere. This dust would block out most sunlight for weeks or months, creating a cold dark world where most plants would die, especially the phytoplankton that are the foundation of the oceanic food chain. (The idea of "nuclear winter" was developed in the early 1980s as a by-product of the computer modeling of the impact-induced end-Cretaceous "winter.") After refining the model and getting it peer reviewed, the Alvarez, Alvarez, Asaro, and Michel paper came out in the journal *Science* in 1980. It has become the most influential paper in the geosciences for the past 44 years, with thousands of scientific papers and numerous books citing it.

The key thing to note here is that the discovery was made by accident while they were looking for something entirely different. Lots of people scorn "basic" or exploratory research in which researchers cannot predict what discoveries they may find. The National Science Foundation and other science funding agencies are very conservative and only fund projects that are "sure bets" and whose results are already mostly known by the time the grant proposal is reviewed. But most of the best science occurs by accident: by scientists exploring and asking basic questions and not knowing what they might find. This discovery by fortunate accident is often called "serendipity," after the old Persian tale of the "Three Princes of Serendip" who make discoveries unexpectedly.

There are hundreds of cases of accidental discoveries in science, especially in chemistry. Alfred Nobel accidentally mixed nitroglycerin and collodium ("gun cotton") and discovered gelignite, the key ingredient for his development of TNT. Hans Von Pechmann accidentally discovered polyethylene in 1898. Silly Putty, Teflon, Superglue, Scotchgard, and Rayon were all accidents, as was the discovery of the elements helium and iodine. Among drugs, penicillin, laughing gas, Minoxidil for hair loss, the "Pill," and LSD were all discovered by accident. Viagra was originally developed to treat blood pressure—not impotence.

Many of the great discoveries in physics and astronomy were unexpected, including finding the planet Uranus, infrared radiation, superconductivity, electromagnetism, and X-rays. Among practical inventions, inkjet printers, corn flakes, safety glass, Corningware, and the vulcanization of rubber were all accidents. Percy Spencer of Raytheon was looking for another use for the surplus magnetrons after World War II ended and accidently discovered they could be used as microwave ovens when one of them melted the candy bar in his lab coat pocket. In 1964 Arno Penzias and Robert W. Wilson were trying to get the noise out of their newly developed microwave antenna when they realized they had discovered the long predicted cosmic background radiation from the Big Bang (see chapter 2). These and many other examples are good reasons for scientists to conduct "pure research." Sadly, many shortsighted and misguided people scorn "pure research" as worthless navel-gazing and demand that every scientist show a practical or useful reason for the research or it will not be funded. This is a sure path to scientific stagnation.

Most of the greatest discoveries in science were not anticipated or planned—they happened by accident. More often than not, scientists who find a crucial new piece of evidence were looking for something else. In the case of science, serendipity works most often when the researcher is prepared to see the implications of some new, unexpected development. Louis Pasteur put it this way, "In the field of observation, chance favors only the prepared mind." As the famous scientist and writer Isaac Asimov said, "The most exciting phrase to hear in science, the one that heralds new discoveries, is not 'Eureka!' but 'That's funny . . .'"

THE IMPACT STORM OF CONTROVERSY

> With most subjects there is a silly season, usually of unpredictable duration and of an intensity correlated with the status of the acceptance of the new idea, [including] proposal of ideas even more far-out than the original one.
>
> **—KEITH STEWART THOMSON**

When the Alvarez, Alvarez, Asaro, and Michel paper came out in 1980, it had a huge impact on the geosciences. Hundreds of scientists began looking for additional evidence to support or refute it, and others suggested alternative explanations for the abundance of iridium. But soon the same iridium abundance was found in other deep-marine sedimentary rocks from the boundary between the Cretaceous and the Paleocene (the KPg boundary).[2] Eventually this abundance was found on land sections as well, ruling out the criticism that it was just a weird geochemical event in the oceans.

Throughout the 1980s, hundreds of researchers were furiously publishing research on every aspect of the KPg extinctions. I vividly remember those years as a grad student and then early in my professional career when every annual meeting of the Geological Society of America featured several different sessions focused on the KPg extinction and related geological phenomena.

Soon there were two distinct camps: the impactors, who argued that the asteroid impact was the only mechanism required to explain the KPg extinctions; and their opposition, who pointed to the gigantic Deccan volcanic eruptions in what is now India and Pakistan (figure 18.2). The Deccan lavas were the second largest eruption in all of Earth's history, exceeded only by the Siberian eruptions that ended the Permian (see chapter 15). The lava flows are more than 2,000 meters (6,600 feet) thick, with a volume of about 1,000,000 cubic kilometers (200,000 cubic miles); and the lava may have covered 1,500,000 square kilometers (600,000 square miles). The eruption may have brought up huge amounts of sulfur dioxide

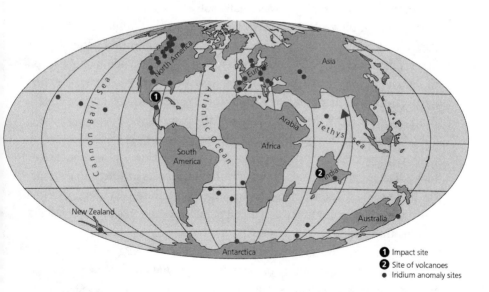

1 Impact site
2 Site of volcanoes
• Iridium anomaly sites

Figure 18.2 ▲

Map showing the location of the Chicxulub crater (1) and the Deccan traps (2), along with the numerous localities where the KPg iridium anomaly has been found. (Drawn out of copyright)

and released dust and ash into the atmosphere, which would have an effect similar to that of the "asteroid nuclear winter" model, which darkened the skies for months. Global temperature dropped 2°C (3.6°F) just 25,000 years before the impact happened. Even more interesting is the fact that mantle-derived volcanoes like Kilauea also contain lots of iridium. (Iridium is extremely rare in the crust but slightly more abundant in the mantle and in space.)

As discovery after discovery was reported in the 1980s, the pendulum swung back and forth. The discovery of impact spherules (blobs of crustal matter from the crater), shocked quartz (only known from impacts and from nuclear explosions), and giant tsunami deposits around the Caribbean and Gulf of Mexico pointed to the impact being real. Yet the improved dating of the Deccan lavas showed that they were a gigantic climate event that happened just 25,000 years before the KPg boundary, with the strongest pulse apparently synchronous with the impact.

The biggest stumbling block was the absence of the "smoking gun"—the KPg impact crater. Several candidates, including the Manson Crater in Iowa, were proposed but rejected later when new dates showed they were the wrong age. The problem was finally solved—again by accident. In the late 1970s, an oil geologist named Glen Penfield had found geophysical data that showed a huge filled-in crater-like feature buried beneath the jungles of the northern Yucatan Peninsula in Mexico. When he published this in an oil company report in 1978, no one was interested in the asteroid impact model (at that time, the Alvarez team had just received the samples from Gubbio). More than a decade later the planetary geologist Alan Hildebrand realized that all the tsunami deposits and impact droplets were scattered around the Caribbean and the Gulf of Mexico, and he began to look for the crater there—he found Penfield's reports in 1990. Since then the buried crater, known by its Mayan name of Chicxulub (CHIK-zoo-loob), has been drilled and studied and dated precisely at the KPg boundary, confirming that it was the impact site.

Complicating and amplifying the furor over the KPg extinction was excitement about explaining other mass extinctions by impact events as well. In the 1980s, the bandwagon for impact explanations had spilled over to many other events. In 1982, David Raup and Jack Sepkoski published a paper about the Big Five mass extinctions (see chapter 11), and in 1984 they caused a huge controversy by claiming that pulses of extinctions occurred every 26 million years. (This idea has since been debunked.) The impact advocates argued that extraterrestrial indicators (such as iridium, shocked quartz, and impact droplets) occur at many of the major mass extinctions, and by extrapolation we can infer that many mass extinctions are impact-induced. I attended the International Geologic Congress in Washington, D.C. in 1989, and Digby McLaren (a Canadian paleontologist who had worked on the Devonian extinctions) told a stunned audience that all mass extinctions were caused by impacts—*whether or not there was evidence of impact in the fossil record!* In 1991 Raup wrote that all extinctions (even normal background extinctions) might be caused by impacts. With statements such as these, why bother gathering data at all? Extinctions occurred and impacts occurred—therefore impacts caused all the extinctions.

WHAT DOES THE FOSSIL RECORD SAY?

The great tragedy of science—the slaying of a beautiful hypothesis by an ugly fact.

—THOMAS HENRY HUXLEY, *BIOGENESIS AND ABIOGENESIS* (1894)

For every problem, there is a solution that is simple, neat, and wrong.

—H. L. MENCKEN

All these scientific papers arguing about geochemical or geophysical or mineralogical evidence might be sufficient to establish that an impact had occurred and that the gigantic Deccan eruptions also had occurred, but the real questions remain:

What effect (if any) do each of these stressors have on the Cretaceous biota?
Which ones were more important?
Does the impact event explain the KPg extinctions, or is there more to the story?

There was a big impact at that time, but there was also the second-largest volcanic eruption in the last 600 million years. In addition, there was a huge drop in sea level as the Western Interior Seaway drained away from the continents, drastically reducing the shallow marine habitat for Cretaceous sea creatures. The impact model predicts that everything should vanish abruptly at the end of the Cretaceous, whereas the volcanism scenario and sea level events predict waves of extinction well before the extraterrestrial impact. The final arbiter of which mechanism is most important is found in the fossil record: Are the species all dying abruptly as if the impact were the only cause, or were a number of groups vanishing before the impact of the rock from space? To answer this question, we need to consider the fossil record for marine and terrestrial extinctions.

MARINE EXTINCTIONS

Let's start with the oceans. They are the most influential ecosystem on Earth. They control Earth's climate and produce most of the oxygen we breathe (figure 18.3). The details of nearly every marine group of fossils were comprehensively reviewed by a distinguished panel of British paleontologists led by Norman MacLeod of the Natural History Museum in London, and published in 1997. The food pyramid of the ocean starts with the phytoplankton, which are the tiniest form of life and are consumed directly or indirectly by all the larger animals up the food chain (see chapter 9). The major group of Late Cretaceous phytoplankton were the coccolithophores, which were responsible for the enormous volumes of Cretaceous chalk (see chapter 17). This group was severely hit as darkness fell over the oceans. However, other major groups of phytoplankton—such as the siliceous diatoms and silicoflagellates and the organic-walled dinoflagellates—show almost no extinctions. This is truly puzzling: If conditions were dark too long, the entire phytoplankton should have suffered more severely. The explanation might be that the oceans were also acidified by climate change, which would dissolve the calcareous coccolithophores but not the siliceous and organic-walled phytoplankton.

The next level up the food chain are the zooplankton, mostly amoeba-like creatures with internal shells. The dominant ones were the planktonic foraminifera, which evolve so rapidly that they are crucial to telling time in the Cretaceous. Their response to the KPg extinction is controversial: some specialists argue that a large number of species vanished right at the boundary, and others claim they dropped out in waves well before the impact event. In contrast, there is almost no extinction in any other group of zooplankton, especially the radiolarians, a group of planktonic amoebas with siliceous skeletons. Again, the acidification of the ocean seems to have been more important than the darkness to these organisms.

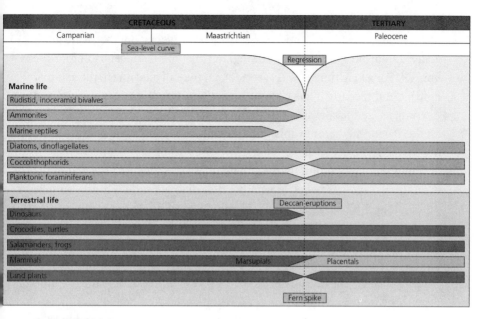

Diagram showing the extinction pattern at the end of the Cretaceous. (Redrawn from several sources)

Among multicellular invertebrates, reef corals lost about 60 percent of their species, but the deeper-water solitary corals that do not depend on light were not affected at all. However, the huge cone-shaped clams known as rudistids (figure 17.3C), which had displaced the tropical reefs during most of the Cretaceous, were already gone long before the impact occurred, and the same is true of the huge flat "dinner plate" clams known as inoceramids (figure 17.3B). So neither of these groups that had dominated the Cretaceous seafloor were alive to see the impact. They were affected by other changes that were already occurring in the Late Cretaceous. The effects on the rest of the clams and oysters were not as dramatic, and marine snails experienced only a modest extinction, with species that feed on organic material in the sediment being the least affected. Shallow-water tropical sea urchins and sea biscuits suffered about a 35 percent extinction, but deeper-water species were unaffected.

The most important marine invertebrate group were the ammonites, which once flourished in huge numbers and high diversity throughout most of the Jurassic and the Cretaceous. They were hardy creatures, surviving the Devonian, Permian, and Triassic extinction events. But they were gradually diminishing in diversity throughout the late Cretaceous before the impact. For a long time, it was asserted that they all died due to the asteroid impact. But recently it has become clear that a handful of genera survived the impact and swam about in the ocean for the first million years or so of the Paleocene before vanishing (possibly in response the acidity of the early Paleocene oceans).

Only about half of the sharks and bony fish died out near the end of the Cretaceous, and it is not clear whether they died out suddenly or gradually throughout the Late Cretaceous. Finally, the top predators of the oceans were the marine reptiles, particularly the gigantic marine monitor lizards known as mosasaurs and the paddling plesiosaurs. There is no evidence that either of these groups ever saw the impact event, but both appear to have been declining during the last part of the Cretaceous, and the dolphin-like ichthyosaurs were gone long before the Cretaceous ended.

Perhaps the best record of the KPg transition in the oceans is found in the thick sections of Cretaceous and Paleocene rocks on Seymour Island in the Antarctic Peninsula. They were first discovered, collected, and studied by William Zinsmeister of Purdue University in the 1980s. Restudied and reanalyzed by Sierra Petersen and others in 2015, these extraordinarily thick, complete, and detailed sections are rich in ammonites and show geochemical evidence of the volcanically triggered climate change causing major extinctions well before the iridium anomaly that marks the impact horizon. These rocks also contain a number of ammonites that survived into the Paleocene, around a million years after the impact. Zinsmeister and coauthors documented these data extensively in the 1980s and 1990s, but the pro-impact faction didn't want to hear about this complete section confirming that the impact wasn't that big of a deal as a cause of extinction.

In summary, the extinctions in the oceans were severe but were not as catastrophic as they are often portrayed. For every group of plankton severely affected (coccolithophores, planktonic foraminifera) there were a number of groups that showed little or no effect (diatoms, radiolarians, silicoflagellates, dinoflagellates). The effects on most of the marine invertebrates (clams, snails, sea urchins) were not nearly as serious as once thought, and some of the most important groups (rudistids, inoceramids, marine reptiles) were apparently gone long before the impact. Even the ammonites, which were once thought to have finally vanished with the asteroid impact, were in decline long before the end of the Cretaceous but survived well into the Paleocene. It appears that many of the victims died long before the impact due to volcanic-induced climate change and sea level drops. Organisms with calcareous shells (coccolithophores, foraminiferans, ammonites) were more affected by ocean acidity than by the darkness and collapse of the marine food chain. The Cretaceous extinctions are not due to the impact alone; climate change effects caused by the Deccan volcanism and the drop of sea level were as much, if not more, important than the impact event.

TERRESTRIAL EXTINCTIONS

The details of the KPg transition on land are much harder to decipher because far less data is available (figure 18.3). Only a handful of land sections that span the KPg boundary have a definite iridium layer, and nearly all are in the Rocky Mountains. The Hell Creek beds of central Montana have the best section across the boundary and lots of fossils as well and these studies were largely confined to that one area, so many scientists have made detailed studies of the stratigraphic sequence there. The Cretaceous beds of the Hell Creek Formation have been legendary for more than a century for producing abundant *Tyrannosaurus rex, Triceratops, Ankylosaurus, Pachycephalosaurus*, and duckbills, but the terrestrial organisms include nearly every group.

At the bottom of the food chain are the land plants. These show the clearest effect, and there are many samples of Cretaceous rocks full of the *Aquilapollenites* pollen typical of the Late Cretaceous. At the boundary itself, the microfossils are mostly fern spores, suggesting a huge spread of ferns during the dark, cold, wet conditions of the "nuclear winter" after the asteroid strike. Remains of fungi are also common, which is typical for a world with cold, damp, decaying vegetation in huge quantities. Samples of the lowest Paleocene rocks yield pollen of completely different plants, with most of the Cretaceous species gone. The same change in typical plants can be documented in the leaves recovered from the Cretaceous and Paleocene beds, especially in the rocks from the badlands of western North Dakota.

Most of the attention has been focused on the nonbird dinosaurs of the Hell Creek Formation. It has been documented that the diversity of dinosaurs was declining throughout the late Cretaceous, so only a few species (*T. rex* and *Triceratops*) were still around to see the rock from space arrive. A few studies claim that lots of species of dinosaurs were present when the impact occurred, but other studies suggest that they were already declining due to the climatic changes triggered by the Deccan volcanic eruptions. In most cases, the dinosaur fossil record is just too sparse for scientists to be confident one way or another.

Putting aside the decline of the nonbird dinosaurs, the evidence regarding other land animals is surprising—almost *nothing else goes extinct* when the asteroid hit. The crocodilians and crocodile-like champsosaurs both survived with only minor extinctions, yet many of them were larger in body size than smaller dinosaurs. If conditions on land were so hellish, why did dinosaurs vanish but crocodilians of the same body size do just fine? Crocodilians are known for being able to hibernate a bit during winters in the subtropics, but they take weeks to prepare for the temperature change and prepare their burrows to survive the cold. The rock from space would have given no warning, and there were no places to hide in the prevailing warm climates of the Late Cretaceous. Most of the other terrestrial animals did just fine during this supposedly hellish world of the impact event. Turtles are common throughout and show no extinction whatsoever. Lizards were somewhat affected, but not as severely as some impact models would predict. Mammals remained abundant throughout the duration of the KPg transition, with only a shift in relative dominance of marsupials in the Late Cretaceous and placentals in the Early Paleocene.

Most revealing of all is that there was *no* extinction in the frogs or salamanders, which have porous skins and tend to be sensitive to slight chemical changes in the water in which they live. Any scenarios that postulate a huge episode of acid rain, which was triggered by the impactor hitting large deposits of gypsum under Chicxulub and created a sulfuric acid atmosphere, are simply false. If there had been such an event, there would be no frogs or salamanders on the planet for the past 66 million years.

Other terrestrial animals don't leave enough fossils, or a detailed enough record in the KPg boundary sections, to determine whether their extinction was sudden or gradual. Pterosaurs vanished completely, and a major extinction wiped out the archaic enantiornithine birds and left only the ornithurines, the group ancestral to living birds. We can tally how many species were represented, but we cannot determine whether they vanished gradually as the Cretaceous came to a close (and the climate was cooling dramatically with the Deccan lavas and their effect on the atmosphere) or suddenly, which is consistent with the impact scenario.

In summary, the terrestrial record shows that land plants were dramatically affected (consistent with the cold and dark of the impact scenario), and the nonbird dinosaurs did vanish (whether gradually or abruptly is still undecided). But *no other group* of land animals was very severely affected. Some people dismiss this inconvenient pattern by arguing that most of the survivors were smaller animals that could hide in aquatic habitats or burrows and survive on decaying matter during the cold dark hell after the impact. But this doesn't explain why the crocodilians did so well. They could not have hibernated, and their body size was in the same size range as many of the dinosaurs. Clearly, conditions on land were not as extreme or hellish as has been suggested, and signs of climate change and cooling was affecting life thousands of years before the impact. If the impact had any real effect at all, it was the coup de grâce for already terrible conditions in the Late Cretaceous.

WHAT IS THE ANSWER?

A detailed analysis of the fossil record does not support a simple "the rock from space did it, end of story" scenario that so many reporters and people have heard (and some scientists also believe). Within the scientific community, the debate continues more than 44 years after it was first begun, and this debate shows no signs of slowing down. Every year I attend the national GSA meeting in different cities in the fall, and every year there are more sessions arguing the details of new data. For a while it appeared that the impact advocates had prevailed, but at the 2014 GSA meeting in Vancouver, the tide was turning back toward the importance of the Deccan eruptions. There were many talks about the Deccan lavas at the 2015 GSA meeting in Baltimore, at the 2016 meeting in Denver, and at the 2021 meeting in Portland.

It is clear that three things happened about the same time: impact, eruptions, and sea level drop. All three may have contributed to these changes, but no one cause is sufficient to explain all of the effects. Nature is complex and defies simplistic models. For such a complex event as the KPg extinctions, there is no simple "right" answer as to the cause.

Meanwhile, the never-ending debate is polarized along the lines of specialization. Geologists and geophysicists and geochemists tend to like simple, clear-cut answers once they get their data out of their machines, and they tend to favor the impact-only scenario. Paleontologists are trained in biology and recognize the complexity of living systems that defy simplistic answers. Vertebrate paleontologists, who study the dinosaurs, reptiles, amphibians, and mammals, were polled in 1985, and only 5 percent agreed that the impact was the cause of the KPg extinctions. In 2004, a survey of vertebrate paleontologists found that only 20 percent accepted the impact as the cause of the KPg extinctions, and 72 percent felt that the extinctions were a gradual process that is consistent with the Deccan eruptions and not the impact. In 1997, a survey of 22 distinguished British paleontologists (specialists in each of the groups that lived in the Late Cretaceous) voted overwhelmingly against the impact being significant in the marine fossil record. In 2010, a paper published in the journal *Science* with multiple authors (few of whom were paleontologists) again pushed the impact as the sole explanation for the KPg extinctions. It was immediately rebutted by a paper authored by 28 paleontologists who demonstrated that the impact was only a minor part of the story. Even Walter Alvarez, in his popular book *T. Rex and the Crater of Doom*, conceded that the KPg extinctions had multiple complex causes.

In short, the battle shows no sign of abating and is largely polarized along the lines of professional specialization. But even more is at stake here. Some people have built their entire careers around promoting one model or the other, so they have a lot to lose: grant money, publications, prestige, and even personal pride. They are unlikely to back down, no matter what evidence is presented.

Sometimes it becomes even more personal. During the heyday of the debate in the 1980s, a lot of name-calling and career-wrecking took place. Luis Alvarez said: "I don't want to say bad things about paleontologists, but they're really not very good scientists. They're more like stamp collectors."[3] On the opposite side, the dinosaur paleontologist Bob Bakker told a reporter:

> The arrogance of these people is simply unbelievable. They know next to nothing about how real animals evolve, live, and become extinct. But, despite

their ignorance, the geochemists feel that all you have to do is crank up some fancy machine and you've revolutionized science. The real reasons for the dinosaur extinctions have to do with temperature and sea level changes, the spread of diseases by migration and other complex events. In effect, they're saying this: we high-tech people have all the answers, and you paleontologists are just primitive rockhounds.[4]

With attitudes like this, and so much at stake, the answer regarding the cause of these events will continue to be the muddled statement: "It's complicated." This debate is unlikely to end until all the original combatants have left the field due to death or retirement. And that day has not yet come.

NOTES

1. Glenn Lowell Jepsen, "Riddles of the Terrible Lizards," *American Scientist* 52 (1964), 231.
2. The geological map symbol for Cretaceous is "K"; "C" stands for "Carboniferous," so the "K" is used here instead; it comes from the German word for "chalk" (*Kreide*), which was the original name for the Cretaceous (which means "chalky"). Before the year 2010, the first 64 million years of the Cenozoic (from 66 to 2 Ma) was called the "Tertiary" ("T"), so the early shorthand for the event was the "KT boundary." In recent years, Tertiary has been formally abandoned as obsolete by the geologic subcommission on stratigraphy; they now prefer the term "Paleogene" ("Pg") for the first three epochs of the Cenozoic: Paleocene, Eocene, and Oligocene (from 66 to 23 Ma) Thus the term "KPg boundary" is now preferred.
3. Malcolm Brown, "The Debate Over Dinosaur Extinctions Takes an Unusually Rancorous Turn," *New York Times*, January 19, 1988, https://www.nytimes .com/1988/01/19/science/the-debate-over-dinosaur-extinctions-takes-an -unusually-rancorous-turn.html.
4. David B. Weinreb, "Catastrophic Events in the History of Life: Toward a New Understanding of Mass Extinctions in the Fossil Record - Part I," *Journal of Young Investigators*, March 23, 2002, https://www.jyi.org/2002-march/2017/10/23 /catastrophic-events-in-the-history-of-life-toward-a-new-understanding-of -mass-extinctions-in-the-fossil-record-part-i.

FOR FURTHER READING

Alvarez, Luis W., Walter Alvarez, Frank Asaro, and Helen V. Michel. "Extraterrestrial Cause for the Cretaceous–Tertiary Extinction." *Science* 208, no. 4448 (1980): 1095–1108.

Alvarez, Walter. *T. Rex and the Crater of Doom*. Princeton, NJ: Princeton University Press, 1997.

Archibald, J. David. *Dinosaur Extinction and the End of an Era: What the Fossils Say*. New York: Columbia University Press, 1996.

——. *Extinction and Radiation: How the Fall of the Dinosaurs Led to the Rise of the Mammals*. Baltimore, MD: Johns Hopkins University Press, 2011.

Archibald, J. David, and Laura J. Bryant. "Differential Cretaceous/Tertiary Extinction of Nonmarine Vertebrates; Evidence from Northeastern Montana." In *Global Catastrophes in Earth History: An Interdisciplinary Conference on Impacts, Volcanism, and Mass Mortality*, ed. Virgil L. Sharpton and Peter D. Ward. Special paper, *Geological Society of America* 247 (1990): 549–562.

Archibald. J. David, and David E. Fastovsky. "Dinosaur Extinction." In *The Dinosauria*, 2nd ed., ed. David B. Weishampel, Peter Dodson, and Halszka Osmólska, 672–684. Berkeley: University of California Press, 2004.

Askin, R. A., and S. R. Jacobson. "Palynological Change Across the Cretaceous–Tertiary Boundary on Seymour Island, Antarctica: Environmental and Depositional Factors." In *Cretaceous–Tertiary Mass Extinctions: Biotic and Environmental Changes*, ed. Norman MacLeod and Gerta Keller, 7–25. New York: Norton, 1996.

Clarke, Julia A., Claudia P. Tambussi, Jorge I. Noriega, Gregory M. Erickson, and Richard A. Ketcham. "Definitive Fossil Evidence for the Extant Avian Radiation in the Cretaceous." *Nature* 433, no. 7023 (2005): 305–308.

Dingus, Lowell, and Timothy Rowe. *The Mistaken Extinction: Dinosaur Evolution and the Origin of Birds*. New York: Freeman, 1997.

Duncan, R. A., and Douglas G. Pyle. "Rapid Eruption of the Deccan Flood Basalts at the Cretaceous/Tertiary Boundary." *Nature* 333, no. 6176 (1988): 841–843.

Gedl, Przemyslaw. "Dinoflagellate Cyst Record of the Deep-Sea Cretaceous-Tertiary Boundary at Uzgru, Carpathian Mountains, Czech Republic." *Special Publications of the Geological Society of London* 230, no. 1 (2004): 257–273.

Henehan, Michael J. "Rapid Ocean Acidification and Protracted Earth System Recovery Followed the End-Cretaceous Chicxulub Impact." *Proceedings of the National Academy of Sciences of the United States of America* 116, no. 45 (2019): 22500–22504.

Hildebrand, Alan R., Glen T. Penfield, David A. Kring, Mark Pilkington, Zuleica Camargo, and Stein Jacobsen. "Chicxulub Crater: A Possible Cretaceous/Tertiary Boundary Impact Crater on the Yucatán Peninsula, Mexico." *Geology* 19, no. 9 (1991): 867–871.

Keller, Gerta. "The Cretaceous–Tertiary Mass Extinction, Chicxulub Impact, and Deccan Volcanism." In *Earth and Life: Global Biodiversity, Extinction Intervals and Biogeographic Perturbations Through Time*, ed. John Talent, 759–793. New York: Springer, 2012.

——. "The End-Cretaceous Mass Extinction in the Marine Realm: Year 2000 Assessment." *Planetary and Space Science* 49, no. 8 (2001): 817–830.

Keller, Gerta, and Andrew C. Kerr. *Volcanism, Impacts, and Mass Extinctions: Causes and Effects*. McLean, VA: Geological Society of America Special Paper 505, 2014.

Li, Liangquan, and Gerta Keller. "Abrupt Deep-Sea Warming at the End of the Cretaceous." *Geology* 26, no. 11 (1998): 995–998.

Macellari, C., and W. J. Zinsmeister. "Sedimentology and Macropaleontology of the Upper Cretaceous to Paleocene Sequence of Seymour Island." *Antarctic Journal of the U.S., Annual Review* 18, no. 5 (1983): 69–71.

Machalski, Marcin. "Late Maastrichtian and Earliest Danian Scaphitid Ammonites from Central Europe: Taxonomy, Evolution, and Extinction." *Acta Palaeontologica Polonica* 50, no. 4 (2005): 653–696.

Machalski, Marcin, and Claus Heinberg. "Evidence for Ammonite Survival Into the Danian (Paleogene) from the Cerithium Limestone at Stevns Klint, Denmark." *Bulletin of the Geological Society of Denmark* 52, no. 2 (2005): 2005–2012.

MacLeod, Kenneth G. "Extinction of Inoceramid Bivalves in Maastrichtian Strata of the Bay of Biscay Region of France and Spain." *Journal of Paleontology* 68, no. 5 (1994): 1048–1066.

MacLeod, Norman. "Impacts and Marine Invertebrate Extinctions." *Special Publications of the Geological Society of London* 140, no. 1 (1998): 217–246.

——. "Nature of the Cretaceous-Tertiary (K–T) Planktonic Foraminiferal Record: Stratigraphic Confidence Intervals, Signor–Lipps Effect, and Patterns of Survivorship." In *Cretaceous-Tertiary Mass Extinctions: Biotic and Environmental Ehanges*, ed. Norman MacLeod and Gerta Keller, 85–138. New York: Norton, 1996.

MacLeod, Norman, and Gerta Keller, eds. *Cretaceous-Tertiary Mass Extinctions: Biotic and Climatic Change*. New York: Norton, 1996.

MacLeod, Norman, Peter F. Rawson, P. L. Forey, F. T. Banner, Marcel K. Boudagher-Fadel, P. R. Bown, J. A. Burnett, et al. "The Cretaceous–Tertiary Biotic Transition." *Journal of the Geological Society London* 154, no. 2 (1997): 265–292.

Marriott, Katherine, Donald R. Prothero, and Alexander Bartholomew. *The Evolution of Ammonoids*. Boca Raton, FL: Taylor & Francis, 2023.

Nichols, Douglas J., and Kirk R. Johnson. *Plants and the K-T Boundary*. Cambridge: Cambridge University Press, 2008.

Officer, Charles, and Jake L. Page. *The Great Dinosaur Extinction Controversy*. New York: Basic Books, 1996.

Petersen, Sierra V., Andrea Dutton, and Kyger C. Lohmann. "End-Cretaceous Extinction in Antarctica Linked to Both Deccan Volcanism and Meteorite Impact via Climate Change." *Nature Communications* 7 (2016): 12079.

Powell, James L. *Night Comes to the Cretaceous: Dinosaur Extinction and the Transformation of Modern Geology*. New York: St. Martin's, 1998.

Retallack, Gregory J. "Greenhouse Crises of the Past 300 Million Years." *Geological Society of America Bulletin* 121, no. 9–10 (2009): 1441–1455.

Retallack, Gregory J., and Giselle D. Conde. "Deep Time Perspective on Rising Atmospheric CO_2." *Global and Planetary Change* 189 (2020): 103177.

Robertson, Douglas S., Malcolm C. McKenna, Owen B. Toon, Sylvia Hope, and James A. Lillegraven. "Survival in the First Hours of the Cenozoic." *GSA Bulletin* 116, no. 5–6 (2004): 760–768.

Thompson, J. B., and S. Ramirez-Barahona. "No Phylogenetic Evidence for Angiosperm Mass Extinction at the Cretaceous-Paleogene (K-Pg) Boundary." *Biology Letters* 19: 20230314.

Ward, P. D., W. J. Kennedy, K. G. MacLeod, and J. F. Mount. "Ammonite and Inoceramid Bivalve Extinction Patterns in Cretaceous/Tertiary Boundary Sections of the Biscay Region (Southwestern France, Northern Spain)." *Geology* 19, no. 12 (1991): 1181–1184.

Zinsmeister, William J., Rodney M. Feldmann, Michael. O. Woodburne, and David H. Elliot. "Latest Cretaceous/Earliest Tertiary Transition on Seymour Island, Antarctica." *Journal of Paleontology* 63, no. 6 (1989): 731–738.

RETURN OF THE GREENHOUSE

The heat of European latitudes during the Eocene period. . . . seem[s]. . . . equal to that now experienced by the tropics.

—CHARLES LYELL, *PRINCIPLES OF GEOLOGY* (1833)

ARCTIC EDEN

The four paleontologists are walking slowly along the barren landscape, heads bent slightly as they focused on the ground just ahead of them. The tundra surface is covered with lichens and crunches slightly when their boots step gingerly across the surface. There is a lot of glare from the low Arctic sun at the peak of the polar summer, but they dare not wear sunglasses for fear of not seeing something important. They all scan the ground slowly as they walk, trying not to miss anything, stooping now and then to pick up something interesting. Most of the time it is just colorful or shiny pebbles, but they are hoping for much more: teeth and maybe even bones of extinct animals. The four paleontologists—Mary Dawson of the Carnegie Museum of Natural History in Pittsburgh, Howard Hutchison of the University of California Berkeley, Robert "Mac" West of the Milwaukee Public Museum, and Paul Raemakers of the University of Toronto—first visited the Canadian Arctic in 1973 (figure 19.1). They found only a few plant fossils in

Figure 19.1 ▲ ▶
The Ellesmere Island expeditions of the 1970s, led by Mary Dawson, Howard Hutchison, Robert "Mac" West, Malcolm McKenna, and Paul Raemakers. (*A*) Dawson and Hutchison studying a gigantic petrified log. (*B*) Dawson and Hutchison searched for tiny fossils by crawling on their bellies, eyes only inches above the ground. (*C*) Unloading all the gear from the Twin Otter aircraft before it leaves. (*D*) Dawson and McKenna inside the cramped seats of the Twin Otter. (*E*) Dawson and Raemakers using portable sieving screens to recover tiny fossils of teeth and bones. (Courtesy of R. M. West)

Figure 19.1 ▲ ▶

(*continued*)

Figure 19.1 ▲
(*continued*)

six weeks of searching, but the geologic maps and the aerial photos seemed promising. Early Canadian geologic surveys had reported abundant petrified wood in the deposits, so the chance of finding land vertebrates seemed very promising.

Working in the Canadian Arctic is challenging, to say the least. They worked during the height of the Arctic summer, above the Arctic Circle, so they had a true "midnight sun" and had to force their bodies to sleep without darkness at night. The temperature was near freezing most of the time, so everyone had to be dressed in warm Arctic parkas and other polar gear. When it warmed up enough for them to shed those layers, hordes of mosquitoes attacked their bare skin. They did not carry guns even though a chance encounter with a huge polar bear or a wolf pack was possible at any time. Eberle and McKenna (2007) described their encounter this way:

> While scaling a steep canyon wall northwest of camp, Dawson and Hutchison sat down to catch their breath and noticed a pack of six white wolves making their way along the shoreline of the fiord. When they caught sight of

the paleontologists, the wolves, seemingly curious, began moving upslope towards them. At first, Dawson and Hutchison sat and watched from their perch, thinking the wolves would move on when they realized their subjects were neither muskoxen nor caribou. However, as the wolves quickly covered the distance between them, coming to within five meters, they stood up, concerned. Dawson and Hutchison were unarmed, save for their backpacks and nearby clods of silt, all of which were heaved downslope at the approaching wolf pack. The fearless wolves made no sound as they continued toward their quarry. The lead wolf fixed on Dawson, and when she had come within about five feet, she leapt and lunged toward her throat. Dawson leaned back, and with arms in front, let out a startled cry. The wolf dropped to the ground, turned, and warily moved away, glancing back at Hutchison and Dawson as they threw clods of silt at her and the other wolves. Although unhurt, Dawson's brush with the lead wolf left traces of wolf saliva on her cheek. The encounter was subsequently published as the first account of a wolf attack in the High Arctic.

The expeditions were costly (roughly $3,000 a day per person), and the logistics were difficult. To get to these remote exposures on Ellesmere Island, they had to fly from civilization to Resolute Bay on Cornwallis Island, where the Canadian Polar Continental Shelf Program maintained a research station. Then they flew in a tiny DeHavilland Otter, a twin-engine propeller plane designed for short takeoffs and landings and carrying only 11 people maximum (including the pilot), to reach their research sites on Ellesmere near Strathcona Fjord (figure 19.1C). As reported by Ebele and McKenna:

Clear days in Resolute were few and far between, but as long as the fog wasn't "down to the deck," the Otters were in the air. Midnight sun that defines the Arctic summer means that you could be placed in camp at any time, day or night, and many can speak about camp moves in the wee hours, when the weather miraculously breaks and a hole the size of an Otter appears above. There was a three-hour Otter ride from Resolute Bay to central Ellesmere that ended in a stomach-churning landing on a tilted frost-heaved terrain where, from above, no one (aside from a well-seasoned bush pilot) would ever dream of landing. When the aircraft lurched to a stop, its occupants gingerly disembarked, as if stepping out onto the moon.

Once all the gear was unloaded, the Otter flew away, and they had to survive for weeks on only what they had brought. Their only contact with the outside world was via a satellite radio for emergency communication with Resolute Bay. On their second field season on Ellesmere Island, Dawson and colleagues finally met with success. As told by Mackenzie Carpenter in a 2000 article in the *Pittsburgh Post-Gazette*:

> In the pale dreaming light of an Arctic summer, a woman is walking along the crunchy surface of the tundra, looking for a place to stop and eat lunch. As usual, Mary Dawson is walking with her head down, silver white hair falling into her face, walking the walk of all fossil hunters—eyes riveted to the ground "just in case there's something that looks a little different." She is not disappointed. There is something that makes her stop. She kneels and picks [up] a dirt-encrusted object, and in a reflexive gesture of someone used to working out in the middle of nowhere, puts it in her mouth. It has to be cleaned somehow. She spits it out and peers at the object, which is indeterminate, foreign. Nothing she had expected to find here. Dawson goes back to camp, washes it off again, this time with some water. And what she sees makes her forget her lunch of crackers and cheese and chocolate bars and get down on her hands and knees for the next several hours, where she will find dozens and dozens of fragments of teeth. Alligator teeth. In the Arctic.[1]

Near Strathcona Fjord, they found alligator teeth and pond turtle shell pieces, indications of a much warmer climate above the Arctic Circle in the early Eocene. Eventually, Dawson and her crew began to find teeth of fossil mammals, and these were even more interesting. They revolutionized our understanding of the Eocene climate, as well as conditions above the Arctic Circle, and how mammals had migrated between Eurasia and North America about 50 Ma.

The leader of these expeditions, Mary Dawson, has a back story almost as remarkable as the fossils she found. Born in 1931, she grew up in Michigan and studied zoology at what became Michigan State University, intending to become a veterinarian—but got hooked on fossil mammals instead. Then she studied under a Fulbright Scholarship at the University of Edinburgh, and finally received her doctorate in vertebrate paleontology at Kansas University in 1958, focusing on fossil lagomorphs (rabbits, hares, and pikas). For almost her entire career, she was the only woman in her college classes and in her graduate program. She encountered nonstop barriers to women at every turn. As she recalled:

> The only thing that bothered me was finding a field party I could go on. . . . They didn't want women along, and I looked and looked and finally found one with a fellow who took his wife along. Isn't that disgusting? It annoyed me, but you go around those annoyances.

After her doctorate, she did a postdoc in Basel, Switzerland, then got a teaching job at the all-female Smith College in Massachusetts in 1957, then took an NSF Fellowship at Yale, followed by a year at the National Science Foundation as the head of the program evaluating grant proposals. Finally in 1962, she was hired at the Carnegie Museum, where she spent the rest of her career working mostly on fossil rabbits and rodents. According to Carpenter:

> Her progress upward at the Carnegie was swift; a year after she was hired, she was made assistant curator, and it was at this time that [museum director] Graham Netting told her that as a woman she could never expect to rise much higher. So why, 10 years later, did he make her curator? Well, it was 1972, and the past decade "had been a real learning experience for him," she laughs.

By 1973, she was chair of the Earth Sciences Division at the Carnegie Museum, and Mary finally retired 30 years later in 2003 at age 74. She continued to work on her research even after she retired, finally passing in 2020 at age 89. She was the only woman in vertebrate paleontology in the United States for most of her career from the late 1950s until more and more women began to break in to the previously all-male profession in the 1970s. She was only the second woman president of the Society of Vertebrate Paleontology (serving a term from 1973 to 1974), and she received all the highest honors a paleontologist can receive over the course of her career—but her pride and joy was her pioneering work in the Canadian Arctic. Between 1973 and 2002, Mary and her teams spent 11 field seasons in the Canadian Arctic, working not only on Ellesmere Island but also on Axel Heiberg, Devon, Ellef Ringnes, Bylot, and Banks islands.

I knew Mary well from the beginning of my career in 1976, especially when I got to work on specimens that she curated at the Carnegie Museum. In addition, many of her expeditions were joined by my own grad advisor,

Malcolm McKenna of the American Museum of Natural History, and Malcolm was a participant in several of her expeditions while I was his student from 1976 to 1982.

So what were some of these amazing discoveries were made by Dawson, West, Hutchison, McKenna, and others? They included fossil relatives of the colugos ("flying lemurs") and primitive tapirs, both denizens of tropical rain forests today, as well early three-toed horses related to *Eohippus*, the extinct hippo-like pantodont *Coryphodon* (figure 19.2), rhino-like brontotheres, raccoon-like primitive carnivorans (*Viverravus, Miacis*), and predators from extinct groups such as the creodonts (*Palaeonictis*) and the hoofed predators known as mesonychids (*Pachyaena*), plus the otter-like pantolestids, insectivorous mammals, and five kinds of very primitive rodents. There was a primitive palaeanodont resembling a small anteater, which was named *Arctiacodon dawsonae* in her honor. They also found fossils of

Figure 19.2 ▲
Reconstruction of the cow-sized early Eocene pantodont *Coryphodon*, living in the lush forests of the Canadian Arctic. (Photo by author)

a variety of fish, turtles, giant salamanders, lizards, snakes, birds, and alligators, none of which live in the polar regions today. They only survive in tropical and subtropical environments.

The petrified tree stumps have wide growth rings, indicating very favorable climate conditions, and included giant redwoods (*Sequoia*), dawn redwoods (*Metasequoia*), and ginkgoes among the gymnosperms, plus members of the walnut family, elms, birch, alders, *Liquidambar, Viburnum*, the bald cypress *Taxodium*, and the Asian katsura tree (*Cercidiphyllum*). Paleobotanists interpreted these as evidence of a polar rain forest with mild temperatures, even though there were 24 hours of darkness during the polar winter. These plants must have grown extremely fast in the 24 hours of midnight sun during the Eocene polar summers. A variety of paleotemperature indicators show that the region above the Arctic Circle was warm and mild, with temperatures averaging about 20°C (68°F) in the summer, with a range of 19°C to 25°C (66°F to 77°F).

Many of these mammals are known from lower Eocene rocks in North America, Europe, and Asia as well. *Coryphodon* were found on all of the northern continents in the early Eocene. Primitive horses and tapirs were also found in Eurasia and North America, as well as some of the other mammals, especially the primitive rodents. The Ellesmere Island fossils show that a wide diversity of mammals lived in a temperate Arctic rain forest and that they could move easily between North America and Eurasia. Some probably took an early version of the Bering Land Bridge from Alaska, but others might have island-hopped across a much narrower North Atlantic Ocean at that time, using Iceland and Greenland and other islands that have since vanished as the North Atlantic grew wider over the last 50 million years.

The early Eocene fossils of Antarctica also show dense forests with temperate vegetation and no sign of freezing in the winter. These forests were inhabited by primitive opossum-like creatures that were related to the pouched mammals of Australia and also to the many kinds of opossums that live in South America today.

THE JUNGLES OF MONTANA

Similar warm and wet subtropical to temperate conditions in the Early Eocene can be found in regions that are now buried under snow each winter. Fossil plants in Wyoming, North Dakota, and Montana demonstrate that conditions warmed to mean annual temperatures as high as 21°C (70°F), and the mean annual cold month temperature could be no lower than 13°C (55°F) because most of these plants are intolerant of freezing. These plants also required a very wet climate, with mean annual rainfall in excess of 150 centimeters (60 inches). Today western North Dakota has a steppe climate. The mean annual temperature is only 5°C (41°F), and the spread between daily extremes ranges over 33°C (90°F). In North Dakota or eastern Montana, it is not at all unusual for temperatures on a hot spring or fall day to start out above 32°C (90°F), then drop below freezing in a matter of hours as an Arctic cold front moves in.

From the evidence of floras in the Bighorn Basin of Wyoming or the Williston Basin of Montana and North Dakota, we can visualize a dense tropical forest much like that found in modern Panama and unlike the barren badlands found there today (figure 19.3). Tall trees formed a dense canopy, with

Figure 19.3 ▲

Lower Paleogene rocks as they look today, a barren badlands, on the slopes of Polecat Bench in the Bighorn Basin of northwestern Wyoming. The line marks the boundary between the last of the Cretaceous Hell Creek Formation (below) and the Paleocene Tullock Formation (above). (Photo by author)

vines and lianas growing all around them. The fossil plants include many tropical groups, including citrus, avocado, cashews, and pawpaw trees. Many of these plant genera are found today only in the jungles of southeast Asia or tropical Central America. In addition to the direct evidence of the plant fossils, there is information in the striking color bands that stripe the badlands slopes. Each band represents an ancient soil horizon, and in many places hundreds of them are stacked on top of each other, representing millions of years of the early Eocene. Each represents another episode of floodplain mud deposition, followed by the development of plants and a soil horizon, and then another episode of flooding, which buried the old soil. These ancient soils were deposited on broad floodplains bordering meandering rivers, much like those of the modern Amazon.

The creatures that roamed the jungles of Montana were unlike anything seen on Earth today (figure 19.4). The warm tropical and subtropical conditions supported a variety of reptiles, including several kinds of crocodilians, snakes, and freshwater turtles. The jungles were inhabited by a wide range of primitive mammals, most of which fit into two broad categories: tree-dwellers and ground-dwelling leaf-eaters. Among the tree-dwellers were the archaic relatives of the lemurs and other primates, which were by far the most common mammals of the Paleocene. Like modern lemurs, they had long snouts, forward-facing eyes, and a more primitive body structure than the more specialized monkeys and apes that came later. Many of them had a squirrel-like body form and occupied the squirrel niche before rodents had appeared. Another tree-dwelling group were the insectivorous mammals, which were the most common mammals of the Cretaceous and continued their diversity in the early Cenozoic. Finally, the jungles of the Paleocene and Eocene were also inhabited by an extremely primitive group of mammals known as multituberculates, which were egg-laying primitive mammals related to the platypus that were built like squirrels.

Moving through the dense jungle undergrowth were many groups of archaic mammals that had simple teeth for eating leaves and fruits. They were not very specialized for running or for any other advanced ecology. These were mostly archaic hoofed mammals of a variety of families, and they ranged in size from that of a small dog to the size of a sheep. Some were closely related to the horse-rhino group of odd-toed hoofed mammals, or perissodactyls. Others were related to the even-toed hoofed mammals, or

Figure 19.4 ▲
Reconstruction of typical animals in the North American jungles of the early Eocene. There are a few distant relatives of rhinos and horses and rodents and primates in that time, but the majority of the mammals shown are extinct and have no living descendants. (Art by Jay Matternes; courtesy of Wikimedia Commons)

artiodactyls. But most were members of groups that vanished in the Eocene without leaving any descendants.

Preying on this wide range of herbivorous mammals were an assortment of mammals and predatory birds and reptiles. Living in dense jungles required predators to rely on stealth and ambush to catch their prey; they did not have to be fast runners or agile tree-climbers. Most of the predatory mammals were members of an archaic extinct group known as creodonts, which vaguely resembled lions and wolverines and weasels but had few specializations for running and tended to be small brained as well. Eventually, they were outcompeted by the mammals of the order Carnivora, or carnivorans, which includes almost all the flesh-eating mammals alive today (dogs, cats, weasels, bears, skunks, raccoons, hyenas, mongoose, seals, sea lions, walruses, and all their extinct relatives). But the earliest Carnivora were quite weasel-like or mongoose-like, and none of these more specialized descendants had yet appeared. There were even two groups of archaic hoofed mammals that switched to a meat-eating diet. A much more diverse array of carnivorous creatures were around in the Paleocene compared to the single order Carnivora today, which includes all meat-eating placental mammals.

Most of these groups, such as the creodonts, small Carnivora, archaic hoofed mammals on the ground, lemur-like primates, insectivores, and multituberculates in the trees, ruled the jungles of the Paleocene and continued to do so in the early Eocene. But a number of new groups appeared that gradually pushed out some of the archaic groups. These included the odd-toed perissodactyls, the earliest relatives of horses, rhinos, and tapirs. These animals are called "odd-toed" because they have one or three fingers or toes on their hands and feet, and the axis of symmetry of their foot runs through the middle finger/middle toe. Their relatives first appeared in Mongolia and India in the late Paleocene, then spread across the northern continents in the early Eocene when the warm Arctic region connected all the landmasses. Also appearing in the early Eocene were the even-toed artiodactyls, including some of the first relatives of pigs, camels, and deer-like creatures, although most were archaic extinct groups with no living descendants. They are called "even-toed" because they have either two or four toes on their feet and hands, and the axis of symmetry of the hand/foot runs through the middle finger/toe and the fourth finger/toe. One of the striking features of the peak of the warming at the beginning of the Eocene is that the horses and several other groups of mammals show an abrupt dwarfing event, possibly in response to the extreme temperatures.

Finally, a major addition to Eocene mammal faunas were the first rodents. They first appeared in China and Mongolia in the late Paleocene and spread across the entire northern hemisphere in the early Eocene. By the end of the Eocene, they may have helped drive the archaic egg-laying multituberculates (that also had a squirrel-like body) to extinction, along with the rodent-like primates.

The same pattern of plants and animals can be seen in other temperate and tropical regions throughout the early Eocene. Even the Pacific Northwest and southern Alaska were relatively warm (25°C or 77°F) and wet, and the land was blanketed with broad-leaved evergreen forests, abundant vines and lianas, and many plants with tropical Asian affinities.

From the London Clay found in the basements of London homes comes an important early Eocene flora. As we saw in Montana, the London flora also included tropical trees and shrubs, lianas, including cinnamon, figs, magnolias, palms, laurels, citrus, pawpaw, cashews, laurels, and vines such as moonseed, icacina, and grapes. About 92 percent of these plants have living relatives in the jungles of southeast Asia. Fringing the coasts of the tropical jungles were mangrove swamps full of *Nypa* palms, also restricted to southeast Asia today. From this evidence, the average temperatures in London were about 25°C (77°F) compared to the modern average of 10°C (50°F). Instead of the cold, foggy London of Sherlock Holmes, London was as warm and tropical as Singapore in the Eocene. And the land mammals found in the London Clay include primitive relatives of horses, plus lots of archaic tapir relatives and different kinds of lemur-like primates, creodont predators, and the ubiquitous hippo-like pantodont *Coryphodon*. The lower Eocene beds of the Paris Basin produce an even richer mammalian assemblage, including multituberculates, opossums, lemur-like primates, archaic hoofed mammals, primitive artiodactyls, horse-like palaeotheres, primitive tapirs, and very primitive rodents. Some of the species of lemur-like primates, horses, and tapirs and other mammals are nearly identical between North America and Europe, suggesting that there was a major travel corridor across the North Atlantic between Wyoming and Paris.

Anywhere we find fossil floras of early Eocene age in temperate or tropical regions around the world, we encounter a similar story. Floras from China, Siberia, India, and southern South America all show the same tropical-subtropical patterns, even though many of these regions were at fairly high latitudes and inland locations. The few floras known from tropical regions, such as Panama, also indicate that conditions were hot and wet there in the early Eocene.

The world of the early Eocene was very different from the icehouse world of today. It was another greenhouse planet, similar to that of the Cretaceous. And the early Eocene was the peak of the warming.

THE BLAST OF GAS FROM THE PAST

What could have warmed the planet so much about 55 Ma? Geochemical evidence from foraminiferans in the deep ocean shows that the extreme greenhouse conditions of most of the Cretaceous had been modified by episodes of cooling during the Paleocene, beginning with the cooling that marked the KPg extinction (see chapter 18). There was much speculation in the 1970s and 1980s about what made the polar regions of the Eocene so warm, but there was no concrete evidence. In the early 1990s, geochemical evidence from the ocean showed that a huge amount of very light carbon-12 was in the sediments and shells of bottom-dwelling foraminiferans, and this had occurred suddenly right around the Paleocene-Eocene boundary. Many explanations were suggested, such as changes in ocean currents and circulation, but the volume of carbon needed and the suddenness of the event ruled most of them out. For a short time, it was suggested that an impact by a carbon-rich comet might have caused the Paleocene-Eocene Thermal Maximum (PETM), but that was debunked shortly after the claim was published. Recently, evidence has emerged that there were huge eruptions of the North Atlantic Magmatic Province which provided the trigger for the warming.

Since the 1990s, the best explanation has been the idea that huge amounts of carbon were released from tiny frozen cages of water ice surrounding methane gas (known as methane hydrates or methane clathrates). This was a relatively recent discovery. Even today methane gas is trapped in the spaces between subfreezing ocean sediments. If the sea bottom warmed quickly enough, it would have thawed many of these ice cages and released enormous amounts of methane, which is an even more powerful greenhouse gas than carbon dioxide. This "blast of gas from the past" explains the peculiar pattern of the PETM: rapid global warming all over the oceans and continents (warm enough to make the polar regions mild enough for alligators) without mass extinctions except in the benthic foraminifera that live in and on top of the seafloor sediment that lost 35 to 50 percent of their species.

THE MIDDLE EOCENE CLIMATE OPTIMUM

After a few thousand years, most of the methane would have oxidized to form carbon dioxide, which continued to warm the planet (figure 19.5). Earth went through a record warm period known as the early Eocene climate optimum (EECO). This event began at the PETM 55 million years ago, and the world kept getting warmer and warmer until it peaked about 50 million years ago. The final pulse of warmth occurred about 40 million years ago and is known as the middle Eocene climate optimum (MECO). Geochemical evidence suggests that the carbon dioxide levels were about 1,000 ppm then, comparable to the peak of greenhouse warming in the Cretaceous and about 50 times higher than our modern levels of 415 ppm.

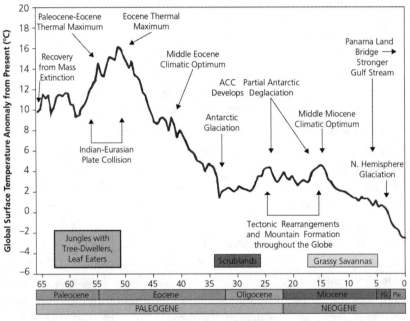

Figure 19.5 ▲

Climate history of the Cenozoic, showing the abrupt warming at the PETM, the gradual warming to the early Eocene climatic optimum, the middle Eocene climatic optimum (MECO), and then cooling until the Eocene-Oligocene transition. (Redrawn from several sources)

The fossils of the middle Eocene reflect a world in the final phase of the greenhouse planet that had begun in the Middle Jurassic. Exposures of middle Eocene rocks are found in many areas of the Rocky Mountains and even in places as far away as San Diego. They are best known from the Bridger Basin in southwestern Wyoming, the Uinta Basin in northeast Utah, and the Washakie Basin in northwest Colorado, and they form the basis for the Bridgerian and Uintan land mammal ages of North America.

The mammals that roamed the middle Eocene jungles were closely related to the animals that roamed the early Eocene jungles of Wyoming and Montana. Extraordinarily complete specimens are known from the Bridger Formation, which were first collected by the pioneers of American paleontology in the early 1870s. Like the jungles of the early Eocene, the forests of the middle Eocene had lots of archaic insectivorous groups that are now extinct. The rodent-like multituberculates were nearly gone, however, apparently crowded out by the huge radiation of rodents that occupied that niche. There were still lots of primates in the trees, but the archaic groups of the Paleocene and early Eocene were nearly gone, replaced by a big radiation of adapid primates (related to lemurs) and omomyid primates (related to tarsiers). The predators were still dominated by lion-sized creodonts such as *Patriofelis*, and the true carnivorans were still primitive groups from the early Eocene.

Among the hoofed mammals, the even-toed artiodactyls had begun to transform from the tiny rabbit-like forms of the early Eocene, and they were becoming more and more specialized. By the end of the middle Eocene, groups such as early camels, early ruminants, and early relatives of pigs were present, especially in Asia. But the odd-toed perissodactyls had evolved even more and had become the most diverse and numerous of all the hoofed mammals. There were numerous relatives of tapirs eating the leaves of the forests, and more advanced three-toed horses like *Orohippus*. But the most diverse group were the rhino-like brontotheres, which were quickly evolving into a diversity of forms and were the largest mammals of the middle Eocene.

The middle Eocene jungles also had some extraordinarily large mammals. Perhaps the most remarkable were the elephant-sized uintatheres, which had huge hoofed bodies and huge heads with three pairs of knobs on the top of the skull and enormous canine tusks protruding below the jaw (figure 19.6). Despite their huge size, they had remarkably small and simple

Figure 19.6 ▲

Reconstruction of the rhino-sized uintathere *Eobasileus*, from the middle Eocene of the Uinta Basin, Utah. (Art by Bruce Horsfall; courtesy of Wikimedia Commons)

leaf-chopping teeth. Their earliest relatives can be traced to the Paleocene of Mongolia, and during the early Eocene they got larger and larger, before finally vanishing at the end of the middle Eocene.

The middle Eocene predators included some enormous birds as well. The most impressive were the huge flightless birds known as *Diatryma* in North America and *Gastornis* in Europe, although most paleontologists think they are the same genus of bird now (figures 19.7A and 19.7B). Over 2 meters (6.5 feet) tall at maximum height, they had huge skulls with extremely robust beaks and powerful legs and necks on their robust bodies. They towered above almost all of the much smaller mammals of the middle Eocene, and they have long been considered as a return to the days when predatory dinosaurs ruled over the mammals. However, recent work on these fossils suggest that their beaks are not really adapted for pure meat-eating, so they may have been omnivores or herbivores.

Figure 19.7 ▲ ▶

(A) Skeleton of the giant flightless predatory bird *Gastornis* (= *Diatryma*); (B) a reconstruction of the bird by R. Bruce Horsfall. ([A] Photo by author; [B] courtesy of Wikimedia
Commons)

Figure 19.7 ▲
(*continued*)

These were the denizens of the middle Eocene jungles, the last of the greenhouse years that spanned the later Mesozoic and ran into the middle Eocene, about 160 to 40 Ma. But a change from a greenhouse to an icehouse world was coming, and the first steps to a colder planet began by the end of the middle Eocene. That is the topic of the next chapter.

NOTE

1. Mackenzie Carpenter, "The Bone Collector," *Pittsburgh Post-Gazette*, March 19, 2000, https://old.post-gazette.com/magazine/20000319dawson1.asp.

FOR FURTHER READING

Aubry, Marie-Pierre, Spencer G. Lucas, and William A. Berggren, eds. *Late Paleocene-Early Eocene Biotic and Climatic Events in the Marine and Terrestrial Records*. New York: Columbia University Press, 1998.
Bowen, Gabriel J., Bianca J. Maibauer, Mary J. Kraus, Ursula Röhl, Thomas Wester-hold, Amy Steimke, Philip D. Gingerich, Scott L. Wing, and William C. Clyde.

"Two Massive, Rapid Releases of Carbon During the Onset of the Palaeocene-Eocene Thermal Maximum." *Nature* 8, no. 1 (2015): 44–47.

Dawson, Mary R. "Early Eocene Rodents (Mammalia) from the Eureka Sound Group of Ellesmere Island, Canada." *Canadian Journal of Earth Sciences* 28, no. 3 (1991): 364–371.

Dickens, Gerald R., Maria M. Castillo Uzcanga, and James C. G. Walker. "A Blast of Gas in the Latest Paleocene; Simulating First-Order Effects of Massive Dissociation of Oceanic Methane Hydrate." *Geology* 25, no. 3 (1997): 259–262.

Eberle, Jaelyn J., and Malcolm C. McKenna. "Early Eocene Leptictida, Pantolesta, Creodonta, Carnivora, and Mesonychidae (Mammalia) from the Eureka Sound Group, Ellesmere Island, Nunavut." *Canadian Journal of Earth Sciences* 39, no. 6 (2002): 899–910.

——. "The Indefatigable Mary Dawson: Arctic Pioneer." *Bulletin of Carnegie Museum of Natural History* 39 (2007): 7–17.

Estes, Richard, and J. Howard Hutchison. "Eocene Lower-Vertebrates from Ellesmere Island, Canadian Arctic Archipelago." *Palaeogeography, Palaeoclimatology, Palaeoecology* 30 (1980): 325–347.

Greenwood, David R., James F. Basinger, and Robin Y. Smith. "How Wet Was the Arctic Eocene Rain Forest? Estimates of Precipitation from Paleogene Arctic Macrofloras." *Geology* 38, no. 1 (2010): 15–18.

Gutjahr, Marcus, Andy Ridgwell, Philip F. Sexton, Eleni Anagnostou, Paul N. Pearson, Heiko Pälike, Richard D. Norris, et al. "Very Large Release of Mostly Volcanic Carbon During the Palaeocene-Eocene Thermal Maximum." *Nature* 548, no. 7669 (2017): 573–577.

Haynes, Laura L., and Bärbel Hönisch. "The Seawater Carbon Inventory at the Paleocene-Eocene Thermal Maximum." *Proceedings of the National Academy of Sciences of the United States of America* 117, no. 39 (2020): 24088–24095.

Jardine, Phil. "Patterns in Palaeontology: The Paleocene-Eocene Thermal Maximum." Palaeontology Online 1, no. 5 (2011): 1–7.

Kennett, James P., and Lowell D. Stott. "Abrupt Deep Sea Warming, Palaeoceanographic Changes and Benthic Extinctions at the End of the Paleocene." *Nature* 353, no. 6341 (1991): 318–322.

McIver, Elisabeth E., and James F. Basinger. "Early Tertiary Floral Evolution in the Canadian High Arctic." *Annals of the Missouri Botanical Garden* 86, no. 1 (1999): 523–545.

Pagani, Mark, N. Pedentchouk, Matthew Huber, Appy Sluijs, Stefan Schouten, Henk Brinkhuis, J. S. Sinninghe Damsté, et al. "Arctic Hydrology During Global Warming at the Palaeocene/Eocene Thermal Maximum." *Nature* 442, no. 7103 (2006): 671–675.

Secord, Ross, Jonathan I. Bloch, Stephen G. B. Chester, Doug M. Boyer, Aaron R. Wood, Scott L. Wing, Mary J. Kraus, et al. "Evolution of the Earliest Horses Driven by Climate Change in the Paleocene-Eocene Thermal Maximum." *Science* 335, no. 6071 (2012): 959–962.

Sluijs, Appy, Stephen Schouten, Mark Pagani, Martijin Woltering, Henk Brinkhuis, J. S. Sinninghe-Damsté, Gerald R. Dickens, et al. "Subtropical Arctic Ocean Temperatures During the Palaeocene/Eocene Thermal Maximum." Nature 441, no. 7093 (2006): 610–613.

Thomas, Ellen, and Nicholas J. Shackleton. "The Paleocene-Eocene Benthic Foraminiferal Extinction and Stable Isotope Anomalies." *Geological Society of London, Special Publications* 101, no. 1 (1996): 401–441.

Wing, Scott L., P. D. Gingerich, Birger Schmitz, and Ellen Thomas, eds. *Causes and Consequences of Globally Warm Climates in the Early Paleogene*. Boulder, CO: Geological Society of America, 2003

Zachos, J. C., Gerald R. Dickens, and Richard E. Zeebe. "An Early Cenozoic Perspective on Greenhouse Warming and Carbon Cycle Dynamics." *Nature* 451, no. 7176 (2008): 279–283.

FROM GREENHOUSE TO ICEHOUSE

I was totally unprepared for that revelation called the Dakota Bad Lands.
What I saw gave me an indescribable sense of a mysterious elsewhere.

—FRANK LLOYD WRIGHT, 1935

BAD LANDS, GOOD FOSSILS

The first thing that comes to mind when I hear the word "badlands" is "good fossils"! The inhospitable bare rock pinnacles of badlands found around the world are indeed a bad place to lose a cow and a bad place to do any agriculture for the farmers who first settled the remote regions of the world (figure 20.1). However, these bare rock exposures are good news for paleontologists and geologists searching for naked outcrops with their fossils weathering out at the surface. The first settlers and farmers often attached hellish or diabolical names to these forbidding landscapes where there was no soil to grow crops, and the maps of the world are littered these place names: Hell's Half Acre, Devil's Backbone, Purgatory Hill, Hell Creek, Devil's Punchbowl, and similar labels. Whenever geologists or paleontologists see names like these on a map, they are drawn to explore them.

Figure 20.1 ▲
The deeply eroded sandstones and siltstones of the White River Group form the dramatic landscape of the Big Badlands of South Dakota. (Photo by author)

But badlands are no picnic in most other ways. During the summer, the temperatures are typically over 40°C (104°F) in the daytime, and the light-colored walls of the pinnacles and ridges and canyons act like a big reflector oven, intensifying the heat. You have to wear loose-fitting light-colored clothing to keep cool, a broad-brimmed hat to keep the sun off your face, and sturdy hiking boots to climb the rough rocks. During my annual summers of research in badlands from 1977 through the late 1980s, I spent weeks in those harsh conditions, climbing steep canyon walls to carve out small oriented blocks of rock samples; by measuring how they record Earth's magnetic field, I could date the exposures. I often walked or even crawled along narrow rocky ridges to reach the top of the outcrop, not realizing that the smart way to collect fossils was to walk in the ravines at the bottom and look for pieces of bone or teeth, then follow the fossil scraps up the cliff face to their origin. Sometimes getting a paleomagnetic sample near the top of a steep cliff required climbing to the top—and I couldn't ask my student field crew to do anything I wouldn't do.

Collecting those fossils was not easy. Invertebrate fossils often occur in dense fossil beds that are easily accessible, but vertebrate fossils tend to be scarce and require days to weeks to years of hard work scouting outcrops of fossil beds in any weather and hoping to find something. I have spent many months of my life walking slowly across the exposed ground in bare rocky ravines and flats, scanning the surface in front of me for something that might be a fossil. If you bend down and pick it up, it's usually just a shiny pebble, but every once in a while I found a bone fragment or a shiny piece of enamel from a mammal tooth. I can't even estimate how many thousands of hours I spent doing fieldwork like this, finding nothing for days at a time and coming home with a only few grubby specimens that are much better represented in the big museum collections.

But the fossils are in those exposures if you know how to look, and they have a bigger story to tell as well. When I began my research in South Dakota and similar-aged beds in Nebraska, Wyoming, Colorado, and North Dakota, the Big Badlands were legendary for their fossil riches. (The South Dakota Big Badlands are capitalized because it is a formal place name, but "badlands" topography is found all over the world.) The Big Badlands were first collected in 1847 by the pioneering geological surveys sent out to explore the wild lands of the West, and their fossils were identified by America's first vertebrate paleontologist, Joseph Leidy. He recognized, named, and described the first camels, horses, rhinoceroses, and other extinct mammals discovered in North America—all were from the Big Badlands.

In the 177 years since 1847, huge collections from those rich fossiliferous deposits of the White River badlands have accumulated in many museums—but the collectors did not recorded the exact map position of each fossil nor the stratigraphic level that would enable us to date it. There was no way to plot the exact range in the rock sequence or time period for each fossil species because the collectors hadn't bothered to record it in the first place. In addition, the age of those beds was still very much up in the air because there were few volcanic ashes that could be dated by potassium-argon methods.

When looking at ideas for a dissertation topic at the American Museum in 1977, I was interested in applying the newly developed method of magnetic stratigraphy first applied to strata bearing vertebrate fossils by previous graduate students Bruce MacFadden and Steve Barghoorn at the Columbia University/American Museum program, where I had begun in 1976 (a second program in magnetic stratigraphy of fossil mammals beds developed independently at the University of Arizona). An additional advantage was that the American Museum of Natural History held the only collection of Badlands fossils with the precise stratigraphic level of each specimen recorded in detail, usually in feet above or below the nearest marker ash. These collections had been built by the legendary paleontologist Morris Skinner and his field crews in the 1950s and 1960s, and the collections (once owned by the millionaire Childs Frick, who had paid the crews to collect them) were now sitting in drawers in the American Museum, virtually unstudied. Using the available data, it would be possible to know exactly when each species vanished or first appeared, and more precise dating could be determined by using the magnetic patterns in the rocks. The prospects of a thesis project on the White River Group fossils and rocks from the Frick collecting areas all across Nebraska, Wyoming, Colorado, and the Dakotas was a very tempting idea. For the first time, we would know exactly when each species first appeared or vanished in these dense fossiliferous beds, and those changes could be correlated with the global geological time scale with great precision (also for the first time ever).

In the summer of 1978, I worked as a volunteer field assistant for Dr. Robert Emry of the Smithsonian and learned all about the White River Group. He was almost the only active expert on these rocks and fossils at that time. The following two summers I went out on my own with a volunteer field crew of two or three undergraduate students from Columbia and Vassar (where I taught part time from 1979 to 1981). I scraped together a makeshift field camp with essential gear and did both field seasons on a shoestring budget with a borrowed 1966 Ford pickup that had once been driven by Skinner. We camped on the land of the ranchers who had known Morris, and they gave us permission to work on the rocks of the old Frick localities that Skinner had so carefully collected and documented. We camped every night, and once or twice a week we'd go into town to restock our food and ice for the coolers, take a shower that we had to pay for in a private campground, and do our week's laundry.

By August the fieldwork was done, and I took my boxes of carefully collected oriented rock samples, trimmed down to one-inch cubes, to the Woods Hole Oceanographic Institute to analyze them on the cryogenic magnetometer. I had no money for lab analysis, so I accepted a graveyard shift when the magnetometer was not being used but was still full of liquid helium so it could measure specimens 24/7 if necessary. I stayed up for 36 straight hours to flip my biological clock ahead 12 hours, and I got used to sleeping away the daylight hours between each night shift. By the fall of 1980, I had the results of dozens of sections from Colorado to Nebraska to Wyoming to South and North Dakota, and I had a great magnetic correlation pattern across the entire White River outcrop. By 1981, I had documented and measured every important mammal fossil from those beds in the American Museum collections and knew precisely when they first appeared—and when they vanished—and I precisely correlated those changes with the global geologic time scale for the first time. It turned out that I had documented the best sequence of rocks and fossils spanning the Eocene-Oligocene transition in North America.

But these data had other surprising implications as well that I had no way of anticipating when I began the project. In the early 1980s, the precise dating of the Eocene-Oligocene time scale was still in doubt, and my dates and magnetic stratigraphy helped resolve that question. The controversy over "punctuated equilibria", and whether fossil species showed gradual evolution over time, was still raging, and my detailed measurements of thousands of well-dated fossils spanning millions of years of Eocene-Oligocene climate change were valuable in the context of that debate as well. The controversy over whether extraterrestrial impacts caused mass extinctions was raging in the 1980s, and my research put me at the center of expertise about another mass extinction during the Eocene-Oligocene transition, which enabled me and my colleagues to correct those who falsely claimed that those extinctions were caused by impacts on the Earth's surface.

Publication of all my magnetic sections and updating the classification of all the White River mammal fossil groups was spread over a decade and a half, but in 1996 Bob Emry and I edited a volume that put nearly all of that information into the same book, which has proven to be the key reference work on the subject. It's been years since I worked on those projects or even returned to my old stomping grounds, but I was fortunate to be in the right place at the right time—and I picked a project with *lots* of useful information that had not previously been compiled or dated. It was a fertile source of expertise for the entire Eocene-Oligocene transition, which I kept on working on well into the 2000s.

PARADISE LOST

During the Cretaceous and the early Cenozoic, warm climates with mild conditions were global, even at the poles (see chapter 19). After the middle Eocene, however, the climate began to shift rapidly into an icehouse world. This can be seen even in the badlands. The White River Group rocks of the Big Badlands of South Dakota preserve quite a bit of detail about the history of many kinds of organisms. In 1983, Greg Retallack of the University of Oregon published a study of the color bands visible in the Badlands sections and found that they were paleosols, or ancient soil horizons (figure 20.2).[1] Those from the upper Eocene Chadron Formation (the lowest exposures of the Big Badlands

Diagram of the faunal sequence and stratigraphy of the Big Badlands, showing the loss of many late Eocene groups (like the brontotheres, labeled "F" here) at the end of the Eocene, and a very different fauna in the lower Oligocene Brule Formation. (Courtesy of G. Retallack)

Cenozoic) were formed under forests with closed canopies of large trees (the huge root casts are particularly conspicuous) with between 500 and 900 millimeters (20 to 35 inches) of rainfall per year. In the overlying lower Oligocene (Orellan) Brule Formation, the paleosols indicate more open, dry woodland with only 500 millimeters (20 inches) of rainfall per year (figure 20.3).

Figure 20.3 ▲

Diorama of mammalian life in the Big Badlands, as drawn by Jay Matternes. A few late Eocene groups, like the brontotheres, lived alongside very primitive relatives of horses (*Mesohippus*), camels (*Poebrotherium*), rhinos (*Subhyracodon*), peccaries (*Perchoerus*), dogs (*Hesperocyon*), rabbits (*Palaeolagus*), and rodents, as well as archaic groups with no descendants, including oreodonts, hyracodonts, and entelodonts. (Courtesy of Wikimedia Commons)

My friend Emmett Evanoff of Northern Colorado University studied the sediments of the White River Group in eastern Wyoming and found that the moist upper Eocene floodplain deposits abruptly shifted to drier, wind-blown deposits by the early Oligocene. In the same beds are climate-sensitive land snails. According to Evanoff, late Eocene land snails are large-shelled taxa similar to those found in wet subtropical regions of modern Central America. Based on modern analogues, these snail fossils indicate a mean annual temperature of 16.5°C (63°F) and a mean annual precipitation of about 450 millimeters (18 inches), similar to the results obtained by Retallack for neighboring South Dakota. In contrast, early Oligocene land snails are drought-tolerant, small-shelled taxa indicative of warm temperate open woodlands with a pronounced dry season. Their living analogues are found today in Baja, California.

The amphibians and reptiles suggest similar trends of cooling and drying in the early Oligocene, according to Howard Hutchison, now retired from the University of California Museum of Paleontology in Berkeley. The latest Eocene was dominated by aquatic species (especially salamanders, pond turtles, and crocodilians) that had been steadily declining in the middle and late Eocene. Crocodiles were gone from the area by the late Eocene, but a few fossil alligators have been recovered from the Chadron Formation. By the early Oligocene, only land tortoises are common, indicating a pronounced drying trend. In fact, these tortoises (*Stylemys nebraskensis*) are so common in the lower Oligocene of the Big Badlands that these beds were originally called the "turtle-oreodon beds" after their two most common vertebrate fossils.

Land plants are not well-preserved in the highly oxidized beds of the Big Badlands (except for the durable hackberry seeds, which are calcified while alive), so we must look to other regions to understand the floral changes. The rest of North American floras show a clear trend (figure 20.4). Based on leaf-margin analysis by Jack Wolfe in 1971 and 1978, the mean annual temperatures in North America cooled 8°C to 12°C (13°F to 23°F) in less than a million years. This is by far the most dramatic cooling event of the entire North American floral record and was the original basis for the popular geological phrase "terminal Eocene event" (even though revised dating now places it in the early Oligocene).

The Rocky Mountains of central Colorado yielded several important floras that span the Eocene-Oligocene transition. The floras of the famous

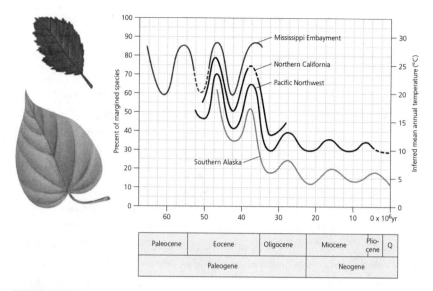

Figure 20.4 ▲

Floral evidence of climate change in the Eocene and Oligocene, as proposed by Jack Wolfe in 1978. Leaves from warm climates have smooth "entire" margins and drip tips, and those from colder climates have jagged margins. Plotting the percentage of entire margins provides a remarkably accurate estimate of ancient temperatures. Notice the large drop in temperatures at the end of the middle Eocene, and especially in the earliest Oligocene. (Redrawn from several sources)

upper Eocene Florissant Formation west of Colorado Springs records the final phase of late Eocene warmth in Wolfe's North American floral climatic curve just before the early Oligocene deterioration. Even though it was at 2,000 to 3,000 meters (6600 to 9800 feet) elevation in the Eocene, the Florissant flora is believed to represent warm temperate climatic conditions of moderate rainfall and a mean annual temperature of 13°C to 14°C (55°F to 57°F), compared to modern mean annual temperatures of 4°C (39°F). Slightly younger than Florissant is the late Eocene Antero flora. In 2008, I used magnetic stratigraphy to date these and several other floras, which showed a gradual cooling in the high-altitude Rocky Mountain floras, from Florissant and Antero to the early Oligocene Pitch-Pinnacle flora to the late Oligocene Creede flora.

The longest and most complete sequence of floras spanning the Eocene-Oligocene transition was found in the Eugene and Fisher formations near Eugene, Oregon. High-resolution lithostratigraphy, biostratigraphy, tephrostratigraphy, ^{40}Ar/^{39}Ar dating, and magnetostratigraphy by Greg Retallack and myself allowed fine-scale dating and correlation of the classic floras and marine invertebrate faunas in this sequence. These floras clearly suggest that the Oligocene deterioration had taken place 31.3 Ma. Thus the Eugene-Fisher floral sequence places climatic change in the earliest Oligocene (consistent with the global record), somewhere between 33.4 and 31.3 Ma. In summary, the cooling event found during the earliest Oligocene in the global climatic record seems to cool gradually over the late Eocene and Oligocene rather than abruptly as Jack Wolfe originally argued in 1978.

Despite these changes in the soils, land plants, land snails, and reptiles and amphibians, the change in the mammalian fauna is not that impressive (figure 20.3). Most of the archaic Eocene mammals (especially the forest dwellers and arboreal forms) were already gone well before the end of the Eocene. A few groups, such as the rhino-like brontotheres, the camel-like oromerycids, the mole-like epoicotheres, and two groups of rodents, died out near the end of the late Eocene, but none were around to witness the early Oligocene climatic deterioration. Most of the mammals that were present before the climatic crash persisted with no change whatsoever, except for a dwarfing event in one lineage. Apparently, the groups that were present in the late Eocene were already adapted to the drier and more open woodlands habitats, so the vegetation change did not make that much difference. It is also possible that the responsiveness of mammals to short-term changes in climate has been oversold and that they are not as sensitive to climate change as we have long believed.

The mammals that lived in what is now the Big Badlands during the early Oligocene were very much like those of the late Eocene, minus the huge brontotheres. Oreodonts (a group of extinct sheep-like artiodactyls that may be distantly related to camels) were by far the most common creatures, with dozens of their skulls discovered in almost every meter of the section. Early gazelle-like camels (*Poebrotherium* and *Paratylopus*) without humps were also common, as were the huge pig-like entelodont *Archaeotherium* and the tiny deer-like artiodactyls with tusks in the males rather than antlers: *Leptomeryx, Hypertragulus,* and *Hypisodus*. Among perissodactyls, the primitive three-toed horses *Mesohippus* and *Miohippus* were common, as were the hippo-like *Metamynodon* rhinos, the long-legged Great Dane–sized running rhino *Hyracodon,* and the true rhinoceros *Subhyracodon*. Tapirs were rare when compared to their great diversity in the middle Eocene. Small mammal faunas were dominated by rodents (especially archaic relatives of modern hamsters, pocket gophers, beavers, and squirrels) and an abundance of rabbits, and a whole suite of insectivorous mammals as well. Preying upon the herbivores was the last of the archaic creodonts, *Hyaenodon*, plus a much greater diversity of true carnivorans: the early dog *Hesperocyon* (which looked more like a weasel), the first members of the weasel family, plus primitive amphicyonids (known as "beardogs," a unique extinct family only distantly related to bears or dogs), and abundant nimravids, which evolved features that resembled those of saber-toothed cats even though they are not closely related to cats at all.

Most of the herbivorous mammals had fairly primitive, low-crowned dentitions for eating leaves and shrubs, and there is no evidence of grasslands or animals with high-crowned teeth for eating them yet. However, animals with extremely primitive low-crowned teeth (such as brontotheres) were gone by this time, and the softest vegetation that they fed on must have disappeared. The absence of tree-dwelling mammals such as primates and multituberculates also indicates that the dense forests must have vanished because almost all of the small mammals (even the squirrels) appear to have been adapted for life on the ground. This "White River Chronofauna" was stable and well established and would remain relatively unchanged from the late Eocene until well into the Miocene.

MAMMALS ON OTHER CONTINENTS

The Eocene-Oligocene transition on other continents has been studied as well. In contrast to the almost complete lack of change in North American mammals, the European record looks very different. In Europe there is a sharp break between the latest Eocene and early Oligocene mammal faunas, which Hans Stehlin in 1910 called the Grande Coupure, the "great cutoff" or "great break." This sharp break in the mammal faunas is not due to climate directly but to its indirect effect on sea level. Middle and late Eocene mammals of Europe are unique to that region, and they evolved in isolation in the archipelago of islands that were Europe. There are unusual families of rodents, horse-like palaeotheres and unique tapiroids, two families of archaic lemur-like primates, and six unique families of even-toed hoofed mammals in particular that are found nowhere else. The early Oligocene glaciation of Antarctica caused a substantial sea level drop worldwide, and suddenly these isolated island faunas were connected to mainland Asia. They were soon invaded by many groups of mammals already established in Asia, such as the pig-like entelodonts, hornless rhinos, early ruminants, the hippo-like anthracotheres, and a whole bunch of Asian rodent families, plus hedgehogs. By the late early Oligocene, nearly all the endemic European groups were gone, and the European mammal fauna looked much like that of Asia, from whence all those invaders came.

Climatic changes in the late Eocene and Oligocene affected plants and animals in North America and Europe. But what occurred in Asia in the late Eocene and Oligocene? The Asian Eocene-Oligocene fossil record is also excellent, not only in China and Mongolia but also in many regions of the former Soviet Union, such as Kazakhstan and Georgia, as well as regions further south and east, such as India, Turkey, the Balkans, and Pakistan. The early Oligocene in eastern Asia is known as the Shandgolian, named after the famous locality of Hsanda Gol in Mongolia. In this unit, the faunas were very different from those of the late Eocene and from those in the rest of the world. In 1998, Meng and McKenna called this dramatic drop in diversity and transformation in the dominant faunas the "Mongolian Remodeling."

It was truly a dramatic change. Many of the groups that dominated the late Eocene (Ergilian land mammal age) landscape were gone, or nearly so. The giant brontotheres, with battering-ram horns known as embolotheres, were extinct, as were the last of the predatory giant mesonychids and most of the tapiroids that dominated in the late Eocene. Both hyracodont rhinos and true rhinos became diverse and common, as were deer and a variety of deer-like artiodactyls. Several new groups occurred for the first time, including a variety of ruminant artiodactyls, several advanced rodents, plus a whole suite of advanced carnivorans: beardogs, weasels, and the first true cats. These shared the predator niche with the hyaenodonts, the cat-like nimravids, and civets, which were already established.

Although this list of characters seems similar to the early Oligocene faunas of other continents, there are important differences. The diversity of late Eocene large mammals is reduced, with only rhinos and entelodonts occupying the large body size niche. The assemblage has almost no midsized mammals, only giant rhinos and entelodonts and a wide array of rodents and rabbits and smaller carnivores. This is typical of the Sahara Desert today, where only a few large mammals (e.g., camels, wild asses, Arabian oryx) can survive and there are few or no medium-sized ground-dwelling or tree-dwelling mammals. The bulk of the mammalian diversity occurred in rodents or rabbits, which can burrow.

The Hsanda Gol fauna is overwhelmingly dominated by rodents and rabbits, many with high-crowned teeth for eating gritty vegetation. According to Meng and McKenna, the overall faunal composition suggests that the environment was very arid, with few tall trees for arboreal species. Almost all the land mammals are either adapted for feeding on the tops of the trees (like the giant indricothere rhinos) or lived close to the ground and within the limited brush cover (most of the deer-like artiodactyls, enteledonts, and their predators, plus the great variety of ground-dwelling and burrowing rodents and rabbits). Studies of the pollen from the Oligocene of China concluded that much of the region was covered by a woody scrubland, with many arid-adapted plants, such as saltbush and *Ephedra*, or Mormon tea, as well an *Nitraria* (nitre bush, a salt-tolerant desert shrub). Trees would have been rare and concentrated in patches where groundwater could be found, such as in riparian habitats near river courses. In Siberia and Kazakhstan, these trees included hardwoods such as oaks, as well as members of the walnut and birch families. In Mongolia and China, the Oligocene tree pollen is mostly from the birch and elm families, as well as oaks and other broadleaf deciduous trees. These would have been the main fodder for the tree-feeding indricotheres.

Above all, Hsanda Gol and other Oligocene rocks of Asia are famous for the gigantic rhinoceros *Paraceratherium*, formerly called *Baluchitherium* and *Indricotherium* (figure 20.5). This beast was 6 meters (18 feet) tall at the shoulder and probably weighed 15 to 20 tonnes. Its head was so high off the ground that it probably browsed in the tops of trees. It was taller even than modern elephants and probably lived a lot like elephants do today. Small herds must have roamed from place to place eating all the vegetation (especially the high vegetation unreachable by small mammals). Like elephants, this animal would have fed almost continuously to fuel its huge bulk and spent some of its day sleeping in the shade and keeping cool in water holes, roaming and eating mostly during the night. The herds would have been small, probably composed mostly of mothers and their calves and their female kin. The adult indricotheres would have had no problems with the wolf-sized predators of the time, such as

Figure 20.5 ▲

Life-sized reconstruction of the gigantic indricothere rhinoceros *Paraceratherium*. Its ancestor, the little running rhino *Hyracodon*, stands beneath it, and a modern elephant provides a sense of scale. (Courtesy of Wikimedia Commons)

hyaenodonts. Also like elephants, they would have lived in herds to protect their young because the young would have feared the local predators. The young would have been born after a two-year gestation, with only one calf every few years.

In summary, although Shandgolian mammals show many families and genera in common with North America (especially the rhinos, entelodonts, anthracotheres, ruminants, plus many of the same weasels, beardogs and nimravids), the fauna was much more arid-adapted than that of the early Oligocene in North America or the early Oligocene of Europe. In Asia, many more mammals had high-crowned teeth, and much of the floral evidence suggests grittier diets and scrubbier vegetation. Although many of these Asian groups migrated to Europe during the Grande Coupure to drive out the native fauna, Europe in the early Oligocene was still a much wetter and milder place than either Asia or North America.

A recent study of African mammals shows that Africa also had an abrupt change in faunas in the earliest Oligocene. The archaic relatives of the elephants and mastodons (such as barytheres and arsinoitheres), many endemic groups of rodents and lemur-like primates, the predatory hyaenodonts, and several other groups vanished after the late Eocene. In the early Oligocene, a much lower diversity of mammals is found in Africa and Arabia, characterized by primates and rodents with teeth better adapted for grinding tougher seeds and vegetation, as well as groups of animals like anthracotheres, which crossed from Asia. The transition in South America is not nearly as well documented, but for some reason many South American hoofed mammals already had high-crowned teeth for eating tough, gritty vegetation in the Eocene, so the cooling and drying effects of climate in the early Oligocene were not as dramatic there.

THE GLOBAL EOCENE-OLIGOCENE TRANSITION

Extrapolating from the Big Badlands, the warm early and middle Eocene greenhouse world, with high carbon dioxide in the atmosphere and temperate forests and alligators on the poles, began to change rapidly at the end of the middle Eocene through the early Oligocene. The records from plants on land and oceanic geochemistry both show this dramatic change. The first pulse occurred at the middle-late Eocene transition (37 Ma), when a dramatic cooling of oceanic bottom waters occurred, resulting in 4°C to 5°C (39°F to 41°F) of global cooling in the world oceans (more change than during any ice age cycle), and a temperature drop of 14°C to 16°C (57°F to 61°F) on some of the continents, as indicated by fossil leaves. There was a mass extinction in the tropical marine plankton, and significant extinction in the subtropical clams and snails that lived in the shallow waters of the U.S. Gulf Coastal Plain.

The next pulse of cooling occurred in the earliest Oligocene (33 Ma). Cores drilled from around the margin of Antarctica show clear evidence of large ice sheets then, proving that polar ice caps had returned for the first time since they vanished from Gondwana in the Permian. The world oceans cooled about 5°C to 6°C (41°F to 43°F) in average temperature, a catastrophic cooling event for tropical and subtropical species of plankton. In fact, the diversity of planktonic foraminifera was at an all-time low, confined mostly to species that were adapted to colder waters. Once again, the tropical snails, clams, and sea urchins so typical of the middle Eocene of the Gulf Coast, the Pacific Coast, and many other shallow seas, were nearly wiped out with a major extinction of most of the tropical species.

In the Oligocene, the archaic Eocene whales were replaced by the two main groups of whales that rule the oceans today: the toothed whales (such as dolphins, porpoises, and sperm whales) and the toothless baleen whales (blue whales, humpback whales, gray whales, and their kin). These whales have a tough, flexible net of baleen in their mouths to screen out and trap their tiny planktonic food (mostly crustaceans known as krill). Baleen whales feed by gulping a huge volume of seawater and plankton and forcing the water out of their mouth cavity while trapping the food in the baleen screen. Many scientists think the explosive evolutionary radiation of baleen whales in the Oligocene and Miocene is due to the large increase in plankton in the Southern Ocean as the Antarctic Circumpolar Current developed. Baleen whales include the largest animals that have ever lived, the blue whale. At 30 meters (100 feet) in length and 190 tonnes (210 tons) in weight, it is larger than any animal that has ever existed, including the largest dinosaurs and the most monstrous of all the marine reptiles.

WHAT CAUSED THIS CLIMATE CHANGE?

The overarching theme in the middle and late Eocene was the rapid decline in atmospheric carbon dioxide, shown by many temperature proxies from sediments in deep sea cores. This trend is well established, but where did all the carbon go? One of the leading ideas is referred to as the "Himalayan weathering hypothesis." The collision of India with Asia began in the early Eocene, and some scientists have argued that the ensuing uplift of the Himalayas would have caused enormous amounts of weathering as the mountains rose into the sky. Increased weathering of soils is one of the mechanisms by which carbon is drawn out of the atmosphere and incorporated into the crust. The principal problem with this idea is that the evidence so far shows that the Himalayan uplift occurred mostly in the Miocene, much too late to explain the cooling at the end of the middle Eocene. More recent analyses have suggested that the northern Tibetan Plateau may have been at high elevations 40 million years ago, but this does not explain why the biggest cooling events occurred 37 and 33 million years ago. Himalayan uplift may have been a major contributor toward the gradual global cooling in the Miocene, but it does not explain the abrupt stepwise cooling events at the Eocene-Oligocene transition.

The other issue is that the slow drawdown of carbon dioxide does not seem to explain the two pulses of cooling through the long interval of changing climate, culminating in an icehouse world by the early Oligocene, and the first polar ice caps since the mid-Permian 290 million years ago. The effect shows up dramatically in land plants where a cooling event of about 15°C to 20°C (59°F to 68°F) is suggested by the leaves as well. Something more abrupt than uplift and weathering is needed. Since the 1970s, the leading candidate has been the enormous Antarctic Circumpolar Current (ACC), which is the main driver of modern ocean temperature and circulation (figure 20.6). It is the largest and fastest of the world's ocean currents, traveling eastward and clockwise around Antarctica (as seen from above the South Pole) at a rate of 25 centimeters per second, with a volume of 233 million cubic meters per second moving past Antarctica, more than 1,000 times the flow of the world's largest river, the Amazon. The ACC has several effects that control the modern oceans. First, this current brings up huge volumes of nutrients from deeper water, producing enormous plankton blooms in the Antarctic waters that are food for their large numbers of baleen whales, as well as penguins, fish, seals, and all other Antarctic marine life.

Another effect is that the cold, oxygen-rich Antarctic surface water then sinks to the bottom of the ocean to become the Antarctic Bottom Water (AABW), the deepest of all the deep ocean water masses. It makes up about 59 percent of the total deep water in the ocean, and it flows along the seafloor all the way to the North Atlantic and North Pacific. This cold oxygen-rich current allows deep sea animals to live in very deep ocean bottoms, even with extremely high pressures and cold temperatures. It also

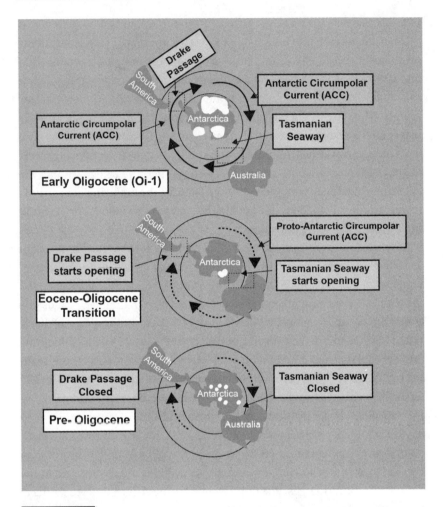

Figure 20.6 ▲

Development of the Antarctic Circum-Polar Current (ACC) in the Eocene and Oligocene. In the pre-Oligocene (*bottom*) warm tropical currents flowed all the way to Antarctica, keeping the climate from getting too cold. Once the ACC had developed (*middle and top*) it isolated Antarctica and surrounded the continent with a barrier of cold water that locked in the cold over the South Pole like a refrigerator door, forming ice sheets in the early Oligocene. (Redrawn by E. Prothero from several sources)

dictates the flow of other deep water currents, such as the North Atlantic Deep Water (NADW) and the Antarctic Intermediate Water (AAIW). The deep water currents of the modern ocean were established in the early Oligocene when the ACC first developed.

But the most striking effect of the ACC is that it acts like a "refrigerator door," isolating the colder waters and air of Antarctica from warmer tropical and subtropical waters. This allowed the cold conditions of the South Pole to be "locked in" and to develop permanently frozen ice caps. During the Paleocene and Eocene, Antarctica and Australia were just beginning to separate, so there was no passage for deep water between them. Likewise, South America was connected to Antarctica, further blocking any Antarctic flow. Warm tropical and subtropical currents in the Eocene could flow all the way to the shores of Antarctica and prevent freezing. But sediments from deep sea cores south of New Zealand clearly show that deep water was passing through the gap between Antarctica and Australia in the early Oligocene, and there is some evidence that it was also passing through the Drake Passage between South America and Antarctica (although the tectonics and timing of this region is complicated). Once the ACC became established, it was the primary reason Antarctica froze over, with the first ice caps appearing 33 million years ago. The rearrangement of ocean currents can explain the drop in temperatures, but this idea is limited because it does not include a mechanism for removing carbon dioxide from the atmosphere. That would require something else, and scientists have not reached a consensus on this yet.

Greg Retallack argued that the evolution of early bunch grassland and scrubland habitats, and their soils, was the new development in the early Oligocene that absorbed all the carbon dioxide. This is supported by the work of Caroline Strömberg of the University of Washington, who found evidence of the tiny silica-rich structures known as phytoliths, which are known from many types of grasses. These grassy habitats were not the full-fledged savanna grasslands that occurred in the Miocene, but grasses and associated floras are huge absorbers of carbon dioxide: they grow very fast, and much of that carbon is deposited in the characteristic soils they leave behind. Eventually these grassy scrublands transformed into savannas as the Miocene climate became cooler and drier (see chapter 21).

The debate over the cause of the Eocene-Oligocene climate change is still unresolved and subject to active debate. Another explanation may come along that no one has yet considered.

DO IMPACTS ALWAYS CAUSE EXTINCTION?

During the peak of the debate about mass extinctions in the early 1980s, the scientific bandwagon rushed to blame *all* mass extinction on impacts. Only two years after publication of the Alvarez and others K/Pg impact hypothesis, two different groups of authors published separate papers claiming that iridium was associated with the Eocene-Oligocene extinctions. People were looking for iridium everywhere, and some geologists were claiming that impacts causing global changes and extinctions were a new scientific revolution in geology (see chapter 18).

In fact, there are several well-dated Eocene impact events, one of which formed the crater that became Chesapeake Bay, and another called Popigai in Siberia (figure 20.7A). They produced iridium abundance spikes in deep sea cores, and tiny droplets of melted crustal rock called microtektites also showed up in those cores. But when these data were scrutinized more carefully, the iridium and microtektites turned out to be evidence *against* the idea that impacts caused most extinctions. They are dated at 35.5 to 36 million years ago, right in the middle of the late Eocene—about 3 million years too early for the Oligocene extinction and 1.5 to 2 million years too late for the middle/late Eocene extinction. Detailed studies of deep sea cores show almost *no* extinction at the time of the impact, except for a few species of plankton.

The size of the craters these impacts produced is enormous, just slightly smaller than the Chicxulub crater blamed for the Cretaceous extinctions. If you fit a "kill curve" to the K/Pg impact event alone (figure 20.7B), it appears that any large impact should cause significant extinction. But when you add the Chesapeake and Popigai impacts to the curve, its shape changes dramatically, suggesting that *only* an event of the magnitude of Chicxulub or larger could cause mass extinction—anything less than that would not produce a mass extinction.

Clearly, impacts do not automatically cause mass extinctions, and only the KPg mass extinction event is associated with impacts. In the 1980s and 1990s, scientists got carried away with the idea that all impacts cause some kind of extinction, but today we know that most impacts have no effect at all. Only the largest collision is capable of triggering a mass extinction. In the late Cretaceous, the huge Deccan lava eruptions and a major sea level drop were both occurring at the same time as the impact, which complicates the story. If there is any common cause for many of the mass extinctions, it appears that gigantic volcanic eruptions, not impacts, are most important—a point that the French volcanologist Vincent Courtillot argued in the 1980s. The Permian extinction was caused by the Siberian eruptions, and the Deccan lavas are correlated with the KPg event. The Triassic

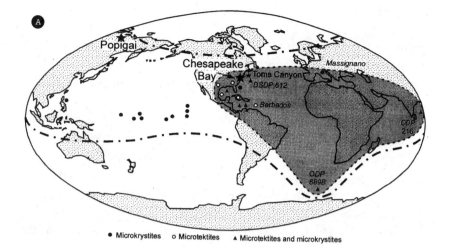

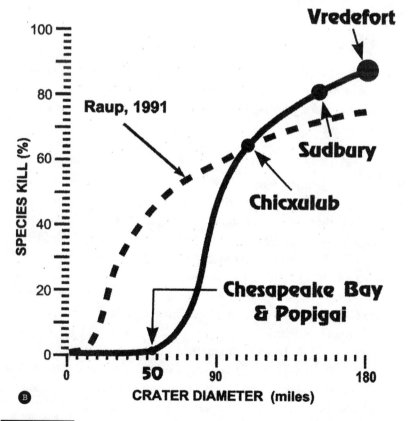

Figure 20.7 ▲

(*A*) Map showing the distribution of late Eocene microtektites from impacts at Popigai, Siberia, as well as Chesapeake Bay, and Toms Canyon. (*B*) Raup's initial "kill curve" (dashed curve) had to be heavily modified due to the lack of any extinction in the late Eocene in response to the Popigai and Chesapeake Bay impacts (solid curve). Clearly, any impact smaller than Chicxulub at the KPg boundary causes almost no extinction. (Redrawn from several sources)

extinction seems to have been caused by the eruption of the Central Atlantic Magmatic Province. Large volcanic events are now associated with the Late Ordovician (see chapter 11) and Late Devonian (see chapter 12) extinctions as well.

Once again, the Eocene-Oligocene is the exception to that rule. Some Oligocene volcanic events have been pointed to by Courtillot and others, but they do not occur precisely at 37 or 33 million years ago, when the largest extinction pulses occurred. The Eocene-Oligocene extinction is the great exception that shot down the fantasies of both impact advocates and volcanologists. Its causes are still unclear, but this extinction seems to have been caused by rapid cooling and changes in oceanic circulation, not impact events or volcanic eruptions.

NOTE

1. G. J. Retallack, "A Paleopedological Approach to the Interpretation Of Terrestrial Sedimentary Rocks: The Mid-Tertiary Fossil Soils of Badlands National Park, South Dakota," *Bulletin of the Geological Society of America* 94 (1983): 823–840.

FOR FURTHER READING

DeConto, Robert M., and David Pollard. "Rapid Cenozoic Glaciation of Antarctica Induced by Declining Atmospheric CO_2." *Nature* 421 (2003): 245–249.

DeVries, D., S. Heritage, M. R. Borths, H. R. Sallam, and E. R. Seiffert. "Widespread Loss of Mammalian Lineages and Dietary Diversity in the Early Oligocene of Afro-Arabia." *Communications Biology* 4, no. 1172 (2021): 1–9.

Hooker, Jerry J., Margaret E. Collinson, and N. P. Sille. "Eocene-Oligocene Mammalian Faunal Turnover in the Hampshire Basin, UK: Calibration to the Global Time Scale and the Major Cooling Event." *Journal of the Geological Society* 161, no. 2 (2004): 161–172.

Hutchinson, David K., Helen K. Coxall, Daniel J. Lunt, Margret Steinthorsdottir, Agatha M. De Boer, Michiel Baatsen, Anna von der Heydt, et al. "The Eocene-Oligocene Transition: A Review of Marine and Terrestrial Proxy Data, Models, and Model-Data Comparisons." *Climates of the Past* 17, no. 1 (2020).

Ivany, Linda C., William P. Patterson, and Kyger C. Lohmann. "Cooler Winters as a Possible Cause of Mass Extinctions at the Eocene/Oligocene Boundary." *Nature* 407, no. 6806 (2000): 887–890.

Kennett, James P. "Cenozoic Evolution of Antarctic Glaciation, the Circum-Antarctic Ocean, and Their Impact on Global Paleoceanography." *Journal of Geophysical Research* 82 (1977): 3843–3860.

Koeberl, Christian, C. Wylie Poag, Wolf U. Reimold, and Dion Brandt. "Impact Origin of the Chesapeake Bay Structure and the Source of the North American Tektites." *Science* 271, no. 5253 (1996): 1263–1266.

Kraatz, Brian P., and Jonathan H. Geisler. "Eocene–Oligocene Transition in Central Asia and Its Effects on Mammalian Evolution." *Geology* 38, no. 2 (2010): 111–114.

Liu, Zhonghui, Mark Pagani, David Zinniker, Robert Deconto, Matthew Huber, Henk Brinkhuis, Sunitar R. Shah, et al. "Global Cooling During the Eocene–Oligocene Climate Transition." *Science* 323, no. 5918 (2009): 1187–1190.

Meng, Jim, and Malcolm C. McKenna. "Faunal Turnovers of Palaeogene Mammals from the Mongolian Plateau." *Nature* 394, no. 6691 (1998): 364–367.

McGowran, B. "Fifty Million Years Ago." *American Scientist* 78 (1990): 30–39.

Molina, Eustoquio, Concepción Gonzalvo, Silvia Ortiz, and Luis E. Cruz. "Foraminiferal Turnover Across the Eocene–Oligocene Transition at Fuente Caldera, Southern Spain: No Cause-Effect Relationship Between Meteorite Impacts and Extinctions." *Marine Micropaleontology* 58, no. 4 (2006): 270–286.

Pagani, Mark, Matthew Huber, Zhonghui Liu, Steven M. Bohaty, Jorijntje Henderiks, Wiiliam Sijp, Srinath Krishnan, and Robert M. DeConto. "The Role of Carbon Dioxide During the Onset of Antarctic Glaciation." *Science* 334, no. 6060 (2011): 1261–1264.

Pearson, Paul N., and Martin R. Palmer. "Atmospheric Carbon Dioxide Over the Past 60 Million Years." *Nature* 406, no. 6797 (2000): 695–699.

Poag, C. Wiley. *Chesapeake Invader: Discovering America's Giant Meteorite Crater.* Princeton, NJ: Princeton University Press, 1999.

Prothero, Donald R. *After the Dinosaurs: The Age of Mammals.* Bloomington: Indiana University Press, 2006.

——. *Greenhouse of the Dinosaurs: Evolution, Extinction, and the Future of Our Planet.* New York: Columbia University Press, 2009.

Prothero, Donald R., and William A. Berggren, eds. *Eocene–Oligocene Climatic and Biotic Evolution.* Princeton, NJ: Princeton University Press, 1992.

Prothero, Donald R., and Robert J. Emry, eds. *The Terrestrial Eocene–Oligocene Transition in North America.* Cambridge: Cambridge University Press, 1996.

Prothero, Donald R., Linda C. Ivany, and Elizabeth A. Nesbitt, eds. *From Greenhouse to Icehouse: The Marine Eocene–Oligocene Transition.* New York: Columbia University Press, 2003.

Retallack, Gregory J. "Global Cooling by Grassland Soils of the Geological Past and Near Future." *Annual Reviews of Earth and Planetary Science* 41, no. 1 (2013): 69–86.

Ruddiman, William J. *Earth's Climate: Past and Future*. New York: Freeman, 2013.

Strömberg, C.A.E. The origin and spread of grass-dominated ecosystems in the Late Tertiary of North America: Preliminary results concerning the evolution of hypsodonty: *Palaeogeography Palaeoclimatology Palaeoecology*, 177 (2002): 59–75.

Sun, Jimin, Xijun Ni, Shundong Bi, Wenyu Wu, Jie Ye, Jin Meng, and Brain F. Windley. "Synchronous Turnover of Flora, Fauna, and Climate at the Eocene–Oligocene Boundary in Asia." *Scientific Reports* 4, no. 1 (2014): 1–6.

Wolfe, Jack A. "A Paleobotanical Interpretation of Tertiary Climates in the Northern Hemisphere." *American Scientist* 66 (1978): 694–703.

Zachos, James C., Terrence M. Quinn, and Karen A. Salamy. "High-Resolution (104 Years) Deep-Sea Foraminiferal Stable Isotope Records of the Eocene–Oligocene Climate Transition." *Paleoceanography* 11, no. 3 (1996): 251–266.

Zhang, Rui, Vadim A. Kravchinsky, and Leping Yue. "Link Between Global Cooling and Mammalian Transformation Across the Eocene–Oligocene Boundary in the Continental Interior of Asia." *International Journal of Earth Sciences* 101, no. 8 (2012): 2193–2220.

AMERICAN SERENGETI

During the late Miocene, North America supported its richest array of large herbivores since the late Mesozoic. Quarries in continental sediments of this age regularly produce 15 or more species of herbivorous vertebrates with body weight exceeding 5 kg. This great assemblage of ungulates included a mixture of apparent browsers and apparent grazers; presumably it lived on a rich savanna composed of trees, grasses, and herbs. The analogy between the North American savanna ungulate fauna of the late Miocene and the present African ungulate fauna has become a fundamental tenet of studies on the subject.

—S. DAVID WEBB, "THE RISE AND FALL OF THE LATE MIOCENE UNGULATE FAUNA IN NORTH AMERICA" (1983)

THE GROWTH OF GRASSLANDS

Almost every nature documentary about the East African savanna in legendary places like the Serengeti plains features its legendary creatures: lions, hyenas, jackals, elephants, giraffes, warthogs, rhinos, hippos, Cape buffalo, zebra, and a great diversity of antelopes. They also show extensive areas of grasslands and small patches of forest and scrub brush, which support a huge diversity of mammals familiar from wildlife documentaries. The modern East African savanna is a surviving remnant of what was once the prevailing habitat across most temperate and dry tropical regions in the middle Miocene. But they were not the only place in the world where such Miocene savanna grasslands and their mammals developed.

Imagine that you could take a time machine back to Nebraska about 10 Ma. The Great Plains of North America once supported its own equivalent of the East African savanna like the Serengeti plains, with different animals playing the roles performed by the groups of mammals found in the modern savanna of Africa (figure 21.1). It's like watching the same play, but with a

Figure 21.1 ▲

Reconstruction of the American savanna by Jay Matternes, showing the major groups of mammals in the middle and late Miocene, which performed the same ecological roles of animals in the modern African savanna, but with different families replacing the modern savanna mammals. Not only are there multiple species of horses, but there are camels performing the roles of giraffes and antelopes, pronghorns in the antelope role, and deer-like dromomerycids. Rhinos include hippo-like *Teleoceras* as well as more normal rhinos like *Aphelops*. Gomphothere mastodonts are in the roles of elephants. (Courtesy of Wikimedia Commons)

different cast or understudies playing the same roles. In North America, the paleontologist S. David Webb named this African-style mammalian assemblage the Clarendonian Chronofauna because it became a homogeneous established entity beginning in the middle Miocene. In North America, the middle Miocene is known as the Barstovian land mammal age (16 to 11.5 Ma), named after deposits near Barstow, California. It continued through the late Miocene (Clarendonian land mammal age, 11.5 to 8.5 Ma, and Hemphillian land mammal age, 8.5 to 4.5 Ma), and then vanished in the early Pliocene (Hemphillian/Blancan boundary in North America, about 4.5 Ma).

Instead of elephants, North America had primitive mastodonts that had arrived from the Old World about 18 Ma. Camels performed the roles of giraffes, with *Aepycamelus giraffinus* reaching 3.5 meters (11 feet) at the shoulder and 6 meters (19 feet) tall and feeding from the tree canopy (figure 21.2A). Six other genera of camels also occupied different habitats, but they were built more like antelopes or gazelles or llamas. The antelope role was also played by a huge diversity (9 genera and dozens of species) of native pronghorns, which are not related to antelopes at all although they are mistakenly called "antelopes" (figure 21.2B). (True antelopes live only in Asia and Africa and are members of the family Bovidae, along with cattle, sheep, and goats. Pronghorns belong to their own family Antilocapridae). Instead of true deer (family Cervidae), North America was dominated by an extinct group distantly related to giraffes known as palaeomerycids or dromomerycids, which sported a variety of bony horns in many shapes rather than antlers (figure 21.2C). North America also had a large radiation of blastomerycids, which may be related to the musk deer (family Moschidae), found only in Asia today.

Even more impressive was the great diversity of horses. In one quarry alone, Railway Quarry A near Valentine, Nebraska, more than 12 species of horses were found that all apparently lived at the same time! Some Miocene horses were browsers that ate leaves with their low-crowned teeth (*Archaeohippus, Hypohippus, Megahippus*), but most had higher-crowned (hypsodont) teeth for eating at least some grasses along with leaves. The last of the camel relatives known as protoceratids behaved like moose or bushbuck in the thick underbrush, and the most extreme protoceratid, *Synthetoceras*, had a peculiar slingshot-shaped branched horn on its nose and two horns over its eyes (figure 21.1D). Instead of hippos, the role of aquatic grazers was performed by the rhino known as *Teleoceras*, which had the short stumpy

Figure 21.2 ▲ ▶

Reconstructions of some of the more unusual American hoofed mammals of the Miocene savanna: (A) the giraffe-like camel *Aepycamelus giraffinus*; (B) the primitive pronghorn *Merycodus furcatus*; (C) the deer-like *Dromomeryx borealis*; (D) the bizarre horned mammal *Synthetoceras tricornatus*; (E) the hippo-like rhinoceros *Teleoceras fossiger*; and (F) the peccary with extremely wide flanges on its cheekbones, *Mckennahyus parisidutrai*. ([A–C and E] Art by R. Bruce Horsfall, courtesy of Wikimedia Commons; [D] art by N. Tamura; [F] art by K. Marriott).

Figure 21.2 ▲ ▶

(*continued*)

Figure 21.2 ▲ ▶

(*continued*)

F

Figure 21.2 ▲
(*continued*)

legs and barrel chest of a hippo and high-crowned teeth for eating grasses (figure 21.2E). The extinct aceratherine rhinos (*Aphelops, Peraceras*) were built much like the modern leaf-eating browsers such as the black rhino. Instead of true pigs (such as warthogs) found in the African savanna, that role was played by peccaries or javelinas (family Tayassuidae), which are only distantly related to pigs (family Suidae). Peccaries flourished in North America for most of their history, and they spread to South America when the Panamanian land bridge opened. True pigs were always an Old World group, and they didn't reach America until European settlers brought them here. Late Miocene peccaries had wicked-looking flanges and ridges on their cheekbones that apparently protected their eyes during combat with other males, or were used to advertise their status in their sounders (figure 21.2F).

The predatory roles were occupied by hyena-like borophagine dogs (figure 21.3A), bears, weasels, and beardogs, instead of cats and hyenas that rule the African savanna today. Although squirrels (sciurids) and beavers (castorids) are common among the rodents, the greatest diversification occurred in the cricetids (New World mice and hamsters), the geomyids (gophers), heteromyids (pocket mice), zapodids (jumping mice), and in the lagomorphs, or pikas and rabbits. These groups are often associated with prairies and grasslands, whereas tree-dwelling squirrels and beavers tend to be denizens of wetter and more forested habitats.

Figure 21.3 ▲ ▶

Some large predatory mammals of the American Miocene savanna: (A) the hyena-like borophagine dog *Epicyon*; (B) the saber-toothed *Barbourofelis*; and (C) the cheetah-like hyena *Chasmoporthetes*. ([A–B] Courtesy of Wikimedia Commons; [C] courtesy of H. Galiano)

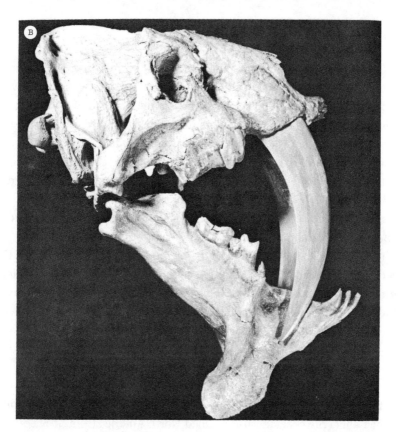

Figure 21.3 ▲

(*continued*)

The Clarendonian Chronofauna was a stable entity with most of its genera persisting through much of the middle and late Miocene, but there were changes as well. The late Barstovian and Clarendonian marked the decline of many groups (especially browsers) that had dominated the earlier Miocene and even the Oligocene. The last of the oreodonts (an extinct group of mammals built like sheep but possibly related to camels) disappeared in the early Hemphillian, about 7 Ma, as did the last of the blastomerycids, or musk deer. The deer-like dromomerycids were very rare after their evolutionary peak during their Barstovian radiation. The protoceratids were also very rare and vanished at the end of the Hemphillian. The last of the horses with low-crowned teeth (anchitheriine horses) died out at the end of the Clarendonian too. Among carnivorans, the last of the beardogs (amphicyonids) disappeared in the Clarendonian, replaced by both bears and dogs, as well as a great radiation of mustelids (weasels, otters, badgers) and true cats (Felidae), plus one last surviving cat-like nimravid ("false cat"), the saber-toothed *Barbourofelis* (figure 21.3B), which persisted until the early Hemphillian. Today hyenas are strictly an African group, but in the late Miocene one hyena, *Chasmoporthetes* (figure 21.3C), reached North America. Unlike most hyenas, it was a long-legged runner built more like a cheetah. (True cheetahs also reached North America from Africa during the Ice Ages.)

Among the small mammals, the hedgehogs vanished from North America in the Clarendonian, along with most of the once-abundant aplodontids (sewellels), as well as the rodents with paired horns on their noses known as mylagaulids, and the archaic eomyid rodents, relicts of the Eocene. The striking change in the Clarendonian Chronofauna through most of the late Miocene was the loss of browsing (leaf-eating) forms with low-crowned or brachydont teeth, as grazers (grass-eaters) with high-crowned or hypsodont teeth gradually took over. Such browsers included the oreodonts, blastomerycids, dromomerycids, protoceratids, the rhino *Peraceras*, the anchitheriine horses, and the archaic rodents (aplodontids, mylagaulids, and eomyids). What remained by the Hemphillian was a fauna dominated by grazers with very high-crowned teeth, including the advanced equine horses, camels, pronghorns, and later species of the rhinos *Aphelops* and *Teleoceras*.

According to paleobotanical data, there were swamps and dense forests in the humid Atlantic and Gulf Coast Plains, and these regions supported a peculiar forest-adapted fauna with abundant protoceratids, weird long-nosed camels, and dwarf rhinoceroses. In the Rockies and the Great Basin, there were abundant pines and other conifers, as well as trees of the walnut, oak, elm, birch, and willow families, and scrub brush dominated by Mormon tea (*Ephedra*), greasewood, sagebrush, and herbs of the family Chenopodiaceae. These plants supported the huge community of grazing mammals of the middle and late Miocene.

CHEMICAL MISMATCHES

For decades, the classic story was that extensive grasslands arose as soon as 16 Ma in the early middle Miocene. This was based on the increasingly high-crowned (hypsodont) ever-growing teeth seen in horses, camels, pronghorns, and rhinoceroses. But middle Miocene fossil grass pollen is relatively rare, and only a handful of grass fossils are known. This striking rarity of grass fossils has always been a puzzle, and it has been further confirmed by the evidence of carbon isotopes.

Most plants (including high-latitude and high-altitude grasses) use the Calvin (C3) photosynthetic pathway, which produces a carbon-isotope signal with lighter carbon values (richer in carbon-12). But most temperate and tropical tall grasses use another photosynthetic mechanism, the Hatch-Slack pathway (C4), which produces much heavier (richer in carbon-13) isotopic values. Studies of the isotopes of carbonate in ancient soils, and also the carbon isotopes in the fossil teeth of herbivorous mammals, revealed that large areas of C4 grassland savanna did not occur until about 7 Ma.

If this is true, what were all those mammals with high-crowned teeth eating between 16 Ma and 7 Ma? There may have been grasslands that used the C3 pathway, but there are no modern analogues for this. So what explains the high-crowned teeth which are more resistant to wearing out? The flora may have been a mixture of grasses and scrub, often with a lot of dirt and grit on it, which would have required higher-crowned teeth. According to Greg Retallack, these sod-forming short grasslands were comprised of mixed scrubland and grassland—not the huge areas of tall grasslands we now see in Africa or North America. If the carbon isotope data are right, true C4 grasslands forming a tall-grass prairie did not emerge until 7 Ma. Caroline Strömberg of the University of Washington has shown that the tiny siliceous fossils of plant cells of grasses (known as phytoliths) show up in the early Miocene, well before there were a lot of mammals with hypsodont teeth. Clearly, the puzzle of the origin of grasslands and grazing mammals is not yet fully resolved.

GLOBAL GRASSLANDS

The patterns of the Clarendonian Chronofauna also apply to Eurasia and Africa, albeit with different casts of characters among the fossil mammals. In Europe, the subtropical evergreen forests of the early Miocene were replaced by seasonal, summer, drought-adapted forests with tough woody vegetation. As the Miocene progressed, this dense forest gave way to more open habitat, although huge plains covered with true grasslands found in North America and Asia never appeared in Europe during the Miocene. An assemblage of mammals adapted for eating this tough vegetation lived in these forests, and true grazers gradually appeared by the late Miocene as patches of grassland developed. They are recorded from localities all over Eurasia, including the famous bone beds in Spain, France, Germany, and Italy; in Pikermi and Samos in Greece; in Maragheh in Iran; in the Siwalik Hills of northern Pakistan; in the Tunggur beds in Mongolia; and in many deposits in China. The hoofed mammals were largely groups that dominate in the Old World today. There was a wide radiation of bovids (antelopes and cattle) including huge 300 kilogram (660 pound) boselaphines (primitive forms related to the modern nilgai antelope), as well as gazelles and several extinct groups. Early Miocene giraffes were built more like antelopes as well, and some had a pair of short unbranched horns or flat sideways-extended horns. By the late Miocene, there were huge giraffes (*Brahmatherium*) with thick, short necks and moose-like horns. The middle Miocene deer (Cervidae) were mostly small primitive forms with no antlers; but the males had large canines, like modern musk deer. By the late Miocene, however, deer really began to diversify, with a wide variety of different types of antlers. The pigs enjoyed a great radiation, including weird forms such as the "unicorn pig" *Kubanochoerus*, which had a straight conical horn pointing out of its forehead.

Both aceratherine and teleoceratine rhinos flourished (as they did in North America), along with the earliest members of the living lineage of Sumatran rhinos (dicerorhinines). There was also a group of rhinos called elasmotheres that had unusually high-crowned molars with convoluted, crenulated enamel on them. Horses were also abundant, primarily the three-toed grazing hipparionine horses, which migrated to Eurasia from North America in several separate immigration events in the Miocene. Gomphothere mastodonts were the largest herbivores, along with the peculiar deinotheres with their downward-pointing lower tusks. Another African immigrant, the hyrax *Pliohyrax*, was found in the streams along with a high diversity of true pigs. As happened in North America, the last of the chalicotheres (the gorilla-like *Chalicotherium*) in Eurasia died in the middle Miocene, and the beardogs vanished from Europe by the late Miocene, replaced by true bears and true cats, as well as civets and mustelids. However, true dogs were not yet found in the Old World, and their roles were played by a diversification of hyenas, many of which were more like cats and dogs than modern hyenas.

As in North America, squirrels and beavers were still common, but there was a huge radiation of cricetids (New World mice and hamsters), jumping mice, and especially murids, the modern family of Old World rats and mice. Finally, mastodonts and hyraxes were not the only African natives to spread to Eurasia after the Arabian land bridge established a connection 18 Ma. By 16 Ma, the primates had made the trip out of Africa too. Beginning with the small anthropoid *Pliopithecus* (one of the first fossil primates described in the history of paleontology by Lartet in 1834), primates soon ranged widely throughout Europe, and by the late Miocene a considerable radiation of dryopithecine apes were found in many places in Europe.

In summary, most of the temperate world mimicked the modern African savanna in the Miocene, although with understudies performing the roles for modern African mammals today.

MIDDLE MIOCENE CLIMATIC OPTIMUM

After the colder climates of the Oligocene, the beginning of the Miocene showed a significant warming event on a global scale. Conditions were not as warm and tropical as they had been during the greenhouse world of the early Eocene, but they were considerably warmer than they had been in the Oligocene. This trend culminated with the "middle Miocene climatic optimum" (MMCO) that occurred from 16 Ma to 14 Ma.

In the oceans, tropical groups of plankton again flourished and diversified, and many of the shelled forms of planktonic foraminifera that had vanished during the Oligocene cooling evolved again in the Miocene. Mollusks also diversified, and middle Miocene shell beds in the Calvert Cliffs of Chesapeake Bay or Sharktooth Hill near Bakersfield, California, are extraordinarily rich in a wide variety of snails and clams. Along with these invertebrates was a great diversification of bony fish in the oceans. Sharks were very diverse as well, and their evolution culminated with the whale-sized predator *Otodus megalodon* (figure 21.4), the largest shark that ever lived, whose teeth are found in the Calvert Cliffs, Sharktooth Hill, and in many other middle Miocene marine deposits. What did such a shark eat? Its main prey was almost certainly whales.

Whales had diversified into a wide range of dolphins and porpoises, as well as huge whale-eating sperm whales known as *Livyatan* and enormous baleen whales. In addition, during the late Oligocene and the Miocene, several other groups of mammals evolved into marine organisms. These

Figure 21.4 ▲

Life-sized model of the gigantic Miocene shark *Otodus megalodon*. (Photo by author)

included the hippo-like desmostylians, found only around the northern Pacific rim; the manatees and dugongs, which had evolved from four-legged terrestrial ancestors in the Eocene; and the pinnipeds (seals, sea lions, and walruses), whose earliest fossils (*Enaliarctos*) can be found in the early Miocene beds of Pyramid Hill, California, not far from Sharktooth Hill and only slightly older.

About 14 Ma, in the late middle Miocene, the middle Miocene climatic optimum came to an end with a sudden global cooling event, sometimes called the "middle Miocene climate transition" (MMCT). Global temperatures dropped 3°C to 5°C (37°F to 41°F), and it is thought that atmospheric carbon dioxide dropped from about 300 ppm to less than 140 ppm. There were significant extinctions in the plankton (especially tropical and subtropical groups), in the mollusks and sea urchins (well documented in the sequence in the Calvert Cliffs), and in many other marine groups. On land crocodilians, chameleons, several other kinds of tropical lizards, and abundant aquatic turtles in North America and Europe during the MMCO quickly vanished from higher latitudes during the cooling of the late middle Miocene, never to return.

The immediate result of this sudden drop in temperature was a dramatic growth of the East Antarctic ice sheet to an extent near its present dimensions. This ice sheet had nearly vanished during the warmer climates of the early and middle Miocene. This was accompanied by an intensification of the flow in the Antarctic Bottom Water (AABW) and North Atlantic Deep Water (NADW), cooling the oceans and changing their circulation patterns. Other factors must have been involved to draw down carbon dioxide so quickly and trigger this global cooling. Several scientists have noted that the orbital variation (Croll-Milankovitch) cycles of Earth around the sun switched from one dominated by Earth's axial tilt with respect to the sun to one dominated by the eccentricity cycle, in which Earth's orbit around the sun goes from nearly circular to more elliptical. Some have pointed to the huge increase in silica production in the Pacific Ocean as gigantic blooms of the siliceous phytoplankton known as diatoms produced thick deposits of organic-rich siliceous shales and diatomites, which trapped enormous volumes of hydrocarbons as oil and gas in the Monterey Formation in California and in similar deposits in Japan. But the most significant cause was probably a major phase of uplift of the Himalayas, which increased the rates of weathering, absorbing carbon dioxide as silicate rocks exchange with the atmosphere and absorb carbon dioxide. The record of this uplift is preserved in the huge volume of river and delta sediments deposited below the Himalayas, especially in the Siwalik Hills of Pakistan, which preserve an excellent record of Asian life and climate through the Miocene and Pliocene. The Himalayan uplift seemed to increase throughout the late Miocene, absorbing more and more carbon dioxide, and pushing Earth's climate to the icehouse world with Arctic glaciers by the Pliocene.

Another consequence of the drop in carbon dioxide is the dramatic change in the floras of the Miocene. In the early Miocene, the temperate regions were still covered by scrublands with limited grasses and small stands of trees, as they had been since the Oligocene. As discussed earlier in the chapter, starting about 7 Ma the carbon dioxide levels apparently were so depleted that new plants using the more efficient Hatch-Slack photosynthetic pathway evolved that could cope with these lower levels. These were the enormous grasslands and prairies inhabited by C4 grasses, which are the dominant types of grasses in temperate and tropical grasslands today. By the late Miocene (7 to 5 Ma), huge grasslands and savannas spread across the temperate regions in North America, Eurasia, and even tropical Africa.

Grasslands, in turn, might have driven climate change. Retallack (2001, 2012) argues that the expansion of grasslands, especially since the middle Miocene, are hugely effective in driving climate change. Just as the explosion of coal swamps in the Carboniferous absorbed the atmospheric carbon dioxide and changed our greenhouse planet to an icehouse planet by the mid-Carboniferous (chapter 13), so does the expansion of grasslands in the later Miocene absorb huge amounts of carbon dioxide in their soils. Retallack argues that this factor, more than the uplift of the Himalayas, was the main agent of late Cenozoic carbon dioxide storage, and with it, the cooling of the climate since the MMCO.

THE MEDITERRANEAN WAS A DESERT

The Straits of Messina, between Sicily and Italy, were the supposed home of the legendary terrors Scylla and Charybdis. In Homer's *Odyssey*, Odysseus and his men were forced to navigate between these two forces that guarded the narrow straits. The monster Scylla lived in a cave high up on the cliff, from whence she would thrust her long necks (she had six heads), with each of her mouths seizing one of the crew of every vessel passing within reach. The other terror, Charybdis, was a gulf nearly on a level with the water. Three times each day the water rushed into a frightful chasm and was then disgorged. Any vessel coming near the whirlpool when the tide was rushing in must inevitably be engulfed; not even Poseidon himself could save it.

Although these mythical monsters never really guarded the straits, the rocks themselves are evidence of an even more terrifying event. From 5 to 6 Ma, the Mediterranean Sea itself dried up and then refilled with frightening waves of water, over and over again. Near the straits are outcrops that are the basis for the final stage of the Miocene, the Messinian Stage (named after the straits). They include gravels of rock salt and gypsum that were formed in massive floods and were of enormous thicknesses (up to 2,000 to 3,000 meters) that only form when huge bodies of salt water dry up (figure 21.5A). Geologists had known about this evidence for more than a century but had assumed the salt and gypsum deposits were produced when a local basin evaporated.

In 1970, Leg 13 of the Deep Sea Drilling Project took a number of drill cores from the sea bottom of the western Mediterranean around the Balaeric Islands between Spain, France, and Italy. They recovered a variety of puzzling clues. As they drilled the shallow margin of the edge of the Balaeric Basin, they found gravels and other evidence of huge desert flash-flood deposits on the floor of the Mediterranean. Then they drilled in the center of the basin, and underneath the Pliocene and Pleistocene marine deposits they found thick deposits of gypsum and rock salt, as well as ancient algal mats that had once formed in shallow pools on the fringe of a giant salty lake, not unlike the Dead Sea in Israel and Jordan. This was an important piece of evidence. The only way such huge quantities of evaporite minerals (over a million cubic kilometers in total) and algal mats could be found beneath the center of the deepest Mediterranean was if the sea had dried up completely and the bottom was exposed to sunlight (figure 21.5B). As more and more cores were drilled, the story became even more spectacular. The Mediterranean had dried up a number of times during the late Miocene, only to refill with catastrophic floods, then dry up again.

How could such an amazing event happen? The rivers that now feed the Mediterranean, such as the Rhône and the Nile, mostly drain out of dry "Mediterranean" climates in Italy and Greece and Spain, or out of the Egyptian desert, so their flow is insufficient to keep up with the huge rate of

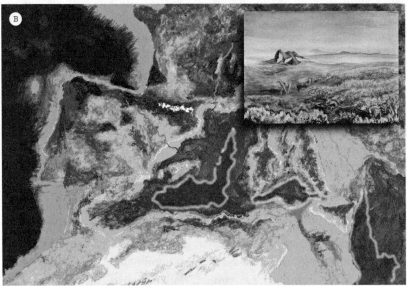

Figure 21.5 ▲

The Messinian salinity crisis: (*A*) outcrops of salt and gypsum near the Straits of Messina in Italy; and (*B*) an artist's reconstruction of the Mediterranean when it dried up completely in the late Miocene. (Courtesy of Wikimedia Commons)

evaporation (3,300 cubic kilometers per year) in this semidesert latitude. With such high evaporation rates, the Mediterranean Sea would vanish in less than a thousand years if it were not constantly receiving cold water from the Atlantic through the Straits of Gibraltar. If some sort of event occurred that cut off this supply of Atlantic Ocean water, the Mediterranean would dry up. Indeed, the geological evidence shows that the Atlas Mountains in Spain and Morocco (through which the Straits of Gibraltar are cut) were rising in the late Miocene as Africa collided with Europe, and apparently they began to form a barrier. In addition to the Gibraltar Straits, there were flows through the Betic Corridor in southern Spain and the Rifian Corridor in northern Morocco. Once these uplifted barriers reached a critical threshold, the Mediterranean was isolated. Each drying episode lasted about 1,000 years and produced about 70 meters of salt in the bottom of the basin, leading to estimates that at least 40 separate episodes must have occurred between 5.8 Ma and 5.6 Ma to produce the 2,000 to 3,000 meters of evaporites now found there.

Once the Mediterranean had dried up, it was only a matter of time before erosion would let the Atlantic water back in, forming a huge waterfall at Gibraltar that was at least 10 times the size of Niagara Falls. The water had to be moving so fast through this gap that it broke the sound barrier! Then the mountains would shift again, the drying resume, followed by another flood through Gibraltar—over and over again 40 times. Finally, 5.2 Ma the Atlantic water rushed in for the last time, the Gibraltar Straits stayed open, and marine waters returned to the Mediterranean permanently. Normal marine sediments were deposited, and the Pliocene had begun.

Confirmation of this startling hypothesis was found in the 1950s when Russian geologists and engineers tried to find hard bedrock beneath the Nile Gorge in southern Egypt so they could provide a firm foundation for the Aswan High Dam. They drilled and drilled and found that beneath the thick blanket of Nile sediments was a canyon almost 200 meters below the present level of the Mediterranean. Surveys were done at the mouth of the Nile and in the Rhône Valley, and buried gorges almost 3,000 meters deep were found! The only way such deep valleys could have formed so far below the present level of the Mediterranean was if the sea had dried up and sea level had dropped 3,000 meters below its present level. Under such conditions, the Nile and the Rhône would have cut down to the level of the bottom of the Mediterranean when it was a dry lakebed. Once the Mediterranean refilled in the early Pliocene, the grand canyons of the Nile and the Rhône would have filled with sediment until they matched the modern level of the Mediterranean Sea.

So what triggered this amazing series of events? The collision of Africa with Europe was an ongoing phenomenon in the Miocene, and the rise of the Atlas Mountains was an inevitable result. But what about the late Miocene glaciation? According to data from drill cores around the Antarctic, the West Antarctic ice sheet (which is unstable because it is held in place by floating ice shelves) developed about this time. In chapter 20, I described how the first Antarctic ice cap formed 33 Ma, and earlier in this chapter I explained that the East Antarctic ice cap became permanent 14 Ma during the Miocene climate disruption. But the West Antarctic was not glaciated in the middle Miocene; glaciation began only in the latest Miocene. This can be seen in glacial sediments in South America dated at 6.75 million years of age. In addition, there were oxygen isotope events 6.3 Ma, which indicates dramatic cooling, and a one part per thousand global carbon isotope shift, which signals a significant turnover in the world's oceans. All of these events seem to be the culmination of a steady global cooling trend throughout the latest Miocene, as the East Antarctic ice cap got larger and was joined by the West Antarctic ice sheet.

The effects of this global cooling were dramatic. When sea level dropped and global temperatures cooled, it triggered the Messinian salinity crisis in the Mediterranean region. The Sea of Japan was also cut off from the Pacific, forming a giant freshwater lake. Huge unconformities were cut in the sedimentary records of most continents and ocean basins at the end of the Miocene, making this global cooling and regression a natural boundary between the Miocene and Pliocene.

The drying of the Mediterranean may have had additional consequences. Bill Ryan, a codiscoverer of the Messinian salinity crisis, calculated that the drying of the Mediterranean and the withdrawal of all that salt would have lowered global oceanic salinity by 6 percent. Decreased salinity raises the freezing temperature of water, so ice could form at a higher temperature, which would have helped the Antarctic ice sheets form more easily once the Mediterranean had dried up. Although William Ryan and colleagues say that this is the cause of the late Miocene glaciation, in fact the evidence shows that the ice sheets were developing 6.75 Ma and the global oceanic change took place 6.3 Ma, so the much later (5.6 to 5.2 Ma) drying of the Mediterranean cannot have caused the first global cooling, but it may have accentuated it.

The latest Miocene global cooling event and the local drying of the Mediterranean had some surprising effects on life. On the shores of the hot dry desert that was once the Mediterranean Sea lived a forest-adapted fauna, including abundant tapirs, pigs, deer with long lyre-like antlers (*Croizetoceros*), diverse antelopes, gomphothere mastodonts with 4.3-meter-long straight tusks (*Anancus*), and the mastodon *Zygolophodon*, saber-toothed cats (*Machairodus* and *Dinofelis*), bears, mustelids, and hyenas. More dramatic was the effect of the drying of the Mediterranean on land corridors. Many African forms, including the hippopotamus *Hexaprotodon*, macaque monkeys, gerbils, and many African antelopes, migrated to Europe across the dry Mediterranean basin during the Messinian. In addition, we see immigrants from even further away, including the first camels and dogs to leave North America and migrate to Eurasia, presumably due to lower seas across the Bering Straits. A number of typically Miocene mammals also vanished by the Pliocene. These included the giant pig *Microstonyx*, the primitive aceratherine and teleoceratine rhinoceroses (leaving only the ancestors of the modern rhinoceros genera), most of the hipparionine horses, and most of the saber-toothed machairodont cats and Miocene hyenas.

Although the drying of the Mediterranean did not affect life so directly in other parts of the world, effects from global cooling included a rapid change in oceanic circulation and a drop in sea level. The remaining forests and leafy vegetation were rapidly replaced by dry grasslands and steppe vegetation. In North America, many of the animals adapted to the mixed grasslands and woodlands of the late Miocene vanished, including the last North American rhinos, the last horned camel-like protoceratids, the last deer-like dromomerycids, and the last archaic aplodontid, eomyid, and horned mylagaulid rodents. In addition to the loss of these entire families, there were extinctions of species and genera *within* families, including many horses (5 genera), peccaries (2 genera), pronghorns (7 genera), camels, gomphothere mastodonts (3 of the 5 genera known), mustelids (3 genera), bears (2 genera), and dogs (3 genera). Among small mammals, major extinctions also occurred in pocket gophers, jumping mice, the cricetids, beavers, pocket mice, and in the shrews and moles. According to Dave Webb, this was the largest extinction event in North American land mammals in the entire Cenozoic, dwarfing even those at the end of the Eocene and during the ice ages.

Africa, too, lost some of its archaic groups, such as the teleoceratine rhinoceros *Brachypotherium*, leaving only the modern lineage of African rhinos. The diversity of antelopes and cattle, pigs, horses, and mastodonts took a hit, requiring the African faunas to evolve new forms in the Pliocene. In Asia, the huge diversity of late Miocene rhinos, horses, mastodonts, pigs, antelopes and cattle, deer, and giraffes was considerably reduced, although all those families persisted, but the pig-like anthracotheres vanished from Asia for good. Likewise, the huge diversity of late Miocene hyenas, mustelids, bears, and cats was greatly reduced in the early Pliocene, but these groups all survived, along with the dogs, which had just immigrated from North America. However, the beardogs, which had straggled on in Asia long after their disappearance elsewhere, finally died out.

Thus the Miocene savannas came to an abrupt end with the Messinian event. Its effects were highly varied, with dramatic responses on the European continent and in North America, and lesser responses in Asia and Africa. The causes for these varied mammalian faunal changes have yet to be fully explained or even discussed. But the stage was now set for the cool, dry world of the Pliocene and the final development of the ice ages.

FOR FURTHER READING

Agusti, Jordi, and Mauricio Anton. *Mammoths, Sabertooths, and Hominids: 65 Million Years of Mammalian Evolution in Europe.* New York: Columbia University Press, 2002.

Böhme, Madelaine. "The Miocene Climatic Optimum: Evidence from Ectothermic Vertebrates of Central Europe." *Palaeogeography, Palaeoclimatology, Palaeoecology* 195, nos. 3–4 (2001): 389–401.

Cerling, Thure E., John M. Harris, B. J. MacFadden, M. G. Leakey, Jay Quade, Véra Eisenmann, and J. R. Ehleringer. "Global Vegetation Change Through the Miocene/Pliocene Boundary." *Nature* 389 (1997): 153–158.

Cerling, Thure E., Yang Wang, and Jay Quade. "Expansion of C4 Ecosystems as an Indicator of Global Ecological Change in the Late Miocene." *Nature* 361 (1993): 344–345.

Flower, Benjamin P., and James P. Kennett. "The Middle Miocene Climatic Transition: East Antarctic Ice Sheet Development, Deep Ocean Circulation, and Global Carbon Cycling." *Palaeogeography, Palaeoclimatology, Palaeoecology* 108, nos. 3–4 (1994): 537–555.

Garcia-Castellanos, Daniel. "Messinian Salinity Crisis Regulated by Competing Tectonics and Erosion at the Gibraltar Arc. *Nature* 480, no. 7377 (2011): 359–363.

Garcia-Castellanos, Daniel, Ferran Estrada, Ivone Jiménez-Munt, Christian Gorini, M. Fernàndez, Jaume Vergés, and R. De Vicente. "Catastrophic Flood of the Mediterranean After the Messinian Salinity Crisis." *Nature* 462, no. 7274 (2009): 778–781.

Edwards, Erika J., Colin P. Osborne, Caroline Strömberg, Stephen A. Smith, and Grasses Consortium. "The Origins of C$_4$ Grasslands: Integrating Evolutionary and Ecosystem Science." *Science* 328, no. 5978 (2010): 587–591.

Eronen, Jussi T., Mikael Fortelius, A. Micheels, Felix T. Portmann, K. Poulamäki, and Christine M. Janis. "Neogene Aridification of the Northern Hemisphere." *Geology* 40 (2012): 823–826.

Figueirido, Borja, Christine M. Janis, Juan A. Pérez-Claros, Miquel De Renzi, and Paul Palmqvist. "Cenozoic Climate Change Influences Mammalian Evolutionary Dynamics." *Proceedings of the National Academy of Sciences* 109, no. 3 (2012): 722–727.

Gargani, Julien, and Christophe Rigollet."Mediterranean Sea Level Variations During the Messinian Salinity Crisis." *Geophysical Research Letters* 34, no. 10 (2007): L10405.

Govers, Rob. "Choking the Mediterranean to Dehydration: The Messinian Salinity Crisis." *Geology* 37, no. 2 (2009): 167–170.

Harris, Judith. "Ecosystem Structure and Growth of the African Savanna." *Global Planetary Change* 8 (1993): 213–248.

Hodell, David A., Kristin M. Elmstrom, and James P. Kennett. "Latest Miocene Benthic δ^{18} Changes, Global Ice Volume, Sea Level and the 'Messinian Salinity Crisis.'" *Nature* 320 (1986): 411–414.

Holbourn, Ann, Wolfgang Kuhnt, Michael Schulz, and H. Erlenkeuser. "Impacts of Orbital Forcing and Atmospheric Carbon Dioxide on Miocene Ice-Sheet Expansion." *Nature* 438, no. 7067 (2005): 483–487.

Hsü, Kenneth J. *The Mediterranean Was a Desert: A Voyage of the Glomar Challenger.* Princeton, NJ: Princeton University Press, 1983.

Janis, Christine, John Damuth, and Jessica M. Theodor. "Miocene Ungulates and Terrestrial Primary Productivity: Where Have All the Browsers Gone?" *Proceedings of the National Academy of Sciences* 97, no. 14 (2000): 7899–7904.

——. "The Origins and Evolution of the North American Grassland Biome: The Story from the Hoofed Mammals." *Palaeogeography, Palaeoclimatology, Palaeoecology* 17, nos. 1–2 (2002): 183–198.

———. "The Species Richness of Miocene Browsers, and Implications for Habitat Type and Primary Productivity in the North American Grassland Biome." *Palaeogeography, Palaeoclimatology, Palaeoecology* 207, no. 3 (2004): 371–398.

Jardine, Phillip E., Christine Janis, Sarda Sahney, and Michael J. Benton. "Grit Not Grass: Concordant Patterns of Early Origin of Hypsodonty in Great Plains Ungulates and Glires." *Palaeogeography, Palaeoclimatology, Palaeoecology* 365–366 (2012): 1–10.

Kennett, James P. *The Miocene Ocean: Paleoceanography and Biogeography.* Boulder, CO: Geological Society of America, 1985.

Krijgsman, Wout, Frits J. Hilgen, Isabella Raffi, Francisco J. Sierro, and D. S. Wilson. "Chronology, Causes, and Progression of the Messinian Salinity Crisis." *Nature* 400, no. 6745 (1999): 652–655.

Morales-Garcia, Nuria M., Laura K. Salla, and Christian. M. Janis. "The Neogene Savannas of North America: A Retrospective Analysis on Artiodactyl Faunas." *Frontiers in Earth Science* 8, no. 191 (2020): 1–24.

Pearson, Paul N., and Martin R. Palmer. "Atmospheric Carbon Dioxide Concentrations Over the Past 60 Million Years." *Nature* 406, no. 6797 (2000): 695–699.

Quade, Jay, and Thure E. Cerling. "Expansion of C_4 Grasses in the Late Miocene of Northern Pakistan: Evidence from Stable Isotopes in Paleosols." *Palaeogeography, Palaeoclimatology, Palaeoecology* 115, nos. 1–4 (1995): 91–116.

Retallack, Gregory J. "Cenozoic Expansion of Grasslands and Climatic Cooling." *Journal of Geology* 109, no. 4 (2001): 407–426.

———. "Global Cooling by Grassland Soils of the Geological Past and Near Future." *Annual Review of Earth and Planetary Science* 41, no. 1 (2012): 1–18.

———. "Neogene Expansion of the North American Prairie." *Palaios* 12, no. 4 (1997): 380–390.

———. "Soil Carbon Dioxide Planetary Thermostat." *Astrobiology* 22, no. 1 (2022): 116–123.

Rössner, Gertrud E., and Kurt Heissig, eds. *The Miocene Land Mammals of Europe.* Munich: Friedrich Pfeil Verlag, 1999.

Ruddiman, W. J. *Earth's Climate: Past and Future.* New York: Freeman, 2013.

Ryan, William B. F. "Decoding the Mediterranean Salinity Crisis." *Sedimentology* 56, no. 1 (2008): 95–136.

———. "Modeling the Magnitude and Timing of Evaporative Drawdown During the Messinian Salinity Crisis." *Stratigraphy* 5, no. 34 (2008): 227–243.

Shevenell, Amelia E., James P. Kennett, and D. W. Lea. "Middle Miocene Southern Ocean Cooling and Antarctic Cryosphere Expansion." *Science* 305, no. 5691 (2004): 1766–1770.

Strömberg, Caroline A. E. "The Origin and Spread of Grass-Dominated Ecosystems in the Late Tertiary of North America: Preliminary Results Concerning the Evolution of Hypsodonty." *Palaeogeography, Palaeoclimatology, Palaeoecology* 177 (2002): 59–75.

———. "Using Phytolith Assemblages to Reconstruct the Origin and Spread of Grass-Dominated Habitats in the Great Plains of North America During the Late Eocene to Early Miocene." *Palaeogeography, Palaeoclimatology, Palaeoecology* 207, nos. 3–4 (2004): 239–275.

———. "Evolution of Grasses and Grassland Ecosystems." *Annual Review of Earth and Planetary Sciences* 39, no. 1 (2011): 517–544.

Wang, Yang, Thure E. Cerling, and Bruce J. MacFadden. "Fossil Horses and Carbon Isotopes: New Evidence for Cenozoic Dietary, Habitat, and Ecosystem Changes in North America." *Palaeogeography, Palaeoclimatology, Palaeoecology* 107, nos. 3–4 (1994): 269–279.

Webb, S. David. "A History of Savanna Vertebrates in the New World. Part I: North America." *Annual Reviews of Ecology and Systematics* 8 (1977): 355–380.

———. "The Rise and Fall of the Late Miocene Ungulate Fauna in North America." In *Coevolution*, ed. M. H. Nitecki, 257–306. Chicago: University of Chicago Press, 1983.

Webb, S. David, Richard C. Hulbert Jr., and W. David Lambert. "Climatic Implications of Large-Herbivore Distributions in the Miocene of North America." In *Paleoclimate and Evolution, with Emphasis on Human Origins*, ed. Elizabeth S. Vrba, George Denton, Timothy C. Partridge, and Lloyd H. Burckle, 91–108. New Haven, CT: Yale University Press, 1995.

Webb, S. David, and Neil D. Opdyke. "Global Climatic Influences on Cenozoic Land Mammal Faunas." In *Effects of Past Global Change on Life*, ed. J. P. Kennett and S. M. Stanley, 184–208. Washington, DC: National Academy of Sciences, 1995.

Weijermars, Ruud. "Neogene Tectonics in the Western Mediterranean May Have Caused the Messinian Salinity Crisis and an Associated Glacial Event." *Tectonophysics* 148, nos. 3–4 (1988): 211–219.

Zachos, James, Mark Pagani, Lisa Sloan, Ellen Thomas, and K. Billups. "Trends, Rhythms, and Aberrations in Global Climate 65 Ma to Present." *Science* 292, no. 5517 (2001): 68.

ICE PLANET

The glacier was God's great plough set at work ages ago to grind, furrow, and knead over, as it were, the surface of the earth.

—LOUIS AGASSIZ

MYSTERY OF THE ERRATICS

In the late 1700s and early 1800s, geology was an infant science, practiced mostly by wealthy gentlemen and clergymen as a hobby. There were no professional geologists who earned their living doing geology, although mining geology was well advanced. There were no geology professors at the universities, although some European universities taught mineralogy and crystallography in the context of finding ore deposits. Few geologists traveled beyond their hometown in Europe, so they had little experience with glaciers or volcanoes or other phenomena not in their own backyard. The intellectual framework of geology was also in transition. The literal interpretation of the first chapters of Genesis still shaped most people's ideas about the age of the earth and the nature of rocks, although the Scotsman James Hutton and other pioneers were writing about natural rather than supernatural processes, and speaking of an Earth with "no vestige of a beginning, no prospect of an end" as early as 1788. Most geologists still thought of the layers of strata known as "Tertiary" as having been laid down by Noah's flood, and they puzzled over the many huge boulders that were found over northern Europe. Some of these boulders were as large as a house and were not made of any of the local bedrock types, so they had come from a distant place (figures 22.1A, 22.1B, and 22.1C). These *erratics*, or "wanderers," were attributed to the immense powers of Noah's flood. Likewise, the poorly sorted loose gravels, sands, and muds found just below the surface nearly everywhere in northern Europe were attributed to flood waters and called "drift" because they supposedly drifted into place as the flood waters receded. (We now recognize that this poorly sorted unstratified collection of sediment was dumped by the retreat of melting glaciers, and we refer to it as "till.")

This Diluvial school of thought (after the old name for Noah's flood, the "Diluvium") was modified and expanded in the early 1800s as geologists with strong religious ties (many of them ministers) used these natural

Figure 22.1 ▲ ▶

Glacial erratics: (*A*) Kummakivi, a balanced glacial erratic in Finland; (*B*) a glacial erratic made of limestone from Scandinavia, sitting on top of glacial sediments in Konin, Poland; and (*C*) the Swiss Patrouille auf Gletschertisch ("patrol of the glacial table") examining an erratic sitting on the toe of a glacier near Zermatt, Switzerland between 1914 and 1918. (*D*) A glacial erratic precariously balanced on a pedestal of a totally different bedrock. ([*A–C*] Courtesy of Wikimedia Commons; [*D*] James Forbes, *Travels Through the Alps of Savoy* [London: Longman, Brown, and Green, 1843]).

Figure 22.1 ▲ ▶
(*continued*)

Figure 22.1 ▲
(*continued*)

features to support their religious ideas. However, evidence was beginning to accumulate that contradicted the Diluvialist dogma. Sailors had reported seeing huge rocks caught in icebergs as they floated down through the North Atlantic shipping lanes, but no one made the connection to the erratics. As early as 1787, the Swiss minister Bernard Friederich Kuhn interpreted enormous boulders in the Swiss Alps as evidence of ancient glaciation (figure 22.1D). In 1794, James Hutton himself visited the Jura Mountains of France and Switzerland and reached the same conclusion. (Ironically, he never recognized the glacial origin of many of the same features in his own backyard in Scotland.) In 1824, the Norwegian Jens Esmark argued that glaciers had once covered most of his homeland and possibly much of Europe. Over the next decade several other geologists amassed evidence that the Alpine glaciers had once extended far onto the lowlands or that the polar ice caps had once spread as far as central Germany.

EISZEIT

The entrenched Diluvial explanation of the erratics and the drift was difficult to uproot, both for religious reasons and due to simple inertia—

Figure 22.2 ▲

Lithograph of a young Louis Agassiz about the time when he was promoting his idea of ice ages. (Courtesy of Wikimedia Commons)

scientists are human and surrender the notions they learn as students only with reluctance. It took the young Swiss paleontologist Louis Agassiz to change their ideas about the glacial explanation (figure 22.2). In 1836, Agassiz visited many of the Alpine glaciers with the Swiss geologist Jean de

Charpentier, who had been documenting evidence of the expansion and retreat of these glaciers for years. Agassiz was convinced, and at the July 1837 meeting of the Swiss Society of Natural Sciences in Neuchâtel he rose and presented a paper that startled his audience. Instead of the expected presentation on fossil fishes, Agassiz argued that the erratics were glacial in origin and that large areas of the globe had once been covered by ice during an *Eiszeit*, or Ice Age.

The meeting was thrown into turmoil, and some geologists never had a chance to present their own papers (including Amanz Gressly, whose idea of "sedimentary facies" later became an important concept in geology). After the meeting ended, a field trip to the glaciers was hastily arranged to convince the most distinguished skeptics in the society—Elie de Beaumont and Leopold von Buch—but they remained skeptical. However, Agassiz's 1837 paper and his monumental 1840 book *Studies on Glaciers* were not written in vain. The glacial theory had made a splash, and it now had to be taken seriously. In 1840, Agassiz visited England and eventually convinced his two most prominent British critics, William Buckland and Charles Lyell, and they became ardent converts and proselytizers for the glacial theory.

Beyond religious and intellectual conservatism, the biggest hurdle was the simple fact that most geologists (except those who had visited the Alps) had never seen a glacier in action, and no European had any experience with a continental-sized ice sheet at that time. This all changed in 1852 when an expedition to Greenland by the American explorer Elisha Kent Kane documented the scale of its continental ice sheets. Kane's books amazed the Europeans and Americans with the sheer magnitude of the Greenland ice. After nine years of battling over the glacial theory with European geologists, Agassiz left Europe and assumed a position at Harvard in 1846, where he happily began documenting the evidence for glaciation in North America as well. When he died in 1873, Agassiz's glacial theory had been universally accepted, and geologists had spent more than 30 years documenting examples of phenomena that could be explained by the great ice sheets.

As studies continued, geologists found that there had not been just one *Eiszeit*, as postulated by Agassiz, but several. In some places in Europe, at least five major sets of glacial deposits were stacked upon one another, suggesting a minimum of five ice ages (figure 22.3): the Donau, Gunz, Mindel,

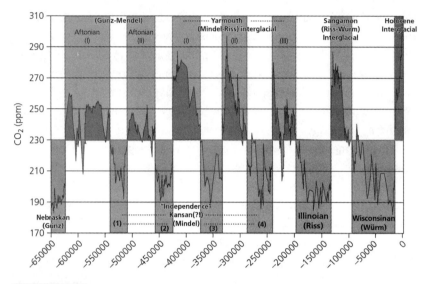

Figure 22.3 ▲

Time scale for the last 650,000 years of the Pleistocene. At one time, only four glacial stages were recognized in North America and five in the Alpine region. In the 1960s and later, deep sea cores and ice cores provided a much more detailed record of the climate changes, and about 25 glacial stages were recognized in the past 2.6 million years. (Redrawn from several sources)

Riss, and Würm (from oldest to youngest). North America, too, yielded evidence of multiple major ice advances, and they were named for the states that recorded their southernmost extent: Kansan, Nebraskan, Illinoian, and Wisconsinan (from oldest to youngest). Deposits indicating warm periods between the glacial tills were also identified; they were called interglacials, and these too had names. In North America, the glacial retreat between the Wisconsinan and Illinoian was known as the Sangamonian; the one between the Illinoian and Nebraskan was the Yarmouthian interglacial; and that between the Kansan and Nebraskan was the Aftonian interglacial. Geologists were unable to date these deposits precisely at that time, so they counted back from the most recent glacial and equated the Wisconsinan with the Würm, the Riss with the Illinoian, the Mindel with the Kansan, and the Gunz with the Nebraskan. Thanks to recent advances in dating methods, we now know that the Riss and Würm are both equivalent to the Wisconsinan, that the Mindel was equivalent to the Illinoian, the Gunz correlates with the Nebraskan, and the Donau correlates with the Kansan.

THE JANITOR AND THE MATHEMATICIAN

In addition to dating problems, geologists were confronted with an even bigger puzzle. Why did Earth's climate fluctuate between glacial and interglacial cycles? And what caused Earth to slip into the glaciated state in the first place? Many ideas were proposed, but good evidence for any explanation was lacking.

One idea being discussed among astronomers was that the amount of sunlight received on Earth's surface (known as solar insolation) might determine the shift from glacial ice advances to interglacial ice retreats. No one had worked out the mathematics of the required orbits or calculated the energy differences involved, and into this breach stepped a remarkable Scotsman named James Croll. He was the perfect example of someone of limited means and education who reached the pinnacle of a profession through sheer effort and intelligence. Born on a farm in Perthshire, Scotland, he had almost no formal schooling and had to go to work before he was 16 years old. He started as an apprentice wheelwright and mechanic, then became a tea merchant, failed in his effort to run a temperance hotel, and then became an insurance agent. In 1859, he was hired as the janitor at the Andersonian University in Glasgow, and using his access he spent many hours in the library teaching himself mathematics, physics, and astronomy.

Building on the work of the astronomer Urbain Le Verrier, who have first shown that Earth's orbit around the sun and its axial tilt were constantly changing, Croll showed his work to Sir Charles Lyell and Sir Archibald Geike, who were impressed. Geike hired him as keeper of the geologic maps and correspondence of the Geological Survey of Scotland in Edinburgh, where he had plenty of spare time to read and do research with all the necessary documents around him. In 1875, he wrote the book *Climate and Time, in Their Geologic Relations*, which laid out the basics of how Earth's orbital motions change the amount of solar radiation received, thus triggering ice ages. This work earned him a university research post, an honorary degree by the University of St. Andrews, and eventually election to the Royal Society.

Croll did calculations for the known cycles of Earth's orbit around the sun to see how much they might explain the ice ages. He pointed out that astronomers as early as Johannes Kepler in 1609 knew Earth's orbit around the sun was not a circle but an ellipse. That ellipse changed shape from nearly circular to more egg-shaped very slowly (we now know it takes about 110,000 years). This is the cycle of the "eccentricity" of Earth's orbit, Another cycle, known since the days of Hipparchos in 130 BCE, was the precession or "wobble" cycle. As the ancient Greeks knew, Earth's axis wobbles like a top with its spin axis pointing in different directions. Today, for example, it points to Polaris, the North Star, but 10,000 years ago it pointed to a completely different star, Vega. As the axis points in different directions, it affects the amount of sunlight the poles receive. This is the precession or "wobble" cycle, which we now know takes about 21,000 to 23,000 years to complete, the fastest of the three cycles.

Croll also pointed out that the ice can grow and melt back rapidly due to the albedo feedback loop (see chapter 8). Albedo is a scientific measure of the reflectivity of a surface. When there is a lot of ice, it is very reflective, or has a high albedo. This tends to bounce more solar energy into space and cool the temperature, further increasing the ice. However, only a little melting of an ice-covered landscape exposes dark, sunlight-absorbing low albedo surfaces (seawater or vegetation), which in turn absorb more heat and accelerate the melting.

Croll's ideas were provocative and worth taking seriously, but no data suitable for testing it was available at that time. Nothing could be reliably dated, and the land record of ice advances was incomplete and too poor to be used to evaluate his ideas. Unfortunately, they languished in the pile of interesting but untestable ideas for decades. Croll himself sustained a serious head injury in 1880 that forced him to retire at age 59. He lived 10 more years before dying in 1890 with no further progress having been made on his pioneering suggestions.

Croll's ideas were nearly forgotten when a Serbian astronomer and mathematician named Milutin Milankovitch revived them. Born in what is now Croatia (part of the Austro-Hungarian Empire in 1879), he was a stellar student and received an engineering degree from the Vienna Institute of Technology in 1904. He became a top civil engineer, building bridges, viaducts, aqueducts, dams, and other structures in Austria, and he received six patents for his inventions. He then moved to Serbia to become chair of applied mathematics at the University of Belgrade.

Despite his day job as an engineer, he was more interested in fundamental research, and by 1912 he was involved in solving the problems of how variations in solar insolation might affect climate. Milankovitch lamented that "most of meteorology is nothing but a collection of innumerable empirical findings, mainly numerical data, with traces of physics used to explain some of them. . . . Mathematics was even less applied, nothing more than elementary calculus. . . . Advanced mathematics had no role in that science."[1] In 1912 and 1913, he published several papers calculating the amount of solar insolation Earth receives at each latitude and how this affects the position of climatic belts. Then in July 1914, Archduke Franz Ferdinand was assassinated in Sarajevo, and the crisis between Serbia and the Austro-Hungarian Empire boiled over to become World War I. As a Serbian, Milankovitch was arrested during his honeymoon in Austria and imprisoned in Esseg Fortress. Milankovitch had powerful connections in Vienna, so he was then interned in Budapest and had access to materials so he could continue his research. Even though he was technically a prisoner, he used his undisturbed time in Budapest and his university library access to make huge advances in mathematical meteorology, which he published in a series of papers from 1914 to 1920. The war finally ended, and Milankovitch and his family returned to Belgrade in March 1919, where he resumed his professorship at the University of Belgrade.

Milankovitch built on this research to calculate exact models for how solar insolation cycles might cause ice ages. He realized that the key factor is how much sunlight Earth's surface receives in the summertime; this determines how much the ice melts or stays frozen. Milankovitch built on Croll's eccentricity cycle and the precession cycle, and added a third cycle that had been discovered by Ludwig Pilgrim in 1904: the obliquity or "tilt" cycle. The rotational axis of Earth is not straight up and down with respect to the plane of its motion around the sun; it is tilted, and today Earth's axis is tilted 23.5° from the plane of Earth's orbit around the sun (figure 22.4). That angle is not always constant; it fluctuates between about 22° and more than 24.5°. When the angle is as steep as 24.5°, the polar regions receive a lot more sunlight and the ice melts; when the angle is as shallow as 22°, the poles receive much less sunlight and ice forms. This is known as the obliquity cycle of Earth's orbit, and it takes about 41,000 years to complete a cycle from 22° to 24.5° and back again. Milankovitch had all the pieces he needed, and he did the painstaking calculations and plots by hand on many reams of paper, with no computers or electronic calculators. After writing dozens of scientific papers and short books on the topic of Earth's solar radiation and climate, by the late 1930s Milankovitch put it all together in

Milankovitch Cycles

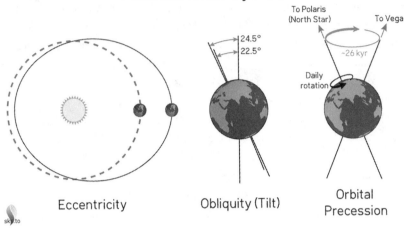

Figure 22.4 ▲

The three cycles of orbital variation described by Croll and Milankovitch control how much solar energy Earth receives, and together these cycles change the ice volume on Earth and cause glacial-interglacial cycles. (*A*) The longest is the cycle of eccentricity, or how long it takes for Earth's orbit around the sun to go from nearly circular to slightly more elliptical and back. This cycle takes about 110,000 to complete. (*B*) The second fastest is the cycle of tilt or obliquity. This cycle takes about 41,000 years to complete. (*C*) The third cycle, the precession or "wobble" cycle, describes how Earth's spin axis is tilted toward a different part of the sky as it "wobbles" like a top. This cycle takes about 21,000 to 23,000 years to complete. (Redrawn from several sources)

one book, *Canon of Insolation of the Earth and Its Application to the Problem of the Ice Ages*.

Once again world events reached a crisis and interfered with Milankovitch's life and work. Four days after he sent the book to the printer in 1941, the Germans invaded Yugoslavia, and the printing house was destroyed during the bombing of Belgrade. Luckily, the printed pages were in another warehouse, undamaged, and eventually bound and published. As the Nazis invaded Serbia in May 1941, two German officers and some geology students came to his home to help him, and he gave them his only bound copy of the book for safekeeping in case something should happen to him or his work. He spent the rest of the war years holed up in his home, writing his memoirs. When the war ended, he once again returned to his duties at the University of Belgrade and also served as vice president of the

Serbian Academy of Sciences. He suffered a stroke in 1958 and died at the age of 79 without ever knowing whether his ideas would be supported by geological evidence.

PLANKTON TO THE RESCUE

The scientific community had been debating Milankovitch's ideas since they were first introduced in 1924. By the late 1950s, they were generally accepted, but the problem Croll had faced still persisted: dating. Although geologists could identify at least four or five glacial-interglacial cycles, they could not date these deposits precisely enough to determine whether they matched the Croll-Milankovitch predictions. In the early 1950s, radiocarbon dating was developed, but many of the dates run on late Ice Age materials did not seem to match Milankovitch's predicted age for the last glacial maximum, or for the warm periods predicted by the astronomical cycles. By the early 1960s, the Croll-Milankovitch theory was in disrepute. The problem remained: dating was still inadequate to test the theory.

Radiocarbon dating only worked for objects less than 40,000 years old, which only allowed a test of the end of the last astronomical cycle; the rest of the events were much older than the limits of radiocarbon dating. Moreover, the dates used came from isolated specimens of wood or shells in the patchy glacial deposits on land. There was no long-term record anywhere on land that continuously recorded changes in climate that would allow a valid comparison with the astronomical cycle theory.

In the late 1950s and early 1960s, a new source of data emerged that did preserve a continuous record of the past few hundred thousand years. Unlike the episodic, irregular deposition of sediment by glaciers on land, sediments on the deep seafloor are deposited in a nearly continuous "rain" of muds and planktonic microfossils from the surface. After World War II ended, the major oceanographic institutions sent out research vessels that began to map and study the ocean floor in detail for the first time. These ships also routinely dropped a long coring device over the side. When retrieved, each core recovered as much as 15 meters of sea bottom muds and fossiliferous oozes, which recorded several hundred thousand years of climatic change in the ocean.

Micropaleontologists such as Fred Phleger and David Ericson soon realized that the abundance of temperature-sensitive planktonic foraminiferans (such as the warm-water indicator species *Globorotalia menardii*) recorded changes in oceanic water temperature and were good indicators of climatic fluctuations in the ocean (figure 22.5). At the same time, Cesare Emiliani developed the oxygen isotope method for measuring the chemistry of the ancient ocean as recorded in the shells of plankton, and we soon had another measure of how Earth's climate had changed in the past several hundred thousand years.

But dating problems remained. These long cores showed climatic cycles, but how old were they? Only the youngest part of the core was within the range of radiocarbon dating. By the late 1960s, however, the thorium/protactinium dating method was successfully used to date ancient coral reef terraces on Barbados, and it demonstrated that the ages of the sea level terraces on Barbados matched the Milankovitch prediction for warm interglacials, when the ice sheets melted and sea level rose. In addition, by the mid-1960s, Neil Opdyke had shown that the deep-sea cores could be dated by paleomagnetism. The final clinching research began in the 1970s, when the multimillion dollar Project CLIMAP (**CLI**mate **MA**pping, and **P**rediction) was launched under the direction of John Imbrie of Brown University. Hundreds of cores from all over the world's oceans were analyzed for both isotopes and changes in the plankton (radiolarians as well as foraminiferans), and dated paleomagnetically. By 1972, CLIMAP project members could show that there were at least 20 glacial-interglacial cycles in the past 2 million years (not the four originally discovered on land) and that they were in good agreement with the Croll-Milankovitch theory (figure 22.6A).

The final breakthrough was the development of a computerized statistical method called filter analysis, which uses spectral analysis to break down the complex climatic curve that was a composite of several different cycles into its component cycles. When the best deep-sea climatic records were analyzed in this way, they showed clearly that the complex oceanic climatic signal was a composite of three different cycles, which interacted in different ways to produce the highs, lows, and intermediate values of the climatic curve (figure 22.6B). The most prominent cycle was the 110,000-year eccentricity cycle predicted by Croll more than 100 years earlier, and there were weaker signals that matched the 43,000-year tilt (obliquity) cycle, and two peaks at 23,000 and 19,000 years that matched the precession cycle. In 1976, Jim Hays, John Imbrie, and Nick Shackleton published the famous "Pacemaker" paper, which finally confirmed that the Ice Age cycles were caused by the orbital variations first predicted by Croll and Milankovitch.

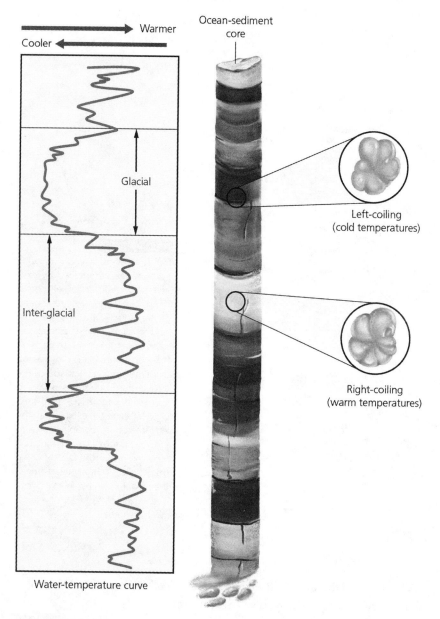

Cooler ← → Warmer

Ocean-sediment core

Glacial

Inter-glacial

Water-temperature curve

Left-coiling
(cold temperatures)

Right-coiling
(warm temperatures)

Figure 22.5 ▲

Certain species of planktonic foraminiferans are very sensitive to changes in temperature in the seawater in which they live. Some increase or decrease in abundance and dominance depending on water temperature, and others change from right coiling to left coiling as temperature changes. By plotting these data against depth in deep sea sediment cores, micropaleontologists were able to reconstruct a detailed history of ocean temperatures and climate. (Redrawn from several sources)

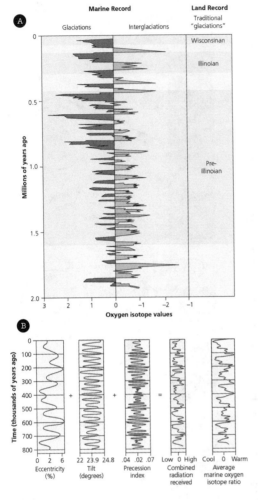

Marine Record

Land Record

Glaciations Interglaciations Traditional "glaciations"

Figure 22.6 ◀

(A) The detailed record of temperature (as determined by oxygen isotopes in deep sea cores) for the last 2 million years. Instead of only four or five ice ages, as suggested by land glaciations, deep sea cores showed at least 20 glacial-interglacial cycles in this time span. The shape of the temperature curve has a jagged saw-tooth pattern. (B) The three Croll-Milakovitch cycles (on the left) interact in complex ways to form a pattern of solar radiation received on Earth (second curve from right). This signal matches the one measured in deep sea cores (right curve). (Redrawn from several sources)

At the end of the decade, this paper was hailed as one of the five most important scientific discoveries of the 1970s.

After the 1976 Pacemaker paper, many developments occurred in the orbital variation theory. Greenhouse gases, such as carbon dioxide and methane, are now thought to play important roles in modifying climate cycles driven by changes in solar radiation. Detailed study of cores covering the Pliocene and the Pleistocene show that during the Pliocene the 41,000-year and 23,000-year cycles of tilt and precession, respectively, were the most dominant, but about 900,000 years ago, the 110,000-year eccentricity cycle became the more dominant, causing major cycles of 110,000 years between the peak warmth of interglacials, or between the coldest part of the glacials. Currently, climatic models suggest that as climate cooled in the Pliocene due to the changes in continents, oceans, and atmospheric composition, growth of the ice sheets exceeded a critical threshold. Once this threshold had been passed, the effects of the direct differences in solar radiation at 41,000 and 23,000 years became less important, and the huge, slowly changing ice sheets responded mainly to the slower (but less variable in solar radiation) 110,000-year cycle of eccentricity.

How did marine life respond to these rapid changes in oceanic temperature? In most cases, the species of Pleistocene marine organisms are still alive today, so we can trace their known temperature tolerances into the past. Some of them, in fact, are so temperature-sensitive that they are good paleoclimatic indicators. Planktonic foraminiferans have been used in this regard by marine geologists for more than 30 years. For example, the presence of foraminiferans with known temperature tolerances tells the micropaleontologist about the water temperature in each centimeter of a deep-sea core. The thick-shelled *Neogloboquadrina pachyderma* is a polar indicator species, whereas *Globigerina bulloides* prefers subpolar waters. *Globotruncana truncatulinoides* occurs in temperate waters. *Globorotalia menardii* and the pink-shelled *Globigerinoides ruber* are found in the tropics and subtropics, and *Globigerinoides sacculifer* is strictly tropical. In other cases, individual species show climate in their shell morphology. *Globorotalia menardii*, *Neogloboquadrina pachyderma*, and *Globotruncana truncatulinoides* from deep-sea cores are predominantly right-coiling during warmer interglacial periods, but switch to left-coiling when the water cools below a certain threshold during glaciations.

Likewise, changes in the molluscan faunas of the Pleistocene are subtle. Relatively few new species appeared or died out during the glacial and interglacial cycles. Instead, they tended to shift their biogeographic ranges as shallow waters along the coast became warmer or cooler with each glacial advance or retreat.

GLACIERS SCULPT THE LAND

The world of the ice ages is much more familiar to us than any other time in the Cenozoic because Ice Age mammals (especially "cave men," mammoths, and saber-toothed cats) are familiar from many movies, television shows, and even cartoons like the Flintstones and the animated movie series *Ice Age*. But the 2 million years of the Pleistocene are not one homogeneous interval with a single fauna that persisted through the many ice age fluctuations. Instead, there were clear changes in land mammal faunas on all the continents throughout the many Pleistocene cycles.

At the peak of the last glacial period 20,000 years ago, more than 18 million square kilometers of ice covered northern North America, and about 11 million square kilometers of ice covered most of northern Europe and nearly all of Siberia. In comparison, the modern Antarctic ice sheet is only 12.6 million square kilometers (it expanded to almost 14 million square kilometers during the ice ages). It is calculated that the North American ice sheet had 27 million cubic kilometers of frozen water, with about 26 million cubic kilometers in Antarctica and 12 million cubic kilometers in Eurasia. The center of the ice sheet was more than 3 kilometers thick, and its weight depressed the underlying bedrock until it was 370 meters below modern sea level. Some regions, such as Scandinavia, are still rebounding from this pressure 12,000 years after all of the ice melted away. This ice took an enormous amount of water out of the oceans, dropping global sea level more than 120 meters. At that time, the outer continental shelf 120 meters below the present sea level was a dry coastal floodplain.

The ice sheets also radically transformed the land. All of the northern regions of North America as far south as Kansas and Nebraska were scoured flat as the giant ice bulldozer scraped over them again and again. Bedrock in many places has long parallel scratches formed by the rasping of rocks embedded in the base of the glacier and dragged along. In other places, the glaciers dumped huge erratic boulders, or left enormous mounds of unsorted unstratified sediment called till. Glacial depressions at the margins of the ice sheets became gigantic lakes, which are now represented by remnants such as Lake Winnipeg (a fraction of the size of glacial Lake Agassiz), the Great Lakes, and the Finger Lakes in New York. In the basins of Nevada and Utah, there were many lakes where there is now desert. These included the immense Lake Bonneville, which once covered most of western Utah but has been reduced to the small remnant known as the Great Salt Lake and Bonneville Salt Flats today. In northern Idaho, an ice dam trapped glacial melt water in the narrow valleys of western Montana. Each time the dam broke, huge floods roared across eastern Washington, producing the "Channeled Scablands" (and Grand Coulee). Giant dune fields of wind-blown glacial dust, or loess, accumulated in the "breadbasket" regions of the world: the Great Plains, the Ukrainian and Polish Plain, and central China. This loess is now a rich soil that is responsible for the growth of much of the world's grain belt.

Advances and retreats of the ice sheets had enormous effects on the vegetation. During glacial maxima, the ice sheets extended across most of the northern United States and completely covered Canada. South of this region was a broad belt of tundra vegetation in Ohio and Iowa, and most of North America was covered by a northern spruce forest like that located today in northern Canada. The southeastern United States had a pine-hardwood forest (like that of the modern Great Lakes), and all the oak forests and prairie vegetation, as well as the swamps of Florida, vanished. Most of Europe was covered by tundra where today there are mixed deciduous forests and grasses, and steppes covered a much larger region of Russia and Siberia than they do now.

ICE AGE: THE MELTDOWN

The last of the 25 or so Pleistocene glaciations that began 2.6 million years ago peaked about 18,000 to 20,000 years ago (figure 22.7). The final 7,000 years of the Pleistocene (from about 18,000 years ago to about 11,000 years ago) is called the last glacial-interglacial transition. The peak of glaciation was explained by the Milankovitch cycles of orbital variation when all three cycles reached a point of minimal solar radiation to Earth. Then the cycles shifted again, and Earth began to slowly warm out of the peak glacial conditions and toward an interglacial phase 10,000 years ago, when all three Milankovitch cycles produced the maximum amount of solar input and planetary warming. It was not a simple steady increase in temperature, however. There were many small-scale fluctuations in climate, with both short-term warming and cooling events, before the climate stabilized into an interglacial mode. Some were due to complex interactions of the Milankovitch cycles, but most were due to events in Earth's oceans and atmospheres.

The most famous of these events was a Heinrich event about 16,000 years ago, when the Arctic ice sheet began to fall apart and released huge volumes of floating icebergs and freshwater into the North Atlantic Ocean. Huge Heinrich events release lots of icebergs. These are detected by finding large pebbles

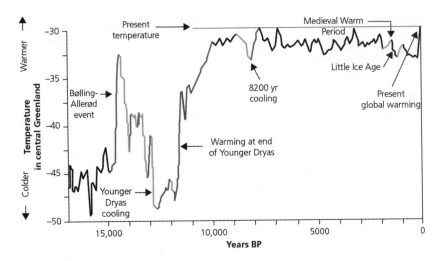

The temperature record of the glacial-interglacial transition as recorded in ice cores, mainly from Greenland. The peak of glaciation ended about 18,000 years ago, and about 16,000 years ago was the last of the Heinrich events releasing Arctic icebergs into the ocean. The detailed record shows numerous short-term warming and cooling events known as Dansgaard-Oeschger cycles. About 14,700 was another warming cycle, called the Bølling-Allerød event, followed by the abrupt cooling event from 12,900 to 11,700 years ago known as the Younger Dryas. After the planet warmed up again at the end of the Younger Dryas, climate has remained relatively stable for the past 11,000 years of the Holocene. (Redrawn from several sources)

and coarse sand grains melted out of icebergs and dropped into the mud found in deep-sea cores. Heinrich events are known from the last seven glacial periods, dating back 640,000 years. However, the Heinrich event 16,000 years ago launched the most rapid phase of warming in the glacial-interglacial transition. About 14,680 years ago, an abrupt warming event increased the moisture of Greenland, and the ice sheet almost doubled in thickness with all the extra snow and ice. This warming episode is correlated with another peak in the eccentricity cycle of the Milankovitch orbital variation cycles.

In addition, throughout this entire interval of climatic fluctuations there were a number of Dansgaard-Oeschger events: rapid warming episodes lasting a decade or more, followed by gradual cooling. There have been at least 25 of these events during the last glacial period and the glacial-interglacial transition, roughly spaced about 1,470 years apart. They show up mainly in Greenland ice cores, but they are not found in Antarctic ice cores.

Based on these records and North Atlantic deep-sea sediment cores, it is thought that these events are caused by rapid pulses of fresh water and ice released from the melting of the Arctic ice sheet. What causes these quasi-periodic events is still debated, although it might be due to the cooling and excessive buildup of Arctic ice, followed by a rapid "purge" of ice afterward, or possibly by changes in the shape and size of northern ice sheets.

Even more dramatic was a rapid cooling event near the end of the Pleistocene, about 12,900 to 11,700 years ago, known as the Younger Dryas event. First discovered in Greenland ice cores with very high-resolution yearly records, and confirmed in cores of Swiss and Scandinavian glacial lake sediments that record every year precisely, it showed just how quickly the entire climate system can flip from warming interglacial back to full glacial—in about 10 years! The interval got its name after the abundance of pollen of *Dryas octopelta*—a European Alpine tundra flower that indicates cold conditions—was found in cores taken from Swiss lake beds. The Younger Dryas was long a mystery to paleoclimatologists. It began with an abrupt cooling lasting a few decades about 12,900 years ago, then kept the world plunged into another glacial episode until it abruptly warmed in less than a decade about 11,700 years ago.

But then deep-sea cores from the North Atlantic led climate modelers to suggest a startling explanation. The flow of ocean waters around Earth are all connected by what is known as the "global ocean conveyer belt" that connects the flow of all the world's oceans in a gigantic looping current so long that it takes a water molecule more than 2,000 years to complete the circuit (figure 22.8). For example, during the middle Pliocene, the Gulf Stream was diverted northward to the North Atlantic when the Panama land bridge closed and shut off the flow from the Caribbean to the Pacific. That "conveyer belt" of the Gulf Stream brought warm moist air to the Arctic, where it was frozen into snow, allowing the Arctic ice sheet to develop. Once the Gulf Stream reaches far enough into the North Atlantic north of Scotland, it runs into cold Arctic waters, where it quickly cools, then sinks and returns to the South Atlantic in a deep water current. That deep current then connects with cold Antarctic currents running all the way to the North Pacific, where it warms again. From there it becomes a warm surface current that flows back into the Indian Ocean, around Africa, and then back into the South Atlantic, eventually joining with the Gulf Stream and completing the conveyer loop.

That is how the global ocean conveyer works today. But what happens if it were to shut off or slow down? Then all the climate systems that

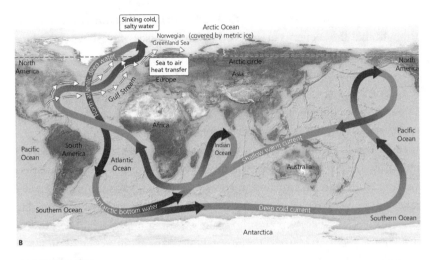

B

The global ocean conveyer belt, showing the warm Gulf Stream coming north from the tropical Atlantic and bringing warm, moist conditions to northern Europe. If the Gulf Stream were disrupted, the Arctic would lose its source of warm water and quickly freeze over again. (Redrawn from several sources)

we now take for granted would be thrown into chaos, and there would be an entirely new regime. Apparently, this is what happened during the Younger Dryas event. Deep-sea cores showed evidence of a sudden change in the water chemistry in the North Atlantic when it began. They traced this change to a sudden flood of water from ice-dammed lakes on the fringes of the glaciers melting in North America (especially glacial Lake Agassiz, which covered much of Manitoba and Saskatchewan as well as northern Minnesota and North Dakota). When the ice dam on the St. Lawrence River broke, it released a gigantic flood of fresh water into the North Atlantic. Fresh water is less dense than salt water, so this freshwater "lid" sat on top of the saltier ocean water and caused the "conveyer" to sink much further south than it does today. This prevented the warmth of the Gulf Stream from reaching the far northern Atlantic, and within a decade ice age conditions had returned to the northern hemisphere. From this we know that the Earth system is very sensitive to slight changes and can switch from warm interglacial to full glacial in just a few years given the right conditions.

The end of the ice ages is well documented by numerous geological records. But what happened to the mammals of the Ice Age? That is the subject of the next chapter.

NOTE

1. Andre Berger, "Milankovitch, the Father of Paleoclimate Modeling," *Climate of the Past* 17, no. 4 (2021): 1727–1733.

FOR FURTHER READING

Blakey, Ronald C., and Wayne D. Ranney. *Ancient Landscapes of Western North America: A Geologic History with Paleogeographic Maps*. Berlin: Springer, 2018.

Bolles, Edmund B. *The Ice Finders: How a Poet, a Professor, and a Politician Discovered the Ice Age*. New York: Counterpoint, 1999.

Childs, Craig. *Atlas of a Lost World: Travels in Ice Age America*. New York: Pantheon, 2018.

Flint, Richard Foster. *Glacial and Pleistocene Geology*. New York: Wiley, 1971.

Gribbin, John, and Mary Gribbin, *Ice Age: The Theory That Came in from the Cold*. New York: Barnes and Noble, 2002.

Guthrie, R. Dale. *Frozen Fauna of the Mammoth Steppe: The Story of the Blue Babe*. Chicago: University of Chicago Press, 1990.

Imbrie, John, and Katherine P. Imbrie. *Ice Ages: Solving the Mystery*. Short Hills, NJ: Enslow, 1979.

Macdougall, Doug. *Frozen Earth: The Once and Future Story of Ice Ages*. Berkeley: University of California Press, 2004.

Pielou, E. C. *After the Ice Age: The Return of Life to Glaciated North America*. Chicago: University of Chicago Press, 1991.

Pyne, Lydia, and Stephen J. Pyne. *The Last Lost World: Ice Ages, Human Origins, and the Invention of the Pleistocene*. New York: Penguin, 2013.

Ruddiman, William F. *Earth's Climate: Past and Future*. New York: Freeman, 2013.

Stanley, Steven M. *Children of the Ice Age. How a Global Catastrophe Allowed Humans to Evolve*. New York: Crown, 1995.

Vrba, Elizabeth S., George H. Denton, Timothy C. Partridge, and Lloyd H. Burckle, eds. *Paleoclimate and Evolution, with Emphasis on Human Origins*. New Haven, CT: Yale University Press, 1995.

Woodward, Jamie. *The Ice Age: A Very Short Introduction*. Oxford: Oxford University Press, 2014.

THE DEATH OF THE MEGAMAMMALS

It is impossible to reflect on the state of the American continent without astonishment. Formerly it must have swarmed with great monsters; now we find mere pygmies compared with the antecedent allied races.

—CHARLES DARWIN, *VOYAGE OF THE H.M.S. BEAGLE* (1839)

WHERE HAVE ALL THE MAMMOTHS GONE?

If you visit the Serengeti Plains of East Africa today, you will see a wide array of large mammals over 46 kilograms (100 pounds) in body size: elephants, hippos, rhinos, giraffes, Cape buffalo, many larger antelopes, and of course, huge predators like lions. A similarly diverse fauna of many large mammals roamed the grasslands of the Miocene around the world and continued to do so throughout the Pliocene and most of the Pleistocene. The fossils of the La Brea tar pits show that Hollywood was once inhabited not by movie stars but by now extinct mammoths, mastodonts, horses, tapirs, camels, ground sloths, and many predators including huge short-faced bears, large lions, and saber-toothed cats, as well as living groups such as bison, deer, and pronghorns. Today bison, deer, moose, and pronghorns constitute the only "large hoofed mammals" of North America, and the cougar and grizzly bear are the largest predators—a pathetic remnant of what once lived on this continent only 10,000 years ago (figure 23.1). Even the Arctic regions

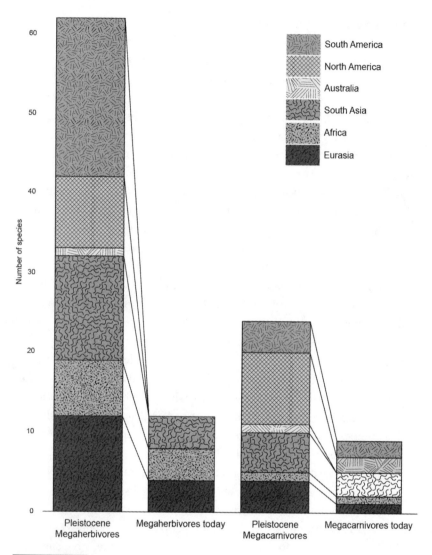

Figure 23.1 ▲

Percentages of extinction of both herbivores (left columns) and carnivores (right columns) from the late Pleistocene until today, shown continent by continent. In every case, the total diversity of megamammals is hugely reduced from its Pleistocene peak. (Redrawn from several sources)

were once covered by a grassland known as the "mammoth steppe," with a diversity not only of mammoth but of horses, bison, muskoxen, and a surprising diversity of hoofed mammals, as well as large cats and bears. But today only parts of Africa and southern Asia support faunas dominated by elephants and rhinos and other large hoofed mammals. What happened?

The realization that this had occurred is a recent discovery in the history of science—barely 200 years old. As early as 1796, the great anatomist Baron Georges Cuvier was the first to argue that the great mammoths, mastodons, and ground sloths were extinct and that our world had been inhabited by huge beasts that were now long gone. In *Voyage of the H.M.S. Beagle* in 1839, Charles Darwin wrote, "it is impossible to reflect on the state of the American continent without astonishment. Formerly it must have swarmed with great monsters; now we find mere pygmies compared with the antecedent allied races."[1] Most of the large mammals of the other continents are gone as well, from the mammoths, hippos, rhinos, giant deer, and saber-toothed cats of Eurasia, to the South American mammoths, mastodonts, horses, camels, saber-toothed cats, to the uniquely South American hoofed mammals known as toxodonts and litopterns, and the giant marsupials of Australia. The latest Pleistocene was a period of widespread extinction in large mammals, but not so much in the medium-sized and small mammals. The one exception to this pattern is Africa, where elephants, rhinos, hippos, giraffes, and a high diversity of antelopes and cattle still live, providing us with a glimpse of the world before the Pleistocene megafaunal extinctions. Darwin's codiscoverer of natural selection, Alfred Russel Wallace, wrote that "we live in a zoologically impoverished world, from which all the hugest, and fiercest, and strangest forms have recently disappeared."[2]

Almost as soon as Cuvier demonstrated that the mammoths and other large Pleistocene mammals were extinct, people began to entertain explanations for their extinction. Most of these explanations have now been discredited due to better evidence, but several still remain. They fall largely into four categories, which (thanks to the long usage of the term "overkill") have been waggishly named:

1. "overkill" hypothesis (effects of humans)
2. "overchill" hypothesis (effects of climate)
3. "overgrill" hypothesis (effects of an extraterrestrial impact)
4. "overill" hypothesis (a global pandemic or "hyperdisease")

I'll quickly deal with the last two explanations, and then I'll focus on the main contenders: the overkill and overchill explanations.

"OVERILL"

In 1997, Ross MacPhee, a mammalogist at the American Museum of Natural History in New York, along with P. A. Marx, published a paper in which they claimed that the megamammals were wiped out by a pandemic of a virulent pathogen, what they called the "hyperdisease" hypothesis. The idea has gotten lots of publicity from media, which prints any catchy idea, and it is promoted in MacPhee and Schouten's 2018 book, *The End of the Megafauna*. But how well does it explain the data?

The first thing most scientists thought when the idea was proposed was "How could you test this hypothesis?" Fossils almost never preserve any trace of a disease or infection of this kind that rapidly kills in large numbers. The evidence of disease we do see in fossils is from slow, bone-altering diseases, such as arthritis or cancer or a massive infection that goes clear down to the bone. Most scientists initially dismissed this idea as untestable speculation and didn't give it much further thought. But a study by Kathleen Lyons and colleagues in 2004 made an attempt to evaluate it. They looked at known examples of recent diseases that have wiped out animal populations and found two examples: the West Nile virus on North American bird populations and the introduction of malaria to Hawaiian bird populations. They concluded that these pathogens never completely wipe out a species, and the victims are restricted to just a few groups across many different size classes. In contrast, the victims of the late Pleistocene extinctions were all of large body size, and there were few or no extinctions in medium-sized or smaller mammals. That speaks strongly against the hyperdisease hypothesis. But the strongest argument is that we know of *no* virulent pathogens that infect animals across a wide spectrum of classes and animals that are not closely related. Even the worst epidemics end eventually when herd immunity is reached, and there are no documented examples of disease wiping out all members of a species in many genera over an entire fauna.

Let's leave this improbable suggestion behind and focus on those ideas that can be scientifically tested.

"OVERGRILL"

The most recent entry on the bandwagon idea that impacts cause extinctions was the claim by Firestone and colleagues in 2007 that the extinction of the Ice Age megamammals was due to the impact of an extraterrestrial object about 12,900 years ago. When this idea was first proposed, the media had a field day, and almost no dissenters or critics were heard at all. Some geology textbooks even inserted this untested idea into their new editions without waiting to see whether it would pan out or not. Just like every other half-baked idea from the impact advocates, the "late Pleistocene impact" scenario has been shot down by a whole range of observations.

The late Pleistocene impact hypothesis emerged from observations by Firestone and colleagues that there was a distinctive "black mat" organic layer in several localities across the southwestern United States, immediately above the last appearance of some Ice Age megamammal fossils. These include not only the huge mammoths and mastodonts but also ground sloths, horses, camels, two genera of peccaries, giant beavers, plus predators such as short-faced bears, dire wolves, and saber-toothed cats—but not bison, deer, pronghorns, and a number of other large mammals still found in North America today. The black mat is also above the first known artifacts of the Clovis culture, which were thought to be the first human arrivals from Eurasia, and they were allegedly responsible for overhunting the megamammals to extinction. Firestone and colleagues also claimed to have found "nanodiamonds," iridium, helium-3, "buckyballs," and a number of other geochemical and mineralogical "impact indicators" in the black mat layer. They painted a variety of different (and sometimes conflicting) scenarios about the impacting object (whether it was a comet or an asteroid) hitting near the Carolina Bays region. This supposedly affected the Laurentide ice sheet in the northeastern part of North America and triggered the Younger Dryas cooling event 12,900 years ago.

The entire scenario has been completely demolished by a number of lines of evidence. As studies by Nicholas Pinter, Scott Ishman, and colleagues showed in 2008 and 2011, there is no evidence that there was an impact in the Carolina Bays, and most of the alleged "impact evidence" such as nanodiamonds, buckyballs, iridium, and other geochemical indicators was questionable when analyzed by other labs. Firestone and

colleagues claimed that the impact was an airburst because there is no crater, no tektites, no shocked quartz, and no other high-pressure minerals,
which are the best indicators of a true impact. Most of the material that was
allegedly impact derived is also consistent with the normal rain of micro-
meteorites, and none of it is abundant enough to be good evidence of an
impact. Some of this material turned out to be fossilized fecal pellets, causing some to joke that the impact hypothesis was "full of sh*t".

The claim that the black mat was an impact layer has been thoroughly
debunked, and it is more likely an indicator of a high water table and wetter conditions associated with the abrupt Younger Dryas cooling event. The
supposed "instantaneous" extinction of megamammals at this horizon has
also been debunked because the extinctions were scattered across a wide
geographic area with different genera going out locally at different times.
The fact that mammoths, mastodons, giant deer ("Irish elk"), ground
sloths, and many other megamammals did not die 12,900 years ago but
survived in most cases until 10,000 years ago or even more recently, is
fatal to the idea that a single impact killed them all off. In fact, *none* of the
well-dated extinctions occurred 12,900 years ago. Most of the extinctions
are either significantly younger than that interval or have no good final
dates for their last appearance—very little appears to have happened to the
megamammals precisely 12,900 years ago.

Particularly striking is the persistence of mammoths and ground sloths
well into the Holocene (as young as only 6,000 years ago), and, of course,
the bison, deer, grizzly bear, cougars, peccaries, and pronghorns are still
with us, and elk and moose came to North America at this time (rather than
being wiped out then). The impact hypothesis does nothing to explain the
selectivity of this extinction. In addition, the South American, Australian,
and Eurasian-African megafaunal extinctions are not synchronous with the
alleged impact, so it does nothing to explain their demise.

The claim that the impact had a severe effect on human cultures has
been completely shot down as well. There is no evidence whatsoever that
human cultures changed dramatically at that time, nor that there was a
major population decline. Clovis culture was gradually transformed into
Folsom, Dalton, and eastern U.S. Paleoindian cultures, and they apparently
spread widely at this time rather than being in decline. In 2009, Jacque-
lyn Gill (now at the University of Maine) published a study analyzing the
details of lake sediments from the Northeast that preserved a high-fidelity

record of that time. She found no evidence of the impact debris that was supposed to be common—and her data were gathered even closer to the alleged impact site than the evidence garnered from the western United States, where the original black mat was found in New Mexico. Nor was there any great shift in vegetation, pollen, spores, or any other biotic signal that would be consistent with the impact hypothesis.

"OVERKILL"

The overkill hypothesis, originally proposed in 1967 by the late Paul Martin of the University of Arizona, postulates that whenever humans with advanced sophisticated hunting technologies invade a new area, the large mammals are unwary or insufficiently protected from the voracious hunting techniques of humans and are hunted out of existence. Martin and other overkill advocates pointed to the apparent coincidence of the arrival of humans with sophisticated Clovis points in North America about 13,000 years ago and the disappearance of the large mammals between 11,000 and 9,000 years ago (figure 23.2). Likewise, the megafauna of South America died out at about the same time, supposedly as humans spread south. The overkill model has also been applied to Australia and allegedly dated to when the aboriginal peoples arrived from Asia. The disappearance of mammals in Eurasia and Africa is less clear-cut because Africa still has its megamammals—but Martin would argue that they evolved side by side with humans, so they were not easily eliminated. In Eurasia, the extinction of the megamammals was also gradual, but Martin argues that it peaked about 32,000 years ago when sophisticated Paleolithic hunting tools appeared. However, the latest dating shows that mammoths, rhinos, and large deer survived in Eurasia until about 14,000 years ago. In fact, a recent study by Cooper and others showed a strong correlation between the extinction of Eurasian megamammals and rapid warming episodes (figure 23.3). Few scientists dispute the overkill hypothesis when it comes to islands like those of the Pacific (especially New Zealand and Hawaii), the Caribbean (especially Hispaniola and Cuba), or the Indian Ocean (especially Madagascar and Mauritius). It is well documented that their peculiar endemic animals were wiped out when humans arrived.

But the explanation is not as simple as this. Although 32 species of large mammals died out in North America between 11,000 and 9,000 years

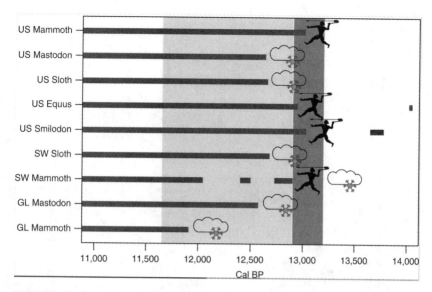

Figure 23.2 ▲
Summary of the chronology for megafaunal population busts (dark lines) by region in relation to the Clovis period (dark column) and the Younger Dryas (light column). Symbols to the right indicate extinction cause(s) suggested by this analysis: clouds and snowflake = climate; human with spear = hunting. (From Jack M. Broughton and Elic M. Weitzel, "Population Reconstructions for Humans and Megafauna Suggest Mixed Causes for North American Pleistocene Extinctions," *Nature Communications* 9, no. 5441 [2018]; by permission of *Nature*)

ago, only a few have been shown to have been hunted by humans (mainly mammoths, mastodonts, bison, horses, and camels). We have no evidence that the rest of the megamammals were extensively hunted, and critics argue that we should find more evidence of hunting if these mammals were hunted to extinction. Moreover, some of the groups that we know were extensively hunted (such as bison and deer) survived, whereas other groups did not. Even more critical for the overkill hypothesis is the fact that many small mammals *did not* go extinct, even though humans find small mammals much easier to catch and kill. Also, dozens of bird species *did* go extinct, even though they were not obvious food for humans. If the Clovis hunters were as voracious and devastating as Martin's "blitzkrieg" model suggests, why did it take several thousand years for them to wipe out the North American megamammals?

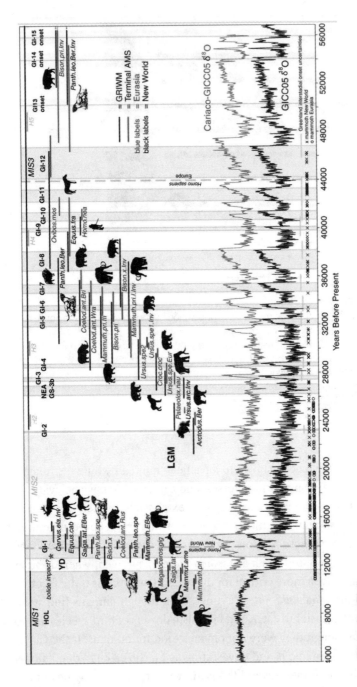

Figure 23.3 ▲

Correlation of extinctions of Eurasian Pleistocene mammals with key warming events during the Pleistocene. (After Alan Cooper, et al., "Abrupt Warming Events Drove Late Pleistocene Holarctic Megafaunal Turnover," *Science* 349, no. 6248 (2015): 602–606; by permission of AAAS)

The crux of this debate is whether or not the megamammals did in fact vanish shortly after humans first appeared in the Americas or Australia. At one time, the first appearance of humans in North America and South America seemed to be closely linked. But more and more sites have now been found that show humans were clearly here thousands of years before the megafaunal extinction. Ironically, for a long time the oldest evidence of humans in the Americas was *not* near the retreating edge of the northern ice sheet in Canada (as the Martin blitzkrieg model predicted) but all the way down to the southern tip of South America. This is the famous Monte Verde site in southern Chile, dated to at least 14,500 years ago, and possibly as old as 18,500 years old (figure 23.4). This suggests that humans didn't arrive in a wave walking across the North American mainland, crossing the Plains from north to south. Instead, it is likely that they canoed their way down the West Coast and inhabited sites on the coast that have now vanished underwater, drowned by the postglacial rise of sea level.

There are no archeological sites in North America close to the ice-free corridor in Canada and Alaska where the first humans supposedly traveled. For a long time, the Meadowcroft Rock Shelter in Pennsylvania was the oldest site in North America at more than 16,000 years old. There is human fecal matter in Paisley Cave in southern Oregon dated to 14,300 years ago. Now there are 20-some "pre-Clovis" sites across the Americas with dates older than 12,000 to 13,000 years ago.

In 2021, the entire overkill argument took several severe blows. Fossilized human footprints found near White Sands National Monument in New Mexico were dated at 22,500 years ago—about 10,000 years before the megafaunal extinctions. This discovery was topped in 2022 by a study by Tim Rowe and others who found fossil mammoths showing clear signs of human butchering—and they were radiocarbon dated at 39,000 years old! Clearly, humans were not overhunting megamammals for thousands of years, and the overkill model proponents have a lot of explaining to do. They usually backtrack and argue that it's the arrival of the people with distinctive Clovis points and a rapacious hunting culture around 13,000 years ago that made the difference, but this is an ad hoc hypothesis, an attempt to salvage an idea that has failed numerous tests.

In 1996, M. W. Beck tested the overkill hypothesis in North America by examining the date of the last occurrence of each extinct species. According to the prediction, they should have died out first in areas near the Bering

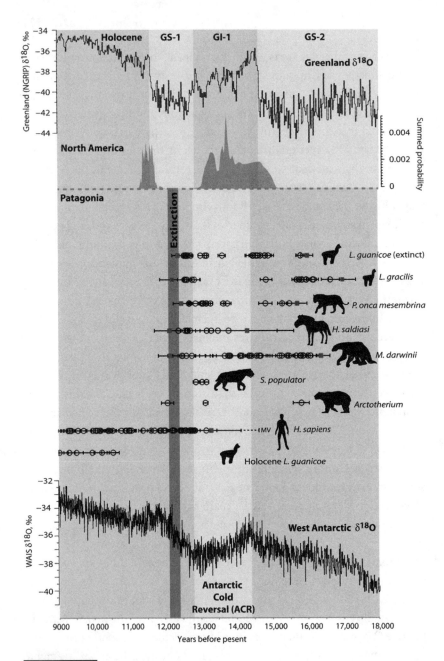

Figure 23.4 ▲

Timing of extinctions in South America and correlation (or lack thereof) with climate events and human arrivals. MV on the human timeline = Monte Verde sites, the oldest evidence of humans in South America. (After J. L. Metcalf, et al., "Synergistic Roles of Climate Warming and Human Occupation in Patagonian Megafaunal Extinction During the Last Deglaciation," *Science Advances* 2, no. 6 (2016): 1–8; by permission of AAAS)

land bridge, and last in the most remote areas. In fact, the opposite pattern was found: the most remote faunas died out long before those closer to the Bering land bridge.

The most devastating argument comes from Australia, where humans were present as early as 46,000 years ago, but the extinction of large mammals is not coincident with human arrival. Studies summarized by Wroe and others in 2013 showed that nearly all the large mammals and giant monitor lizards (like goannas and Komodo dragons) were gone long before human arrival around 46,000 years ago; most were gone 100,000 years ago (figure 23.5). The only animals that seemed to die out with the arrival of humans were a few giant wombat relatives known as diprotodonts and giant kangaroos. In fact, some of the large, slow-moving marsupials survived until as recently as 16,000 years ago. In addition, Australian archeologists have long pointed out that almost none of their sites show much evidence of hunting or butchering of the large Pleistocene marsupials. Finally, the late Pleistocene giant marsupials show a dwarfing trend toward

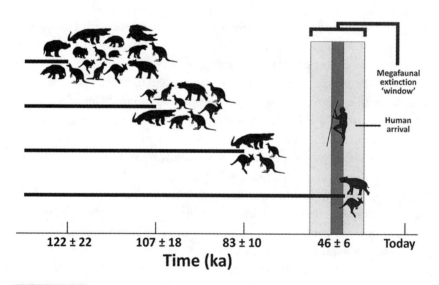

Megafaunal
extinction
'window'

Human
arrival

| 122 ± 22 | 107 ± 18 | 83 ± 10 | 46 ± 6 | Today |

Time (ka)

Figure 23.5 ▲

Waves of extinction in Australian Pleistocene megafauna, most of which occurred long before the first humans reached the region. (From S. W. Wroe, et al., "Climate Change Frames Debate Over the Extinction of the Megafauna in Sahul (Pleistocene Australia-New Guinea)," *Proceedings of the National Academy of Sciences* 110, no. 22 (2013): 8777–8781; by permission of National Academy of Sciences)

the end of their history, consistent with the idea that they were adapting to climatic change before their final demise.

Another variation of the idea that humans impacted the fauna is the "keystone herbivore" hypothesis of Norman Owen-Smith. In his studies of living African elephants, Owen-Smith points out that they are critical to maintaining a diverse habitat and fauna in eastern Africa. Elephants are capable of pruning back and destroying trees and brush at an alarming rate. They act as gardeners, modifying the brush and preventing it from becoming overgrown, so diverse habitats are available for many kinds of antelopes and other mammals. When they are poached out of an area, the brush quickly becomes overgrown and diversity drops. If mammoths performed the same role in the Pleistocene, their extinction might have had a critical effect, causing habitat change that affected most other mammals as well. Thus the extinction of mammoths (either by human overkill or by abrupt climatic change) may be a key factor.

"OVERCHILL"

An alternative school of thought blames most of the late Pleistocene extinctions on climatic change. The period around 11,000–10,000 years ago was the end of the glacial-interglacial transition from the end of the last Pleistocene glacial (around 18,000 years ago) to the current Holocene interglacial episode, and indeed it was a period of rapid climatic change. Critics argue that there were many previous glacial-interglacial transitions during the Pleistocene, so they cannot see how the events 10,000 years ago were any different from earlier transitions. But recent paleoclimatic studies have shown that the last glacial-interglacial transition was amazingly abrupt, taking place in a few decades or less, and it was characterized by extreme swings in climate before interglacial warming took hold. Apparently it became much colder and drier than ever before, and there was a great expansion of deserts. Russ Graham and Ernie Lundelius point out that the late Pleistocene fauna consisted of a mosaic of small habitats, with a wide variety of animals overlapping in ranges. In contrast, the Holocene is characterized by a few monotonous habitats (desert, grasslands, tundra, taiga) that extend over long distances. Many living animals are now living in narrow niches at the edge of their former range, and they no longer overlap with animals that once shared their habitat.

A rapid climate change would have disrupted the habitat mosaic and led to many extinctions.

Critics of the climate change hypothesis have argued that all of the major extinctions are correlated with the appearance of humans, not with the end of the Pleistocene about 10,000 years ago. But the extinction of North and South American land mammals now more closely match the end-Pleistocene climatic signal, which is much later than the arrival of humans 39,000 years ago. Likewise, the disappearance of Australian megamammals more than 100,000 years ago and the Pleistocene extinctions in Eurasia 14,000 years ago better correspond with the end of glacial-interglacial transitions than with the occurrence of humans on those continents, as Cooper and others pointed out.

For the longest time, the debate focused on the mass extinctions at the end of the Pleistocene 10,000 years ago, with the assumption that the megamammals were gone from Eurasia, Africa, and the Americas after that date. But recent discoveries have shown that both the woolly mammoth and the giant deer *Megaloceros* (the misnamed "Irish elk," which is not an elk, nor is it exclusively Irish) survived well into the Holocene. The giant deer died out in most places by the beginning of the Holocene, but they survived in Siberia until about 7,700 years ago. The woolly mammoth, too, survived in Siberia through much of the Holocene and vanished from Wrangel Island in the Siberian Arctic as recently as 4,000 years ago, where it had become a dwarfed species. The American mastodon was still alive in North America until possibly 6,000 years ago, and the ground sloth did not vanish from the Caribbean until about 5,000 years ago. In 2004, Stuart and colleagues pointed out that these variable last occurrences due to differences in climate and geography do not support the "overkill" blitzkrieg hypothesis (human hunters were present in all of these regions long before the extinction of giant deer or woolly mammoths) and show that extinction can be due to complex ecological factors that vary from region to region.

And there's another twist: the fossil bone record may not be complete enough to provide a reliable answer. In 2009, scientists led by James Haile analyzed samples of soil from Alaska's deeply buried permafrost. They found DNA from extinct horse and mammoth in permafrost that was 2,000 years younger (only 7,000 years old) than the youngest bones of these animals! Even though there are no fossilized bones to confirm it, horses and mammoths survived for thousands of years after human hunters arrived, long after the major changes in climate and the supposed blitzkrieg by human hunters—and the last known fossil bones of a species do not reliably tell us when they actually vanished.

In short, there is a tremendous diversity of explanations, with no consensus after more than 55 years of the overkill vs. climate debate (figure 23.6). In 1984, Larry Marshall listed these diverse explanations along a continuous spectrum, with human causes at one extreme and climate at the other. With so many lethal events, some scientists have remarked that it's a miracle that *anything* survived! In all likelihood, the end-Pleistocene extinctions were a complex series of events, with both human and climatic components. Indeed, a distinguished panel of scientists led by Tony Barnosky and who were not affiliated with either the climatic change or overkill camps supported this conclusion: Both human hunting and extreme climatic change contributed to the Pleistocene extinctions. Scientists, like most people, prefer simple answers in black and white, but nature is rarely so cooperative

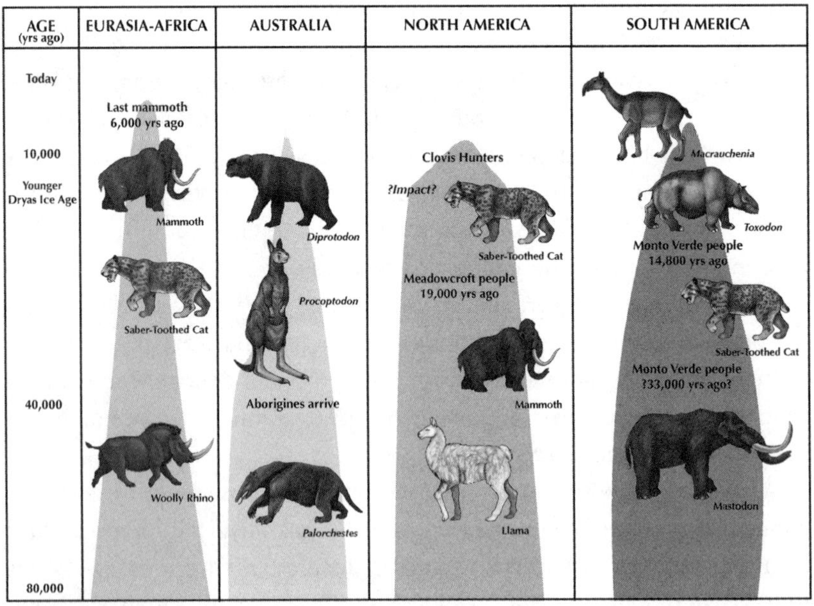

Figure 23.6 ▲

Summary of the patterns of megafaunal extinctions on four continents.

and more often the truth lies in a nexus of complex causes. When someone asks, "Did humans or climate wipe out the Ice Age mammals?" the answer (as it often is in science) is: "It's complicated." The media may prefer simplistic answers that fit in a sound bite, but in the real world almost nothing in complex biological and geological systems fits that description.

Whatever the cause, by the beginning of the Holocene about 10,000 years ago, most of the Ice Age giants were gone. Humans had spread across the entire world, which was largely denuded of its large mammals, and they were facing rapidly changing environmental hostility. Similar changes continued throughout the Holocene and still face us today.

NOTES

1. Charles Darwin, *Voyages of the Adventure and Beagle, Volume III, Journal and Remarks* (London: Henry Colburn, 1839), 116.
2. Alfred Russel Wallace, *The Geographical Distribution of Animals* (London: MacMillan, 1876), 150.

FOR FURTHER READING

Barnosky, Anthony D., Paul L. Koch, Robert S. Feranec, Scott L. Wing, and Alan B. Shabel. "Assessing the Causes of Late Pleistocene Extinctions on the Continents." *Science* 306, no. 5693 (2004): 70–75.

Beck, Michael W. "On Discerning the Causes of Late Pleistocene Megafaunal Extinctions." *Paleobiology* 22, no. 1 (1996): 91–103.

Bennett, Matthew W., David Bustos, Jeffrey S. Pigati, Kathleen B. Springer, Thomas M. Urban, Vance T. Holliday, Sally C. Reynolds, et al. "Evidence of Humans in North America During the Last Glacial Maximum." *Science* 373, no. 6562 (2021): 1528–1531.

Broughton, Jack M., and Elic M. Weitzel. "Population Reconstructions for Humans and Megafauna Suggest Mixed Causes for North American Pleistocene Extinctions." *Nature Communications* 9, no. 5441 (2018).

Cohen, Claudia. *The Fate of the Mammoth: Fossils, Myth, and History.* Chicago: University of Chicago Press, 2002.

Cooper, Alan, Chris Turney, Konrad A. Hughen, Barry W. Brook, H. Gregory McDonald, and Corey J. A. Bradshaw. "Abrupt Warming Events Drove Late Pleistocene Holarctic Megafaunal Turnover." *Science* 349, no. 6248 (2015): 602–606.

Fariña, Richard A., and Sergio F. Vizcaíno. *Megafauna: Giant Beasts of Pleistocene South America.* Bloomington: Indiana University Press, 2013.

Firestone, R. B., A. West, J. P. Kennett, L. Becker, T. E. Bunch, Z. S. Revay, P. H. Schultz, T. Belgya, D. J. Kennett, J. M. Erlandson, O. J. Dickenson, et al. "Evidence for an Extraterrestrial Impact 12,900 Years Ago That Contributed to the Megafaunal Extinctions and the Younger Dryas Cooling." *Proceedings of the National Academy of Sciences of the United States of America* 104, no. 41 (2007):16016–21.

Graham, Russell W., and E. L. Lundelius Jr. "Coevolutionary Disequilibrium and Pleistocene Extinctions." In *Quaternary Extinctions: A Prehistoric Revolution*, ed. P. S. Martin and R. G. Klein, 223–249. Tucson: University of Arizona Press, 1984.

Grayson, Donald. *Giant Sloths and Sabertooth Cats: Archeology of the Ice Age Great Basin.* Salt Lake City: University of Utah Press, 2016.

Guthrie, R. Dale. *Frozen Fauna of the Mammoth Steppe: The Story of the Blue Babe.* Chicago: University of Chicago Press, 1990.

Haile, James, Duane G. Froese, Ross D. E. MacPhee, Richard G. Roberts, Lee J. Arnold, Alberto V. Reyes, Morten Rasmussen, et al. "Ancient DNA Reveals Late Survival of Mammoth and Horse in Interior Alaska." *Proceedings of the National Academy of Sciences* 106, no. 52 (2009): 22352–22357.

Haynes, Gary. *American Megafaunal Extinctions at the End of the Pleistocene.* Berlin: Springer, 2009.

Koch, Paul, and Anthony D. Barnosky. "Late Quaternary Extinctions: State of the Debate." *Annual Reviews of Ecology, Evolution, and Systematics* 37, no. 1 (2007): 215–250.

Kurtén, Björn. *Before the Indians.* New York: Columbia University Press, 1988.

——. *Pleistocene Mammals of Europe.* New York: Columbia University Press, 1968.

Kurtén, Björn, and Elaine Anderson. *Pleistocene Mammals of North America.* New York: Columbia University Press, 1980.

Lange, Ian. *Ice Age Mammals of North America.* Missoula, MT: Mountain, 2002.

Levy, Sharon. *Once and Future Giants: What Ice Age Extinctions Tell Us About the Fate of the Earth's Largest Animals.* Oxford: Oxford University Press, 2011.

Lister, Adrian. *Mammoths: Giants of the Ice Age.* London: Chartwell, 2015.

Lundelius, E. L., Jr. "Climatic Implications of Late Pleistocene and Holocene Faunal Associations in Australia." *Alcheringa* 7 (1983):125–149.

Lyons, S. Kathleen, Felisa A. Smith, Peter Wagner, Ethan P. White, and James Brown. "Was a 'Hyperdisease' Responsible for the Late Pleistocene Megafaunal Extinction?" *Ecology Letters* 7, no. 9 (2004): 859–868.

MacPhee, Ross D. E. *Extinctions in Near Time: Causes, Contexts, and Consequences.* New York: Kluwer Academic/Plenum, 1999.

MacPhee, Ross D. E., and P. A. Marx. "Humans, Hyperdisease, and First-Contact Extinctions." In *Natural Change and Human Impact in Madagascar*, ed. Steven M. Goodman and Bruce D. Patterson, 169–221. Washington, DC: Smithsonian Institution Press, 1997.

MacPhee, Ross D. E., and Peter Schouten, *The End of the Megafauna: The Fate of the World's Hugest, Fiercest, Strangest Animals.* New York: Norton, 2018.

Marshall, Larry G. "Who Killed Cock Robin? An Investigation of the Extinction Controversy." In *Quaternary Extinctions: A Prehistoric Revolution*, ed. P. S. Martin and R. G. Klein, 785–806. Tucson: University of Arizona Press, 1984.

Martin, Paul S. "Prehistoric Overkill." In *Pleistocene Extinctions: The Search for a Cause*, ed. P. S. Martin and H. E. Wright Jr., 75–120. New Haven, CT: Yale University Press, 1967.

——. "Prehistoric Overkill: The Model." In *Quaternary Extinctions: A Prehistoric Revolution*, ed. P. S. Martin and R. G. Klein, 354–403. Tucson: University of Arizona Press, 1984.

Martin, Paul S., and Richard G. Klein, eds. *Quaternary Extinctions: A Prehistoric Revolution.* Tucson: University of Arizona Press, 1984.

Metcalf, Jessica L., Chris Turney, Ross Barnett, Fabiana Martin, Sarah C. Bray, Julia T. Vistrup, Ludovic Orlando, et al. "Synergistic Roles of Climate Warming and Human Occupation in Patagonian Megafaunal Extinction During the Last Deglaciation." *Science Advances* 2, no. 6 (2016): 1–8.

Pinter, Nicholas, and Scott E. Ishman. "Impacts, Mega-Tsunami, and Other Extraordinary Claims." *GSA Today* 18, no. 1 (2008): 37–38.

Pinter, Nicholas, Andrew C. Scott, Tyrone L. Daulton, Andrew Podoll, Christian Koerberl, R. Scott Anderson, and Scott E. Ishman. "The Younger Dryas Impact Hypothesis: A Requiem." *Earth-Science Reviews* 106, no. 3 (2011): 247–264.

Owen-Smith, Norman. "Pleistocene Extinctions: The Pivotal Role of Megaherbivores." *Paleobiology* 13, no. 3 (1987): 351–362.

Rowe, Timothy D., Thomas W. Stafford Jr., Daniel C. Fisher, Jan J. Enghild, J. Michael Quigg, Richard A. Ketcham, James C. Sagebiel, et al. "Human Occupation of the North American Colorado Plateau About 37,000 Years Ago." *Frontiers in Ecology and Evolution* 10 (2022): article 903795.

Stanley, Steven M. *Children of the Ice Age. How a Global Catastrophe Allowed Humans to Evolve.* New York: Crown, 1996.

Stuart, Anthony J., P. A. Kosintsev, T. F. G. Higham, and Adrian M. Lister. "Pleistocene to Holocene Extinction Dynamics in Giant Deer and Woolly Mammoth." *Nature* 431, no. 7009 (2004): 684–689.

Sutcliffe, Antony J. *On the Track of Ice Age Mammals.* Cambridge, MA: Harvard University Press, 1985.

Trueman, Clive N., Judith. H. Field, Joe Dortch, Bethan Charles, and Stephen W. Wroe. "Prolonged Coexistence of Humans and Megafauna in Pleistocene Australia." *Proceedings of the National Academy of Sciences* 102, no. 23 (2005): 8381–8385.

Vrba, Elizabeth S., George H. Denton, Timothy C. Partridge, and Lloyd H. Burckle, eds. *Paleoclimate and Evolution, with Emphasis on Human Origins.* New Haven, CT: Yale University Press, 1995.

Wroe, Stephen W., Judith H. Field, Michael Archer, Donald K. Grayson, Gilbert J. Price, Judith Louys, Tyler Faith, et al. "Climate Change Frames Debate Over the Extinction of the Megafauna in Sahul (Pleistocene Australia-New Guinea)." *Proceedings of the National Academy of Sciences* 110, no. 22 (2013): 8777–8781.

THE HOLOCENE

Civilization exists by geological consent, subject to change without notice.

—WILL DURANT

PROJECT ICEWORM

During the height of the Cold War, the United States was looking for ways to deter the possibility of a Soviet nuclear attack. They had airbases and missiles stationed in Alaska and in northern Canada that were poised to retaliate by the shortest direct route across the Arctic Circle to the Soviet Union. Then the U.S. Army tried a different approach. They built a secret base on the northwestern tip of Greenland beneath the glacial ice sheet and called it Camp Century, a "city under the ice." Its official codename was "Project Iceworm," and it was powered by its own small nuclear reactor. Consisting of a grid pattern of dozens of tunnels in the ice some 20 feet wide, it was designed to hold military and support personnel and supplies and equipment. If it worked well, they would build chambers to house nuclear missiles that could be launched from beneath the ice sheet and could not be detected by the Soviets.

The base was first occupied in 1959, but by 1961 it began to have problems. The designers of Camp Century failed to fully account for the movement of glacial ice, which is always flowing and deforming even on ice sheets on relatively flat lowlands, which they had chosen for Camp Century. The tunnels began to warp out of shape in less than two years, and soon many parts of the base were useless because of collapsing tunnels. By 1967, the entire experiment was abandoned, long before missiles were to be put inside. Some 65 years later, the ice has completely crushed the old tunnels, and any traces of the old base are forever buried in the ice. However, the Greenland ice is melting so rapidly than in another century, Camp Century's hardware might come to light again.

Camp Century was not a complete waste of time and resources, however. To give the base some official excuse for its peace-keeping mission, a few scientists were allowed to work there and to conduct scientific experiments on the Greenland ice sheet. This was the official "cover" for the existence of the base, giving it the appearance of a regular scientific station doing pure research. From 1963 to 1966, they ran a coring station in one of the larger tunnels on the base, allowing them to drill through the entire 4,560 feet of the Greenland ice sheet in that location (figure 24.1). Once they reached the bottom, they drilled another 12 feet to examine the glacial sediments below the ice sheet and determine what the landscape was like before the ice. The ice cores were brought to the surface in short segments, wrapped and labeled, and stored in the frozen tunnels of Camp Century. Later they were moved into freezers in the United States, eventually ending up in freezers at the University of Buffalo, where the chief scientist, Chester Langway, was a professor.

The Camp Century coring program was one of the first successful large-scale attempts to core through any thick ice sheet, and these cores eventually yielded an incredible amount of information. One of the scientists at the base was Willi Dansgaard, a Danish scientist who had a long history of polar research. (Greenland is a Danish territory, so the U.S. base was there by permission of the Danish government, but they were not told of its nuclear reactor or its ultimate military purpose.) Dansgaard had developed one of the first methods of determining oxygen-18 levels, carbon dioxide levels, and deuterium (heavy water) levels trapped in gas bubbles within the ice. He was the first person to measure the actual composition and temperature of the atmosphere through thousands of years of snowfalls in Greenland. Each year can be dated by the layers in the ice core, and

Figure 24.1 ▲

Photograph of the ice coring rig inside one of the tunnels in Camp Century, Greenland. (Courtesy of Wikimedia Commons)

samples of the actual atmosphere year by year can be examined to understand changes in the atmosphere over time. His work was described in this tribute to Dansgaard:

> Dansgaard set out to test this hypothesis by examining the oxygen isotope ratio in molecules from various precipitation samples. In 1952, a long

spell of rain in Copenhagen provided him with an excellent opportunity. He collected rainwater in a beer bottle in his garden, emptying the rainwater every so often as cold and warm fronts passed over Copenhagen over a two-day period. In his analysis, Dansgaard saw that a higher amount of heavier ^{18}O was present in the rainwater that fell from warmer clouds than in the rainwater from colder clouds. This is due to the fact that heavier water molecules with an atom of ^{18}O have a 10 percent higher tendency to condense from gas to liquid more readily than lighter ones with an atom of ^{16}O. In the years that followed, Dansgaard confirmed his findings by analyzing water collected from around the world, including samples collected by the International Atomic Energy Agency (IAEA), as well as samples he collected himself in the United States and samples of ice and snow collected from glaciers and icebergs in Greenland. Having samples from all over the world allowed Dansgaard to see that the colder the region where the precipitation sample came from, the lower the amount of ^{18}O present in the sample.[1]

When Dansgaard analyzed the entire Camp Century core, he learned that it went back about 100,000 years, to the middle of the previous interglacial epoch. The core recorded the entire glacial-interglacial cycle from its beginning after the last interglacial, through its peak about 20,000 to 18,000 years ago, through the glacial-interglacial transition, and through the last 10,000 years. From this core, he discovered the Dansgaard-Oeschger cycles (see chapter 22), found some of the first evidence of the Younger Dryas event, examined the warming of the last 16,000 years of the glacial-interglacial transition, and uncovered details of climate in the past 10,000 years.

Numerous drill cores have been recovered from Greenland since Camp Century, culminating in the Greenland Ice Sheet Project 2 core. It took five years to drill and recover 3,050 meters (10,000 feet) of cores, the deepest core ever recovered from Greenland. The Northern Greenland Ice Core Project has drilled in the center of the Greenland ice sheet, its thickest part, which is also about 3,000 meters (10,000 feet) thick and reaches back 140,000 years.

Even more ambitious projects have been attempted on the much thicker Antarctic ice sheet. In the 1970s and 1980s, the Soviets built and operated the Vostok Station on the East Antarctic ice sheet and produced a core that went down more than 3,600 meters (12,000 feet), with a climate record that goes back more than 400,000 years and records almost four complete glacial-interglacial cycles. In the early 2000s, the European Project for Ice Coring in Antarctica (EPICA) drilled nearby, on Dome C of Dronnig Maud Land in East Antarctica, and retrieved a core that goes back almost 800,000 years—more than seven complete glacial-interglacial cycles. This core is now the basis for much of our research on past and future climates (see chapter 25).

Ice cores were the first truly detailed piece of direct evidence of climate in the geological past that enabled us to look at Earth's temperature and atmosphere year by year during the Pleistocene and the Holocene. In the 1960s and 1970s, deep sea cores full of sediments and plankton were also yielding detailed records of the geological past (see chapter 22). In addition, tree rings provide an independent measure of climate from the land perspective. Together these studies have helped us understand Earth's climate in the last few hundred thousand years in ways we had never dreamed was possible.

THE HOLOCENE

The last 11,000 years of the present interglacial is known as the Holocene ("completely recent" in Greek) epoch, or just known to geologists as the "Recent" (figure 24.2). It encompasses all of human history and most of human civilization. Human history and civilization could not have happened without the relatively stable climates of the past 10,000 years, especially compared to the wild swings of climate during the peak of the last glacial and over the glacial-interglacial transition from 16,000 years ago until about 10,000 years ago. Climate has had a huge influence on human history, and it continues to affect us as we look toward our future.

Full interglacial conditions stabilized about 11,700 years ago after the world warmed out of the final glaciation of the Younger Dryas (figure 24.3A). Huge masses of ice melt slowly, so the great ice sheets that once covered Canada and Scandinavia and Siberia started melting about 18,000 years ago but did not finally vanish until about 6,500 years ago, well into the Holocene. Indeed, huge continental ice sheets remain on Antarctica and Greenland, and large glaciers are found in

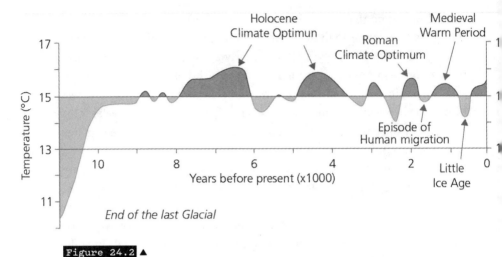

Figure 24.2 ▲
Average surface temperature and climate history of the last 11,000 years in the northern hemisphere, showing the small-scale fluctuations in climate during the relatively stable conditions of the Holocene. (Redrawn from several sources)

the high mountains around the world (although nearly all are vanishing rapidly now).

The beginning of the Holocene was marked by the last phase of the postglacial rise of sea level (figure 24.3B). Between 11,700 and about 9,000 years ago, sea level rose about 35 meters (115 feet) to the steady level that persisted through much of the Holocene. Rising sea levels continued flooding old glacial fjords and drowned river valleys like Chesapeake Bay. The English Channel did not form until the early Holocene, when sea level rose to flood it all the way from the Atlantic to the North Sea. Large areas of low-lying northern Europe were also flooded at that time. In particular, the floodplains and river mouths to the north off the coast of the Netherlands, Denmark, and Germany once extended much further into the North Sea, forming a broad area scientists call Doggerland, named after a submarine plateau in the North Sea called the Dogger Banks (figure 24.4). Today these floodplains are drowned beneath the North Sea. Dredging for human artifacts in the North Sea and imaging the seafloor has turned up evidence that large populations of humans once lived there.

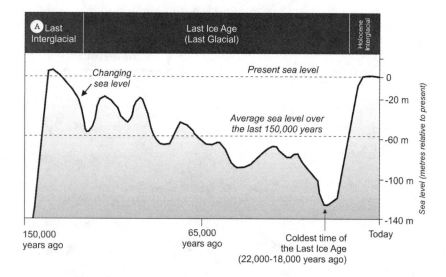

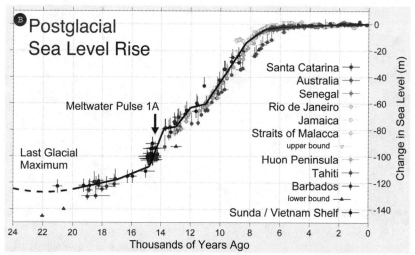

Figure 24.3 ▲

(A) Simplified sea level history of the past 150,000 years from the previous interglacial, through peak glacial, and then to the Holocene interglacial. (B) Details of the postglacial rise in sea level. (Courtesy of Wikimedia Commons)

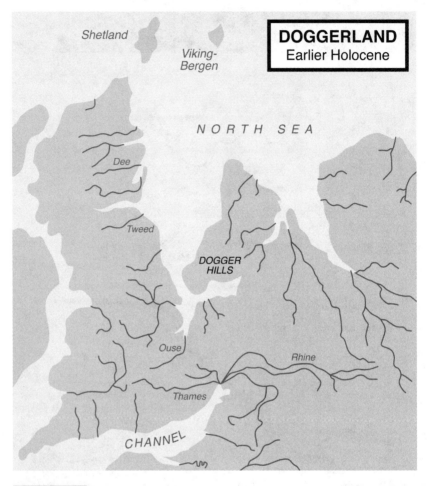

DOGGERLAND
Earlier Holocene

Figure 24.4 ▲

During the low sea level stands of the peak of the last glaciation 18,000 years ago, most of the floor of the North Sea was an exposed upland called Doggerland, and the English Channel cutting Britain off from the continent did not exist. As sea level rose in the past 17,000 years, all of this region was eventually flooded. (Courtesy of Wikimedia Commons)

CLIMATE AND HUMAN HISTORY

THE CLIMATIC OPTIMUM AND THE RISE OF CIVILIZATION

The most rapid and important Holocene changes on Earth were climatic fluctuations. Even though Earth was in a relatively warm stable interglacial episode through the Holocene, small fluctuations of warming and cooling continued in the past 10,000 years. As the planet warmed out of the last Ice Age, the early Holocene is typified by a period of warmer than average temperatures as the peak of solar insolation from the coincidence of the warmest phases of the three Milankovitch cycles occurred. About 9,000 years ago, the axial tilt of Earth reached 24°, maximizing heat on the poles, and Earth was at its closest approach to the sun during the northern hemisphere summer, which substantially increased the amount of solar radiation it received. This early Holocene warming from about 9,000 to 5,500 years ago has been called the Hypsithermal or the Holocene climatic optimum. Average global temperatures were about 0.7° Celsius warmer than typical for the Holocene, and the polar regions were as much as 4°C warmer. Temperate North America and Europe were about 2°C warmer.

The most striking effect of the Holocene climatic optimum is that the desert regions of Central Asia and especially the Sahara Desert were not only a bit warmer but also much wetter than they are today. Civilizations began to develop in many parts of Africa and Asia simultaneously. In fact, the Sahara was covered by a huge network of lakes and rivers dominated by crocodiles and hippos, and a large human population thrived in the rich dense forests. Using satellite radar we can "look beneath" the Sahara sand dunes and see the buried river valleys and lake beds that are now covered completely by sand (figure 24.5). It is thought that extensive overuse of the land and overgrazing by domesticated animals may have contributed to the destruction of the Green Sahara, along with changes in climate around 5,500 years ago.

None of this would have been possible without the earlier invention and spread of agriculture, often called the Neolithic Revolution. It apparently arose somewhere in the Zagros Mountains of Iran about 14,000 years ago during the late Pleistocene. Archeological sites in that region show the first evidence of domestication of wheat and barley, along with the mortars and grinding stones used to grind the grain into flour, as well as bones of domesticated sheep and goats. By 9,000 years ago, the earliest cultures in the Tigris and Euphrates valleys below the Zagros Mountains were living in the open in small houses rather than in caves, and the people were raising sheep, goats, pigs, and dogs. By 8,000 years ago, sites in Greece show that agriculture and domesticated animals were spreading into Europe and reached the recently deglaciated landscapes of Britain and Scandinavia about 6,000 years ago. Around 7,000 years ago cattle were domesticated

Figure 24.5 ◄

During the Holocene climatic optimum, the Sahara was a well-watered region with large rivers draining forested plains—all of which are now buried under sand dunes. This radar view was shot from the space shuttle *Columbia* in 1981 and the swath running diagonally across the photo is the radar image from below the sand dunes. It reveals a network of drainages and large rivers that have vanished beneath the dunes. (Courtesy of the U.S. Geological Survey)

and spread among these early European cultures. All of this expansion of agriculture, domestication of animals, and early civilization was made possible by the warmer, wetter conditions of the early Holocene climatic optimum.

It is no accident that the most ancient civilizations first developed in this same area during the latter part of the Holocene climatic optimum. These regions include the Tigris-Euphrates valleys in what is now Iraq (Ubaid Period, about 8,500 years ago, followed by the Sumerians, about 6,000 years ago), the Nile Valley (Gerzean-period Egyptians, about 5,500 years ago), the Indus River Valley in what is now Pakistan (Harappan Civilization, about 4,500 years ago), and the earliest Chinese civilization in the Yellow River and Yangtze River valleys about 4,200 years ago. The oldest known cities on Earth, such as Jericho, were first settled about 10,500 years ago. All of these early civilizations arose during the warm, wet conditions of the Holocene climatic optimum, when Asia and northern Africa were much wetter and greener and many crops could be grown without advanced irrigation techniques. The Bible describes the Middle East as "the land of milk and honey" and "the promised land," and many have described the Middle East as "the Fertile Crescent." This region was certainly fertile 5,500 years ago, but thanks to climate changes during the later Holocene, it now has the harsh desert climate we associate with modern Egypt, Israel, and Iraq.

THE NEOGLACIAL

The last 5,500 years of climate history is often called the Neoglacial Period because the second half of the Holocene was on average much cooler than the Holocene climatic optimum of the early Holocene. Just as climate changes and warming episodes of wetter climate helped spur the rise of agriculture and eventually civilization, changing climate that made the world cooler and drier had effects on human history as well. About 1177 BCE (or 3,177 years ago) an abrupt and severe drought along with a cooling event across much of the eastern Mediterranean wiped out many civilizations, including the archaic Greek cultures of Mycenae and the Minoan civilization of the island of Crete, as well as the Hittites of what is now Turkey, and the New Kingdom of Egypt. This major crisis in world history, called the Late Bronze Age cultural collapse, caused the entire civilized world to go into a "Dark Ages" for more than a century (figure 24.6). Its causes remain controversial, but the well-documented extreme drought and rapid cooling are certainly part of the explanation.

After this episode, climate began to warm again, and the great civilizations of classical Greece rose, peaking about 500 BCE (2,500 years ago). They were later absorbed into the empire of Alexander the Great, followed by the Roman Empire. In fact, the period around 500 BCE to 200 CE (2,500 to 1,800 years ago) was another episode of warmer wetter weather in Europe, which is called the Roman Warm Period. Many scholars have speculated about the causes of the rise and fall of the Roman Empire, but

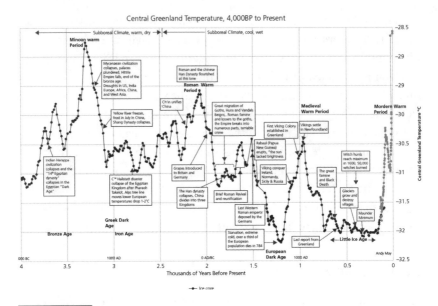

Central Greenland Temperature, 4,000BP to Present

Figure 24.6 ▲

Details of climatic fluctuations and their effect on various events in human civilization. (Redrawn from several sources)

certainly among the underlying factors were the dramatic shifts in climate that promoted agriculture and mild conditions. During the decline of the Roman Empire, climate went into the long cooling episode beginning about 500 CE to 1000 CE (1,500 to 1,000 years ago), which coincided with colder, harsher, drier climates in the Middle Ages. In 536 CE, during the darkest period of the Dark Ages, an extreme cooling event caused famine and mass death, shocking cultures across Europe. Its causes remain controversial, but global cooling from the volcanic dust injected into the stratosphere by an Indonesian volcano was probably a factor (possibly an earlier eruption of Krakatau before its most recent 1883 eruption).

THE MEDIEVAL WARM PERIOD AND THE LITTLE ICE AGE

The cold climates of the Middle Ages gave way to another warming period between 950 CE and about 1250 CE, known as the Medieval Warm Period. Temperatures around the North Atlantic region rose by about 0.5°C, enough to melt much of the ice in the North Sea and North Atlantic and bring warmer climates to Europe and North America. The Vikings from Scandinavia used this opportunity to sail far and wide, invading northern Europe repeatedly. They even sailed to Iceland and then reached Greenland and eventually the Labrador Coast of North America. Greenland was warm enough on its southern tip so it was possible to grow a few crops, raise sheep and cattle, and establish small outposts. However, the warming was not so favorable to people in other regions. In the southwestern deserts of the United States, the Medieval Warm Period caused prolonged droughts that wiped out the Anasazi cultures that built the amazing pueblos found throughout that region.

However, no one should mistake the warming of the Medieval Warm Period as being comparable to the current global warming of Earth. First, the warming was localized to the North Atlantic region, and global average temperatures actually cooled, not warmed, on a worldwide basis. Second, the warming was tiny—only a fraction of a degree Celsius. In contrast, the warming of the past 50 years is almost an order of magnitude bigger, rising more than a full degree Celsius globally in just a few decades. Although the causes of the Medieval Warm Period are not fully understood, there was a maximum of solar radiation at that time and an unusually long interval without volcanic eruptions that block solar radiation and cool the planet. In addition, changes in oceanic circulation brought unusually warm conditions to the North Atlantic, but that did not mean that the rest of the world got warmer. These causes are totally different from the human-caused climate changes we are now experiencing.

About 1500 CE, the Medieval Warm Period gradually began to cool down, and sea ice returned to Greenland, along with expanding ice sheets. These events wiped out the tiny Viking outposts in Greenland and marked the transition to what is known as the Little Ice Age. In 1650, 1770, and 1850, three particularly cold intervals dramatically changed the climates of northern Europe and North America. Regions of Europe that had not frozen in centuries routinely experienced harsh winters. Many famous paintings and historical records of the time depict the freezing of the Thames River in London and the freezing of canals in Holland. There were longer and more frequent snowfalls, colder summers, and more severe deep freezes. This led to numerous crop failures and outright famine during this period. In North America, the effects were similar. The famous painting of Washington crossing the Delaware through ice floes is a good example.

On Christmas Eve 1776, when Washington made the crossing to surprise the British and Hessian troops in the Battle of Trenton, the Delaware River routinely froze over; today it never gets close to freezing. Likewise, the stories of Washington's armies suffering through harsh winters in Valley Forge remind us that those winters were much colder than the winters we experience today.

The causes of the Little Ice Age are not fully understood. The Milankovitch cycles were reaching a point where the next glacial interval should have been starting, so that may be one of the reasons. There was also very weak solar activity at the time, known as the Maunder Minimum, which meant that Earth received less solar warming. Finally, there were far more large explosive volcanic eruptions during this period, blocking the sun with the ash that rose into the stratosphere. The largest of these were the eruption of Laki in Iceland in 1783 and the 1815 eruption of Tambora in Indonesia (figure 24.7). The Tambora eruption was the largest explosive volcanic event in the last 74,000 years (since the eruption of Toba volcano, which nearly wiped out humans completely), ejecting huge amounts of volcanic dust into the stratosphere that cooled the planet. The effect was so extreme that 1816 was called "the Year without a Summer." Summer temperatures were so far below normal that there were deep freezes and snowfalls in June and July in New York and in many European cities known for

Figure 24.7 ▲

Map of Indonesia showing some major volcanoes. (Redrawn from the U.S. Geological Survey)

their summer warmth. Crops froze and farm animals died, causing widespread famine in Europe just as the world was recovering from the shock of the Napoleonic Wars that ended with the 1815 Battle of Waterloo. In Lord Byron's villa on Lake Geneva, his guests Percy and Mary Shelley were spending their summer holiday huddled by the fireside instead of enjoying a warm summer by the lake. They told ghost stories to pass the time, one of which became the basis for Mary Shelley's *Frankenstein*. The Little Ice Age came to an end not long after "the Year without a Summer." The extreme winters of the 1880s were the last gasps of truly cold climates in northern Europe and North America.

One of the factors that may have contributed to the record freezes of the late 1880s was the 1883 eruption of Krakatau, in the Sunda Straits just west of Java and east of Sumatra. The peak of the eruption on August 27, 1883, generated a huge pressure wave that spread from the volcano at 1,086 kilometers (675 miles) per hour. Its volume was measured at 310 decibels, loud enough to be heard in Perth, Australia, more than 3,500 kilometers (2,200 miles) away, and even on the island of Mauritius north of Madagascar 4,800 kilometers (3,000 miles) away, where they sounded like nearby cannon fire. The pressure wave was so strong that it ruptured the eardrums of sailors in the Sunda Strait. The barometric pressure gauges measuring normal air pressure jumped off the scale and were shattered. This aerial shock wave continued around the globe and was registered on barographs around the world up to seven days after the initial explosions. The biggest killer was a huge tsunami that washed steel warships miles onshore and killed 36,000 people in coastal villages around Java and Sumatra. The ash shot into the stratosphere and blocked sunlight and dropped global average temperatures 1°C to 2°C for over a year. Weather patterns were erratic for years, and temperatures did not return to normal until 1888. The skies were dimmed, even darkened for months after the eruption, and the high amount of particulate matter in the stratosphere changed the color of the sky, producing spectacular orange-red sunsets that are shown in Edvard Munch's famous 1893 painting *The Scream*. Munch described the sky over Norway after the eruption as follows: "suddenly the sky turned blood red. . . . I stood there shaking with fear and felt an endless scream passing through nature."[2] Rare atmospheric effects, such as a blue moon, a Bishop's ring around the sun in daytime, and volcanic purple light at twilight were also seen around the world.

How Much Magma Erupts?
Comparison of Eruption Volumes
Dense Rock Equivalent (DRE)*

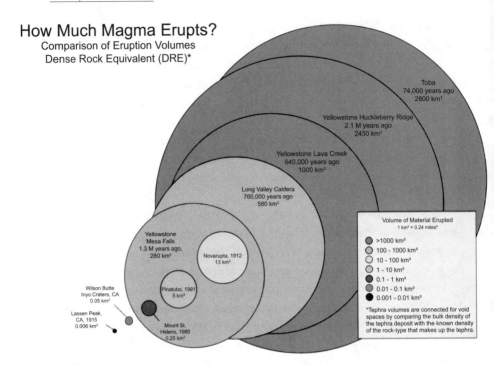

Toba
74,000 years ago
2800 km³

Yellowstone Huckleberry Ridge
2.1 M years ago
2450 km³

Yellowstone Lava Creek
640,000 years ago
1000 km³

Long Valley Caldera
760,000 years ago
580 km³

Yellowstone
Mesa Falls
1.3 M years ago,
280 km³

Novarupta, 1912
13 km²

Wilson Butte
Inyo Craters, CA
0.05 km³

Lassen Peak,
CA, 1915
0.006 km³

Pinatubo, 1991
5 km³

Mount St.
Helens, 1980
0.25 km³

Volume of Material Erupted
1 km³ = 0.24 miles³

- ⬤ >1000 km³
- ⬤ 100 - 1000 km³
- ◯ 10 - 100 km³
- ◯ 1 - 10 km³
- ⬤ 0.1 - 1 km³
- ⬤ 0.01 - 0.1 km³
- ⬤ 0.001 - 0.01 km³

*Tephra volumes are connected for void
spaces by comparing the bulk density of
the tephra deposit with the known density
of the rock-type that makes up the tephra.

Figure 24.8 ▲

Diagram showing the relative size of eruptions based on the Volcanic Explosivity Index; the circles represent the volume of material ejected by each eruption. (Redrawn from the U.S. Geological Survey)

But the eruptions of Krakatau in 1883 or Tambora in 1815 are tiny compared to the biggest eruption in the last 28 million years: the eruption of the Toba volcano on Sumatra, about 74,000 years ago (figures 24.7, 24.8). It was so big that it was over a thousand times larger than the catastrophic eruptions of Mount Tambora in Indonesia or the Krakatoa volcano between Java and Sumatra in Indonesia, both of which changed the climate for several years after the eruption. Toba caused global temperature to drop 3°C to 5°C (5°F to 9°F), further amplifying the cold of the ongoing ice ages. The tree line and snow line dropped 3,000 meters (9,000 feet) lower than today, making most high elevations uninhabitable. Global average temperatures dropped to only 15°C (59°F) after three years and took a full decade to recover. Ice cores from Greenland show evidence of this dramatic cooling in the trapped ash and ancient air bubbles.

What happened to the people and animals during this terrible time? Geneticists and archeologists have found evidence that the Toba catastrophe nearly wiped out the human race. Only a few thousand humans survived worldwide and passed through the genetic bottleneck. Another study found a similar genetic bottleneck in the genes of human lice, and in our gut bacterium *Helicobacter pylori*, which causes human ulcers. Both of these events date to the time of Toba, according to their molecular clocks, which shows how long ago a genetic change took place. The same is true for the genes of a number of other animals, including tigers, orangutans, several monkeys, gorillas, chimpanzees, and pandas—just about every large mammal in southern Asia or Africa that has been sequenced so far. In short, Toba is the largest eruption to occur since humans have been on Earth, and it came very close to wiping us out altogether, along with many other animals.

In summary, historians emphasize factors such as the power of politicians or the effects of wars, famine, disease, and other calamities as the major influences on history. But the role of climate must not be underestimated, and this factor is becoming better and better understood today.

NOTES

1. Joseph Cheek, "Willi Dansgaard: Pioneer of Paleoclimate Research," Science-Poles, February 28, 2011, http://www.sciencepoles.org/article/willi-dansgaard -pioneer-of-paleoclimate-research.
2. Zuzanna Stańska, "The Mysterious Street from Edvard Munch's *The Scream*," *Daily Art Magazine*, May 18, 2023, https://www.dailyartmagazine.com/the -mysterious-road-of-the-scream-by-edvard-munch/.

FOR FURTHER READING

Cline, Eric H. *1177 B.C.: The Year Civilization Collapsed*. Princeton, NJ: Princeton University Press, 2014.

Davies, Jeremy. *The Birth of the Anthropocene*. Berkeley: University of California Press, 2016.

Diamond, Jared. *Guns, Germs, and Steel: The Fates of Human Societies*. New York: Norton, 2005.

Fagan, Brian. *The Great Warming: Climate Change and the Rise and Fall of Civilizations*. London: Bloomsbury, 2009.

——. *The Little Ice Age: How Climate Made History, 1300–1850*. New York: Basic Books, 2001.

——. *The Long Summer: How Climate Changed Civilization*. New York: Basic Books, 2003.

Gore, Al. *An Inconvenient Truth*. Emmaus, PA: Rodale, 2006.

Harari, Yuval Noah. *Homo Deus: A Brief History of Tomorrow*. New York: Harper, 2017.

Klingaman, William K., and Nicholas P. Klingaman. *The Year Without a Summer: 1816 and the Volcano That Darkened the World and Changed History*. New York: St. Martin's, 2013.

Lamb, Hubert H. *Climate, History, and the Modern World*. London: Methuen, 1992.

Prothero, Donald R. *When Humans Nearly Vanished: The Catastrophic Eruption of Toba Volcano*. Washington, DC: Smithsonian, 2007.

THE FUTURE GREENHOUSE PLANET

There's no debate about the greenhouse effect, just like there's no debate about gravity. If someone throws a piano off the roof, I don't care what Sarah Palin tells you, get out of the way because it's coming down on your head.

—JIMMY KIMMEL

THE GREENHOUSE EFFECT

Earth cycles back and forth between having a "greenhouse" climate (when there is a high level of carbon dioxide in the atmosphere) and having an "icehouse" climate (when carbon is trapped in coals and limestones and oil or by deep weathering when mountains are rising and eroding). In addition, carbon is taken up in the tissues of animals and plants. A huge expansion of coal swamps caused icehouse conditions to form in the Carboniferous, and weathering of the rising Himalayas and Andes ranges and deep weathering of grassland soils during the Oligocene returned the planet to an icehouse state that culminated in the Pleistocene ice ages. Human activity is rapidly pushing the planet back into a greenhouse state, mostly by burning the coal and oil that had been locked up in Earth's crust for millions of years, releasing more carbon dioxide into the atmosphere.

But the idea of the greenhouse effect is not a new one. In 1824, Joseph Fourier wrote that "the temperature [of Earth] can be augmented by the interposition of the atmosphere, because heat in the state of light finds less resistance in penetrating the air, than in repassing into the air when converted into non-luminous heat." And in 1836, Claude Pouillet wrote: "the atmospheric stratum . . . exercises a greater absorption upon the terrestrial than the solar rays."[1] These statements were based on physical calculations and were not confirmed by experiment at that time. In 1856, an American scientist, Eunice Foote, did experiments proving that an atmosphere rich in either carbon dioxide or water vapor would absorb heat rather than let it pass through, but her work was difficult to interpret and was not widely known.

In 1859, the Irish chemist John Tyndall performed experiments decisively proving that greenhouse gases such as carbon dioxide and water vapor could absorb heat and could trap heat from the sun's rays that came through Earth's atmosphere, blocking the heat from radiating back into space. As Tyndall wrote in 1859:

Thus the atmosphere admits of the entrance of solar heat; but checks its exit, and the result is a tendency to accumulate heat at the surface of the planet . . . if, as the above experiments indicate, the chief influence be exercised by aqueous vapour, every variation of this constituent must produce a change of climate. Similar remarks would apply to the carbonic acid [carbon dioxide] diffused through the air. . . . Such changes may in fact have produced all the mutations of climate which the researches of geologists reveal.[2]

A generation later in 1896, the brilliant Swedish scientist Svante Arrhenius (who received the Nobel Prize in Chemistry for his work) showed that the average surface temperature of Earth is about 15°C because of the infrared absorption capacity of water vapor and carbon dioxide. This is the natural greenhouse effect. Arrhenius suggested that doubling the CO_2 concentration would lead to a 5°C temperature rise. Arrhenius and Thomas Chamberlin calculated that human activity could warm the earth by adding carbon dioxide to the atmosphere.

The greenhouse effect and global warming is not some "hoax from China," as some climate deniers would have you believe. More than 150 years of research has established its reality, and by the 1970s scientists—like my former professor Wally Broecker—could see the handwriting on the wall, and they successfully predicted how conditions would be in this century if we continued with "business as usual" (as we are still doing in 2024).

Why do we call them greenhouse gases? In a greenhouse, the temperature inside remains very warm because energy from the sun (almost entirely in the visible part of the electromagnetic spectrum) easily passes through the glass walls, and when it hits surfaces the energy is converted into heat (infrared wavelengths). But the glass prevents the heat from escaping, just like greenhouse gases absorb the heat and prevent the energy from leaving Earth (figure 25.1). Another analogy is the inside of a parked car. Just as in a greenhouse, the interior heats up as the visible spectrum rays of sunlight pass through the windows and the solar energy is absorbed by the interior of the car and is radiated back as heat. It is hotter inside the car than outside because the glass windows let light pass through but block heat from escaping.

Because more heat arrives on the surface of the planet than can leave it, Earth's atmosphere is warming. Arrhenius had calculated that doubling the level of atmospheric carbon dioxide would cause global temperatures to

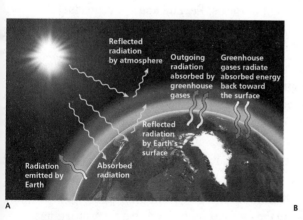
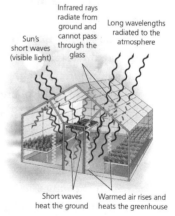

A   B

Figure 25.1 ▲

The greenhouse effect occurs when Earth's atmosphere lets sunlight pass through, but when that solar energy is absorbed by Earth's surface and is converted to a longer infrared wavelength, greenhouse gases in the atmosphere (like the glass panes in a real greenhouse) prevent it from escaping. (Redrawn from several sources)

rise 5°C or 6°C. This is remarkably close to estimates reported by scientists in 2007, and it is consistent with current estimates.

THE CHANGING PLANET

ATMOSPHERIC CHANGES

The next major discovery occurred when Charles David Keeling of Caltech invented one of the first devices for measuring atmospheric carbon dioxide. He tested it in a number of locations, including the Big Sur in California and the Olympic rain forest in Washington, to see how reliably it measured carbon dioxide and what the influence of local plants might be on its measurements. In 1958 he began to take measurements in two places most isolated from major cities (thus minimizing local effects of pollution from cities or industry) and also isolated from the effects of plants absorbing carbon dioxide nearby: the top of Mauna Loa volcano on Hawaii and Antarctica. After a few years of data showing the rapid increase in carbon dioxide, his Antarctic funding ended because the National Science Foundation foolishly decided that he had proved his point, and the potential for collecting a long-term dataset was lost. But the Mauna Loa observatory has been running continuously for more than 65 years, producing one of the longest sets of atmospheric data ever collected. By the 1960s, Keeling and his colleague, the oceanographer Roger Revelle, could see the dramatic increase in carbon dioxide in a steady upward trend (figure 25.2). Superimposed on the overall upward trend are the annual cycles of decreasing carbon dioxide when the northern hemisphere spring plant growth takes in CO_2 and increasing carbon dioxide in the fall when the trees in the north lose their leaves and carbon dioxide is released as the leaves decay. (Most of the world's land vegetation is in the northern hemisphere, so its effects are much stronger in the northern spring and fall than in the southern spring and fall).

From Keeling's initial data and every dataset that has been collected since then, the trend is clear: carbon dioxide in our atmosphere has increased at a dramatic rate in the past 150 years. It was barely 315 parts per million (ppm) when the experiment began in 1958, and in 2023 it's more than 418 ppm. Every dataset of temperature or carbon dioxide collected over a long time span confirms this. A compilation of temperature data from tree rings, ice cores, corals, and direct measurements over the past 900 years

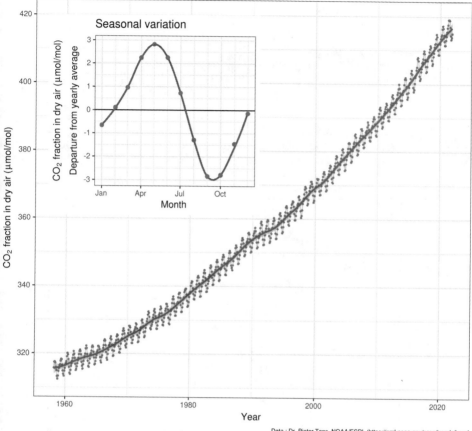

Monthly mean CO_2 concentration

Mauna Loa 1958 - 2021

Seasonal variation

Data : Dr. Pieter Tans, NOAA/ESRL (https://gml.noaa.gov/ccgg/trends/) and
Dr. Ralph Keeling, Scripps Institution of Oceanography (https://scrippsco2.ucsd.edu/). Accessed 2021-12-16
https://w.wiki/4ZWn

Figure 25.2 ▲

In 1958, Charles David Keeling used his newly invented instrument to measure atmospheric
carbon dioxide at the top of Mauna Loa Observatory on the Big Island of Hawaii. That experiment clearly shows the steady and rapid increase in carbon dioxide in the atmosphere. The
Keeling curve has an annual cycle of decrease in the northern hemisphere spring (when
most of the world's plants are growing after the winter and absorbing carbon dioxide) and
an increase in the fall (when those same plants lose their leaves and release that carbon
dioxide back to the atmosphere). (Redrawn from several sources)

Global Average Temperature Change

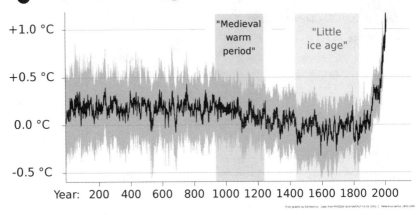

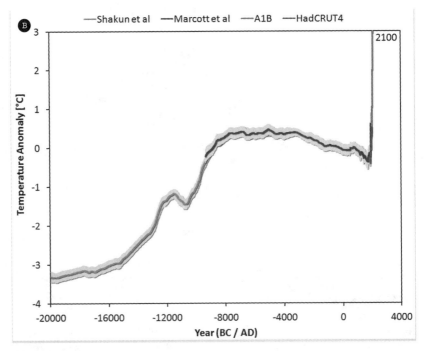

The last thousand years of temperatures for the northern hemisphere. (A) The warming trend begins at the end of the Medieval Warm Period around 1000 CE and reaches its lowest point with the Little Ice Age from 1600 to 1850 CE. Then the effect of human-caused global warming kicks in, and the temperatures rise dramatically and rapidly to a level not seen since the Eocene. (B) The last 20,000 years of climate change, from the peak glacial at 20,000 to 18,000 years ago, to the warming of the glacial-interglacial transition from 18,000 to 10,000 years ago, to the Holocene stability of the past 10,000 years. Even at this scale, the extreme magnitude and rapidity of the heating of the past 150 years stands out as unprecedented, with more and faster warming than any time in the geologic past. ([A] Redrawn from Michael Mann, *The Hockey Stick and the Climate Wars: Dispatches from the Front Lines* [New York: Columbia University Press, 2012] and the Intergovernmental Panel on Climate Change [IPCC]; [B] redrawn from several sources)

shows the sudden increase of temperature of the past century standing out like a sore thumb (figure 25.3A). This famous graph is nicknamed the "hockey stick" because it is long and relatively straight through most of its length, then bends sharply upward at the end like the blade of a hockey stick. Other graphs show that climate was very stable within a narrow range of variation throughout the past 1,000, 2,000, or even 10,000 years, since the end of the last Ice Age, and even to the end of the last glacial maximum about 18,000 years ago (figure 25.3B). The graph clearly shows the minor warming events during the climatic optimum about 7,000 years ago, the Medieval Warm Period, and the slight cooling of the Little Ice Age from the 1700s and 1800s. But the magnitude and rapidity of the warming represented by the last 200 years is simply unmatched in all of geologic history. More revealing, the timing of this warming coincides with the Industrial Revolution, when humans first began massive deforestation and released carbon dioxide by burning coal, gas, and oil.

MELTING POLAR REGIONS

If the data from atmospheric gases were not enough, we are now seeing unprecedented changes in our planet. The polar ice caps are thinning and breaking up at an alarming rate. In 2000, for the first time ever, my graduate advisor Malcolm McKenna was flying over the North Pole in summertime and saw no ice, just open water. His report of the sighting and photographs made the front page of the *New York Times*! So much for Santa's workshop! The Arctic ice cap has been frozen solid for at least the past 3 million years and maybe longer, but now the entire ice sheet is breaking up so fast that by 2030 (and possibly sooner) less than half of the Arctic will be ice covered in the summer (figure 25.4). This is an ecological disaster for everything that lives in the Arctic, from polar bears, to seals and walruses and whales, to the animals they feed upon. The Antarctic is thawing even faster. In February-March 2002, the Larsen B ice shelf, more than 3,000 square kilometers (the size of Rhode Island) and 220 meters (700 feet) thick, broke up in just a few months, a story typical of nearly all of the ice in Antarctica. The Larsen B shelf had survived all the previous ice ages and interglacial warming episodes for 3 million years, and even survived the warmest periods of the last 10,000 years. Now it and nearly all the other thick ice sheets on the Arctic, Greenland, and Antarctic are vanishing at a

 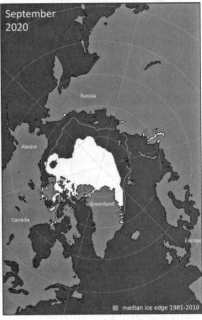

Disappearance of the ice caps in the Arctic over the past decades. Gigantic amounts of ice vanish every summer and fall and are not replaced by the freezing of the following winter. (Courtesy of the National Oceanic and Atmospheric Administration [NOAA])

rate never before seen in geologic history. In 2017, the Larsen C ice shelf began to break apart.

SEA LEVEL RISE

Many people don't care about the polar ice caps, but there is a serious side effect worth considering: all that melted ice eventually ends up as more water in the ocean, causing sea level to rise as it has many times in the geologic past. At the moment, sea level is rising about 3 to 4 millimeters per year, more than 10 times the rate of 0.1 to 0.2 millimeters per year that has occurred over the past 3,000 years (figure 25.5). Geological data show that sea level was virtually unchanged for 10,000 years since the present interglacial began. A few millimeters of sea level rise here or there doesn't impress people, until you consider that the rate is accelerating and that most scientists predict it will rise 80 to 130 centimeters in just the next century. A

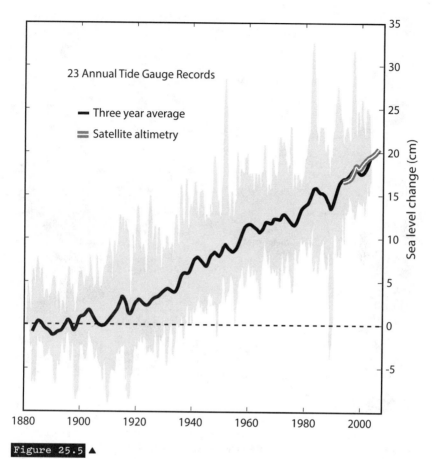

23 Annual Tide Gauge Records

— Three year average
≡ Satellite altimetry

Figure 25.5 ▲

Data showing the rise in global sea level in the past 140 years, after 10,000 years of relative stability of the global sea level. (Redrawn from several sources)

sea level rise of 130 centimeters or 1.3 meters (almost 4 feet) would drown many of the world's low-elevation cities, such as Venice and New Orleans, and low-lying countries including the Netherlands and Bangladesh. A number of tiny island nations (for example, Vanuatu and the Maldives) that barely poke out above the ocean now are already vanishing beneath the waves. Their entire population will have to move someplace else. Low-lying coastal cities such as Miami and Venice are now routinely being flooded during very high tides, but they never used to have this problem. If sea level rose by just 6 meters (20 feet), nearly all of the world's coastal plains and low-lying areas (such as the Louisiana bayous, Florida, and most

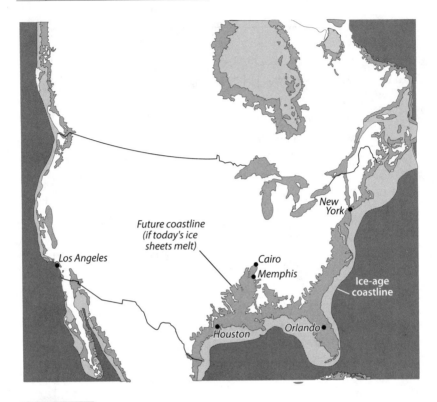

Figure 25.6 ▲

Melting all the world's ice caps and glaciers would cause sea level to rise 65 meters (215 feet) and would completely flood all the coastal cities of North America. Florida and Louisiana would virtually disappear, and a similar effect would be seen on the Low Countries of Belgium and the Netherlands, northern Germany and Poland, Denmark, and much of Great Britain. (Redrawn from several sources)

of the world's river deltas) would be drowned. Most of the world's population lives in coastal cities—New York, Boston, Philadelphia, Baltimore, Washington, D.C., Miami, Shanghai, and London—and all of those cities would be partially or completely under water with such a sea level rise. If all the glacial ice caps melted completely (as they have several times in greenhouse worlds in the geologic past), sea level would rise by 65 meters (215 feet)! The entire Mississippi Valley would flood, and you could dock your boat in Cairo, Illinois (figure 25.6). Such a sea level rise would drown nearly every coastal region under hundreds of feet of water and inundate New

York City, London, and Paris. Only the tall landmarks would remain, such as the Empire State Building, Big Ben, and the Eiffel Tower. You could tie your boats to these high spots, but most of the rest of these drowned cities would be deep under water.

OTHER EFFECTS OF CLIMATE CHANGE

Climates around the world are being affects by climate changes. About 95 percent of the remaining mountain glaciers left over from the Pleistocene are retreating at the highest rates ever documented—or they have already vanished. Many of those glaciers, especially in the Himalayas and Andes and Alps and Sierras, provide most of the freshwater that the populations below the mountains depend on—yet this fresh water supply is vanishing. Permafrost that once remained solidly frozen even in the summer is now thawing, destroying Inuit villages on the Arctic coast and threatening all of our pipelines to the North Slope of Alaska. A more serious issue is that permafrost stores huge amounts of methane from the decay of vegetation in the soil, and methane is being released as the permafrost thaws. Methane is a much more powerful greenhouse gas than carbon dioxide. If methane in the atmosphere increases dramatically, global warming will be even more out of control than it is now.

In recent years we have seen record heat waves over and over again, killing thousands of people, as each year joins the list of the hottest years on record. The year 2020 almost tied 2016 as the hottest year on record, 2021 was in the top seven hottest years on record, and 2022 was the fifth hottest year on record. As I write this, July 3 and 4, 2023, were the hottest days ever recorded, and an extreme heat wave is gripping the world throughout July 2023, so it is clear that 2023 will break the record of the previous warmest year, 2016.

Natural animal and plant populations are being decimated all over the globe as their environment changes. Many animals respond by moving their ranges to formerly cold climates. Places that once did not have to worry about disease-bearing mosquitoes are now infested as the climate warms and mosquitoes move further north.

If you have seen any of recent movies or videos about climate change, the long litany of "things we have never seen before" and "things that have never occurred in the past 3 million years of glacial-interglacial cycles" is staggering. Still, many people are not moved by the dramatic images of vanishing glaciers or by the forlorn polar bears starving to death. Many of these people have been fed misinformation by the powerful lobbies and political organizations and fossil fuel companies that want to cloud or confuse the issue.

HOW DO WE KNOW HUMANS ARE THE CAUSE?

Some people doubt that climate change is happening (despite the overwhelming evidence), and others admit it's real but don't believe humans are the cause. But the evidence is overwhelming that humans are to blame. How do we know that climate is changing in an unusual manner and not just having normal "climate fluctuations"?

"IT'S JUST NATURAL CLIMATIC VARIABILITY."

As we have seen in this book, geologists and paleoclimatologists know a lot about past greenhouse worlds and the icehouse planet that has existed for the last 33 million years. We have a good understanding of why the Antarctic ice sheet first appeared at that time and how the Arctic froze about 3.5 million years ago, beginning the 24 glacial and interglacial episodes of the ice ages that have occurred since then. We know how variations in Earth's orbit (the Milankovitch cycles) controls the amount of solar radiation Earth receives, triggering shifts between glacial and interglacial periods. Our current warm interglacial has lasted for 10,000 years, the duration of most previous interglacials. If it were not for global warming, we would be headed into the next glacial any time now. Instead, by pumping greenhouse gases that had long been trapped in Earth's crust into our atmosphere, our world has been pushed into a "superinterglacial," and the world is already warmer than it was in any previous warming period. We can see the big picture of climate variability most clearly in the EPICA-1 core from Antarctica, which show the details of the last almost 780,000 years of glacial-interglacial cycles (figure 25.7). *At no time during any previous interglacial did the natural carbon dioxide levels exceed 280 ppm, even at their very warmest.* Our atmospheric carbon dioxide levels are already close to 420 ppm today. The atmosphere is headed to 600 ppm within a few decades, even if we stop releasing greenhouse gases immediately. This is decidedly *not* within

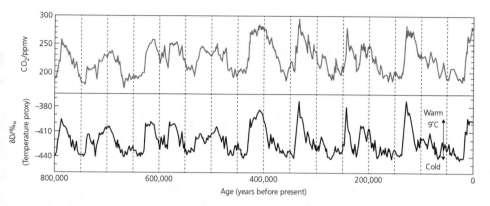

Figure 25.7 ▲

The record of carbon dioxide in the EPICA-1 core from Antarctica, the longest and deepest core ever taken. Natural climate variability is clearly shown by the core, and the highest levels of carbon dioxide (top curve) during the warmest interglacials was only 280 ppm—today it is already over 418 ppm. The bottom curve displays deuterium (hydrogen-2), which is an indirect proxy of paleotemperature recorded in the core. (Redrawn from several sources)

the normal range of "natural climatic variability" and is unprecedented in human history. If you believe "climate changes all the time and humans aren't causing these changes," I urge you to look at the huge amount of paleoclimatic data that show otherwise.

"IT'S JUST THE SUN OR COSMIC RAYS OR VOLCANIC ACTIVITY OR METHANE."

The amount of heat provided by the sun that reaches Earth has been decreasing since 1940, the opposite of what some people claim (figure 25.8). If the sun were causing global warming, it should show an increase in activity, not a decrease. Second, cosmic radiation causes an increase in cloud cover, so increased cosmic radiation would cool the planet, and decreased cosmic radiation would warm it. There are lots of measurements of cosmic radiation, and the result is clear: in the last 40 years, cosmic radiation has been increasing (which should cool the planet) while the temperature has been rising—the opposite of the effect expected if cosmic radiation contributed to recent warming. Third, there is no evidence that large-scale volcanic events (such as the 1815 eruption of Tambora

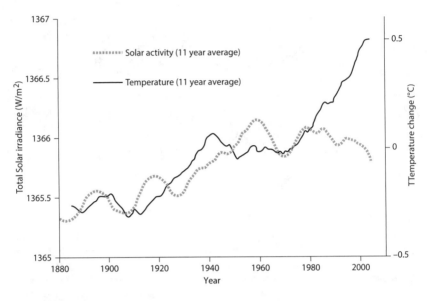

The trends of incoming solar radiation and temperature over the past century. If increasing solar activity explained global warming, there should be an upward trend in the solar curve—but solar input has actually decreased in the past 50 years. (Redrawn from several sources)

in Indonesia, which changed global climate for about a year) have any long-term effects that could explain 200 years of warming and carbon dioxide increase. Volcanoes push only 0.3 billion tonnes of carbon dioxide into the atmosphere each year, but humans emit more than 29 billion tonnes of carbon dioxide a year—about 100 times more. Clearly, we have a bigger effect than volcanoes have on the amount of carbon dioxide in the air. Finally, methane is a more powerful greenhouse gas, but there is 200 times more carbon dioxide than methane, so carbon dioxide is still the most important agent. Every alternative has been examined, and the only clear-cut answer that fits the data is that human-caused carbon dioxide increases are responsible for this era's global warming. We just can't deflect the blame on this one.

"THE CLIMATE RECORDS SINCE 1998 SHOW COOLING."

People who make this argument are cherry-picking the data to make it show what they want it to show, and they point to a slight cooling trend from 1998 to 2000 (figure 25.9A). This trend was caused by a record-breaking El Niño event in 1998, and the next few years were cooler when compared to 1998. In science, cherry-picking data is considered dishonest and unethical—but it is a common tactic in the political sphere and in public relations and

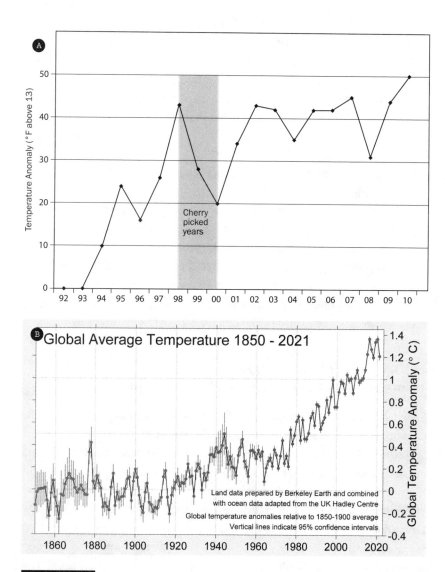

Figure 25.9 ▲

(A) The effect of cherry-picking the anomalously warm El Niño year of 1998 as a starting point to falsely suggest that global temperature has been cooling or steady since 1998. (B) The long-term average temperature on Earth over the past century. (Redrawn from several sources)

propaganda, where lies are the norm. As Mark Twain said, "A lie can travel around the world and back again while the truth is lacing up its boots." Since 2002, the overall long-term trend of warming is unequivocal (figure 25.9B). Saying that there was a "climate pause" or a "cooling trend" since 1998 is a clear-cut case of using out-of-context data in an attempt to distort and deny the evidence. All of the 20 hottest years ever recorded on a global scale have occurred in the last 25 years (only 1996 failed to make the list), and the 9 hottest years on record have occurred since 2013. Almost every year since 2000 is on the top 25 hottest years list, and the 9 hottest years on record (hottest first) are: 2016, 2020, 2019, 2015, 2022, 2017, 2018, 2021, 2014. As pointed out above, 2023 is on track to break the record of 2016. More significant, the record-breaking temperatures of the last few years (2015, 2016, and 2020) each broke or tied the record of the previous year and rocketed upward at a rate never seen in any climatic record, ancient or modern. All of these temperatures are well above the hot year of 1998, when warming allegedly "paused." And it's no surprise that in recent years, the climate deniers have been claiming that global temperatures have "paused" since 2016, because just like in 1998, the years after the El Niño peak are obviously going to be slightly cooler. This is just as dishonest as the "1998 pause" except they cherry-picked the most recent record year. But 2023 will probably be the hottest year on record, shattering the record of 2016, so that specious argument will also be debunked.

"WE HAD HEAVY SNOW AND FREEZING TEMPERATURES LAST WINTER."

So what? This is a classic example of how the scientifically illiterate public cannot tell the difference between *weather* (short-term seasonal changes) and *climate* (the long-term average of weather over centuries to millennia and longer). Weather is what happens from hour to hour or day to day; climate is the average of weather events over long time spans. Climate is what you might look up in an almanac or on a website about the expected average temperature for a given day in the future; weather is what you can see when you look outdoors. Winter doesn't just vanish as the globe gets warmer; winters just become *on average* a bit warmer and shorter than usual. But long cold spells, snowstorms, and freezing events will still happen. More important, local weather tells us nothing about the next continent or the global average; it is only a local effect, determined by short-term atmospheric and oceanographic conditions.

In addition, meteorologists are *not* climate experts; they are trained only to analyze the short-term changes in the weather. (I happen to teach both meteorology and climate science at the college level, and I drive home this distinction). Most have no formal training or published research in climate science, so their opinions about climate change are no better informed than anyone else's opinion. If you see one denying climate change on some climate-denying media, their opinion is worthless. They are *not* the experts they sometimes claim to be.

In fact, warmer global temperatures can actually mean *more moisture* in the atmosphere, which increases the intensity of normal winter snowstorms. In recent years, people living in the northern part of North America have experienced many unusually mild winters, which they conveniently forget about when it finally does get cold. When a long cold spell from the polar vortex (sometimes called a "bomb cyclone") lingers over the northeastern United States (where all the media are headquartered, and all policy and politics is dictated in Washington or New York), it is actually a result of climate change. The strong temperature difference between the pole and the equator, which powers the jet stream, gets weaker and weaker when the North Pole warms and melts. The jet stream slows down and develops a lazy looping path (called Rossby waves) over North America, which can remain stalled for weeks. If the loop down from the Arctic brings a mass of cold air with it, the freezing will stick around for a while—but somewhere else in North America a warm loop of the jet stream will be stalled over the northern Midwest or Plains or the West Coast, bringing abnormally warm weather to that area while the East Coast freezes.

"CARBON DIOXIDE IS GOOD FOR PLANTS, SO THE WORLD WILL BE BETTER OFF."

The people who promote this idea clearly don't know much global geochemistry, or they are trying to exploit the fact that most people are ignorant of science. The Competitive Enterprise Institute (funded mostly by money from oil and coal companies) ran a series of misleading ads that insult the intelligence of any educated person, concluding with this tag line: "Carbon dioxide: they call it pollution, we call it life." Anyone who knows the basic science of Earth's atmosphere can spot the deceptions in this ad. Yes, plants take in carbon dioxide that animals exhale, as they have been doing for billions of years. But the whole point of the global warming evidence (as shown from ice cores) is that the *delicate natural balance* of carbon dioxide has been disrupted by our production of too much of it—far in excess of what plants or the oceans can handle. At the same time, humans are cutting down huge areas of rain forest every day, so there are fewer plants to absorb the gas and these slash and burn practices release more carbon dioxide than plants can absorb. There is much debate about whether increased carbon dioxide might help agriculture in some parts of the world, but that has to be measured against the fact that other traditional "breadbasket" regions (such as the American Great Plains) are expected to become too hot to be as productive as they are today. The latest research shows that increased carbon dioxide actually *inhibits* the absorption of nitrogen into plants, so plants (at least those that we depend on today) are *not* going to flourish in a greenhouse world. Anyone who tells you otherwise is either ignorant of basic atmospheric science or is trying to fool the public who don't know science from bunk.

THE FINGERPRINTS ON THE MURDER WEAPON

How do we know humans are to blame? The lines of evidence for human causes form a very long list and include a wide spectrum of different kinds of data:

- We can directly measure the amount of carbon dioxide humans are producing, and it exactly matches the amount of increase in atmospheric carbon dioxide measured in ice cores and other data sets (figure 25.10). This is one of many lines of evidence that shows our burning of fossil fuels explains the rise in carbon dioxide; there are no other trends in nature that match the carbon dioxide increase.
- Through carbon isotope analysis we can show that the carbon dioxide in the atmosphere is coming directly from our burning of fossil fuels, not from natural sources (figure 25.11). The atmosphere has many other chemical fingerprints proving that the greenhouse gases were produced by humans, such as the ratio of carbon-12 (more abundant due to burning of fossil fuels) versus carbon-13.

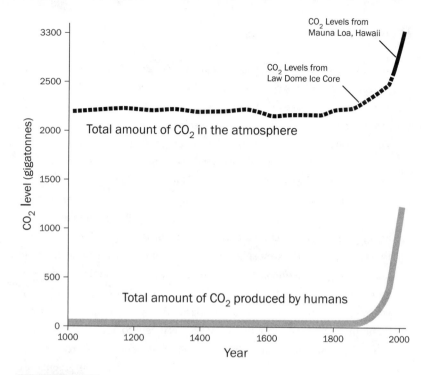

The measurements of carbon dioxide in the atmosphere (as determined from ice cores and direct measurements since the Keeling experiment in 1958) exactly matches the amount of carbon we have released into the atmosphere by burning fossil fuels, starting at the same time and showing the same rate of increase. (Redrawn from several sources)

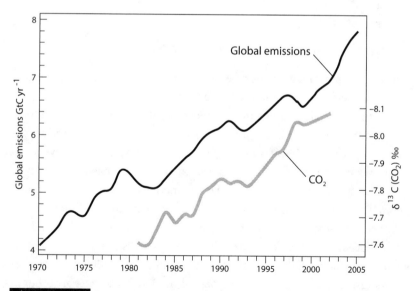

The greenhouse gases accumulating in our atmosphere have distinctive geochemical "fingerprints on the murder weapon" that prove humans are the source of the carbon dioxide, not natural sources. This is demonstrated by many different geochemical indicators. (Redrawn from several sources)

- We can also measure oxygen levels that drop as we produce more carbon that then combines with oxygen to produce carbon dioxide.
- Satellites out in space are measuring the heat released from the planet and can actually *see and measure* the atmosphere getting warmer in real time. They can also see the carbon dioxide emerging from power plants and cities, and it's clear that it comes mostly from those sources.
- The crucial test that warming is coming from sources on the ground and not from solar radiation from space is to measure the relative changes in temperature in the different levels of the atmosphere. Recent climate models of the greenhouse effect predict there should be cooling in the stratosphere (the upper layer of the atmosphere above 10 kilometers [6 miles] in elevation) but warming in the troposphere (the bottom layer of the atmosphere below 10 kilometers [6 miles]) where the human-produced gases originate. If the warming was due to an increase in solar radiation, the stratosphere would warm first and the troposphere would be relatively cool. Indeed, our space probes have measured stratospheric cooling and upper troposphere warming, and just as climate scientists had predicted, the warming is due to greenhouse gases, not to the sun (figure 25.12).
- Finally, we can rule out other culprits: solar heat has been decreasing since 1940, not increasing, and there are no measurable increases in cosmic radiation, methane, volcanic gases, or any other potential cause.

CLIMATE CHANGE AND THE FUTURE

The most important geologic events that we can predict for the future involve the effects of climate change. Many important short-term geological and ecological issues are related to climate change: hotter global temperatures and more heat waves; vanishing ice caps and glaciers; diminishing supplies of fresh water as the glaciers melt away; rising sea levels that will drown most of the coastal regions of the world; increasing intensity of storms, especially hurricanes; and mass extinctions of species driven out of their habitat by human interference and loss of their normal climate.

But the scariest of all is the death of the oceans as overfishing and too much heat kills marine life (especially the coral reefs), and too much carbon dioxide makes the waters more acidic, killing nearly all marine species that require a calcareous shell. The oceans are Earth's biggest absorber of

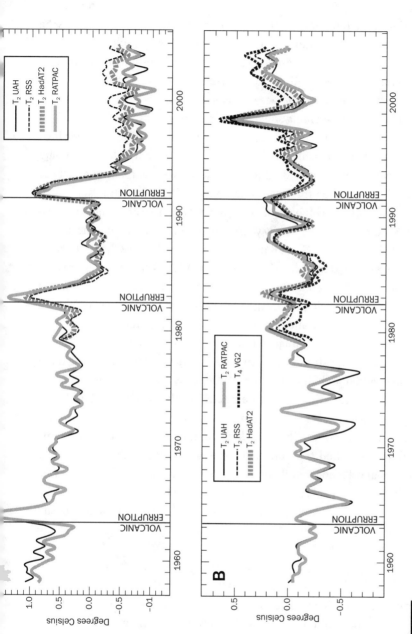

Figure 25.12 ◀

If warming came from space, then the stratospheric layer high above us would warm and the troposphere (the layer in which we live) would be relatively cool. If warming came from below, the troposphere would warm and the stratosphere would remain relatively cool. Satellite measurements show definite stratospheric cooling and tropospheric warming, confirming it comes from ground sources, namely, the burning of fossil fuels. This plot also shows spikes in cooling shortly after volcanic eruptions (dashed vertical lines) due to the blocking of sunlight by volcanic ash and dust blowing in the stratosphere for several years after big eruptions. (Redrawn from several sources)

heat and atmospheric gases, and this enormous stable mass of water has not changed much in temperature or acidity over millions of years. As late as the 1980s, the oceans were thought to be highly stable and could absorb most of what we dumped into them. However, startling rises in ocean temperatures, increasing acidity, and mass deaths of the world's coral reefs illustrate that we have far exceeded the oceans' capacity to absorb our pollution and excess heat. We are seeing a shocking decline in coral reefs ("bleaching") and extinctions in many marine ecosystems that can't handle too much heat and acid. Many of my colleagues in marine biology and paleontology are veteran divers who have studied reefs for 30 or 40 years, and they all say that they have watched this destruction happen in their lifetimes. All over the world, the reefs are in bad shape or dying, and there are almost no pristine reefs left on the planet. A few years ago, scientists found that the largest ecosystem on Earth, the Great Barrier Reef of Australia, was dying. There is strong scientific evidence that the "Mother of All Mass Extinctions" (which wiped out 95 percent of marine species about 250 million years ago) was due to excess carbon dioxide (hypercapnia) in the oceans. This ocean chemistry not only dissolves shells and corals but also suffocates marine life, as happened during the great Permian extinction (see chapter 15).

Once ice caps vanish and oceans warm, it's extremely hard to bring them back unless a major change in climate produces an extreme ice age. What humans have done to the planet far exceeds the ability of the Milankovitch cycles to return us to cooler conditions. It looks like we are headed toward a super greenhouse world of high sea levels and no ice anywhere—the same world that dominated the early Paleozoic and most of the Jurassic and Cretaceous and Paleogene. By burning the remains of prehistoric life (the plants that became coal and plankton which have been turned into oil), we are pushing our planet toward a greenhouse world similar to one that hasn't existed for 55 million years.

A PERSPECTIVE

We assume that our civilization is a permanent feature on Earth and that some form of our culture will persist indefinitely. As any archeologist or historian knows, however, this is not a given. We have many examples of cultures that have vanished, leaving only a few durable artifacts behind, and often we know very little about how they lived—or why they failed. One needs to look only at the mysterious Etruscans, Minoans, Mycenaeans,

Mayans, or Anasazi. The list of failed societies goes on and on. In his 2004 book *Collapse: How Societies Choose to Fail or Succeed*, Jared Diamond points out that when we do know why a society vanished it is truly humbling. For example, the Easter Island culture vanished completely before the European settlers had witnessed their civilization, leaving behind only the famous stone heads, or *moai*, dotting the island. Diamond shows that the extinction of the Easter Islanders was, to a large extent, self-inflicted: too many people, too much overexploitation of their environment when they cut down all the trees on what had been a densely forested island, followed by starvation, disease, and warfare that wiped out the survivors. Diamond suggests that a similar fate could end our world civilization if we overpopulate this planet or damage our environment or overexploit our resources. After all, 99 percent of all species that have ever lived are now extinct. There is no good biological reason to believe that our fate will be different, especially given our accelerated pace of self-destruction.

NOTES

1. Joseph Fourier, "Remarques générales sur les températures du globe terrestre et des espaces planétaires," *Annales de Chimie et de Physique* 27 (1824): 136–167; Claude Pouillet, "Mémoire sur la chaleur solaire, sur les pouvoirs rayonnants et absorbants de l'air atmosphérique, et sur les températures de l'espace," *Comptes Rendus de l'Académie des Sciences* 7, no. 2 (1838): 24–65.
2. John Tyndall, "On Radiation Through the Earth's Atmosphere," *Philosophical Magazine* 4, no. 25 (1863): 200–206.

FOR FURTHER READING

Diamond, Jared. *Collapse: How Societies Choose to Fail or Succeed*. New York: Penguin, 2004.

——. *Guns, Germs, and Steel: The Fates of Human Societies*. New York: Norton, 2005.

Fagan, Brian. *The Great Warming: Climate Change and the Rise and Fall of Civilizations*. London: Bloomsbury, 2009.

Flannery, Tim. *The Weather Makers: How Man Is Changing the Climate and What It Means for Life on Earth*. New York: Atlantic Monthly, 2006.

Gore, Al. *Earth in the Balance: Ecology and the Human Spirit*. New York: Penguin, 1992.

——. *An Inconvenient Truth*. Emmaus, PA: Rodale, 2006.

Harari, Yuval Noah. *Homo Deus: A Brief History of Tomorrow*. New York: Harper, 2017.

Henson, Robert. *The Thinking Person's Guide to Climate Change*. Washington, DC: American Meteorological Society, 2014.

Kolbert, Elizabeth. *Field Notes from a Catastrophe: Man, Nature, and Climate Change*. London: Bloomsbury, 2015.

Mann, Michael. *The Hockey Stick and the Climate Wars: Dispatches from the Front Lines*. New York: Columbia University Press, 2012.

Mann, Michael, and Lee Kump. *Dire Predictions: Understanding Climate Change*. London: DK, 2015.

Mann, Michael, and Tom Toles. *The Madhouse Effect: How Climate Denial Is Threatening Our Planet, Destroying Our Politics, and Driving Us Crazy*. New York: Columbia University Press, 2018.

McKibben, Bill. *Eaarth: Making a Life on a Tough New Planet*. New York: Times, 2010.

——. *The End of Nature*. New York: Random House, 1989.

McNeill, J. R., and Peter Engelke. *The Great Acceleration: An Environmental History of the Anthropocene Since 1945*. Cambridge, MA: Belknap, 2016.

Oreskes, Naomi, and Erik Conway. *Merchants of Doubt: How a Handful of Scientists Obscured the Truth on Issues from Tobacco Smoke to Climate Change*. London: Bloomsbury, 2010.

Pearce, Fred. *With Speed and Violence: Why Scientists Fear Tipping Points in Climate Change*. Boston: Beacon, 2007.

Prothero, Donald R. *Reality Check: How Science Deniers Threaten Our Future*. Bloomington: Indiana University Press, 2013.

Purdy, Jedidiah. *After Nature: Politics for the Anthropocene*. Cambridge, MA: Harvard University Press, 2015.

Rees, Martin. *Our Final Century? Will the Human Race Survive the Twenty-First Century?* New York: Heinemann, 2003.

——. *Our Final Hour: A Scientist's Warning: How Terror, Error, and Environmental Disaster Threaten Humankind's Future in This Century—on Earth and Beyond*. New York: Basic Books, 2003.

Romm, Joseph. *Climate Change: What Everyone Needs to Know*. Oxford: Oxford University Press, 2015.

Scranton, Roy. *Learning to Die in the Anthropocene: Reflections on the End of Civilization*. New York: City Lights, 2015.

Weiner, Jonathan. *The Next One Hundred Years: Shaping the Future of Our Lives on Earth*. New York: Bantam, 1990.

Wilson, Edward O. *The Future of Life*. New York: Vintage, 2003.

INDEX